Crop Wild Relatives

Issues in Agricultural Biodiversity

This series of books is published by Earthscan in association with Bioversity International. The aim of the series is to review the current state of knowledge in topical issues in agricultural biodiversity, to identify gaps in our knowledge base, to synthesize lessons learned and to propose future research and development actions. The overall objective is to increase the sustainable use of biodiversity in improving people's well-being and food and nutrition security. The series' scope is all aspects of agricultural biodiversity, ranging from conservation biology of genetic resources through social sciences to policy and legal aspects. It also covers the fields of research, education, communication and coordination, information management and knowledge sharing.

Published Titles:

Crop Wild Relatives:
A Manual of *in situ* Conservation
Edited by Danny Hunter and Vernon Heywood

The Economics of Managing Crop Diversity On-farm:
Case Studies from the Genetic Resources Policy Initiative
Edited by Edilegnaw Wale, Adam G. Drucker and Kerstin K. Zander

Forthcoming in 2011:

Plant Genetic Resources and Food Security
Stakeholder Perspectives on the International Treaty on Plant Genetic Resources for Food and Agriculture
Edited by Christine Frison, Francisco López and José T. Esquinas

Farmers' Crop Varieties and Farmers' Rights:
Challenges in Taxonomy and Law
Edited by Michael Halewood

Crop Genetic Resources as a Global Commons:
Challenges in International Law and Governance
Edited by Michael Halewood, Isabel López Noriega and Selim Louafi

Crop Wild Relatives

A Manual of *in situ* Conservation

Edited by Danny Hunter and Vernon Heywood

London • Washington, DC

First published in 2011 by Earthscan

Earthscan Ltd, Dunstan House, 14a St Cross Street, London EC1N 8XA, UK
Earthscan LLC, 1616 P Street, NW, Washington, DC 20036, USA

Earthscan publishes in association with the International Institute for Environment and Development

For more information on Earthscan publications, see www.earthscan.co.uk or write to earthinfo@earthscan.co.uk

ISBN: 978-1-84971-178-4 hardback
ISBN: 978-1-84971-179-1 paperback

Typeset by MapSet Ltd, Gateshead, UK
Cover design by Adam Bohannon

A catalogue record for this book is available from the British Library

Library of Congress Cataloging-in-Publication Data

Crop wild relatives : a manual of in situ conservation / edited by Danny Hunter and Vernon Heywood. – 1st ed.
p. cm.
Includes bibliographical references and index.
ISBN 978-1-84971-178-4 (hardback) – ISBN 978-1-84971-179-1 (pbk.) 1. Crops–Germplasm resources. 2. Germplasm resources, Plant. 3. Genetic resources conservation. I. Hunter, Danny. II. Heywood, V. H. (Vernon Hilton), 1927-
SB123.3.C769 2010
333.95'3416–dc22

2010023826

At Earthscan we strive to minimize our environmental impacts and carbon footprint through reducing waste, recycling and offsetting our CO_2 emissions, including those created through publication of this book. For more details of our environmental policy, see www.earthscan.co.uk.

Printed and bound in the UK by CPI Antony Rowe.
The paper used is FSC certified.

Contents

Part IV: Other Major Issues

Annexes

Acknowledgements and Contributors

This book is the result of the combined efforts of many people who were closely involved with the UNEP/GEF Crop Wild Relatives Project since its beginning in 2004. It has involved significant contributions from individuals representing national and international organizations who collaborated closely with the project. Their dedication and commitment to the implementation of the project and the realization of this manual is gratefully recognized and appreciated.

Editors

The compilers and editors of the book were:

Danny Hunter, senior scientist and global project coordinator of the UNEP/GEF Crop Wild Relatives Project, Bioversity International, Rome, Italy

Vernon Heywood, Professor Emeritus, Centre for Plant Diversity and Systematics, School of Biological Sciences, University of Reading, United Kingdom

Chapter authors

Danny Hunter was the lead author for Chapters 4, 5, 15 and 16; Vernon Heywood was the lead author for Chapters 1–3, 6–11, 13 and 14; Ehsan Dulloo was the lead author for Chapter 12.

The editors would like to thank the following chapter co-authors for their contributions to the chapters indicated:

Per G. Rudebjer, scientist, education and capacity development, Bioversity International, Rome, Italy (co-author, Chapter 15)

Elizabeth Goldberg, head, capacity development unit, Bioversity International, Rome, Italy (co-author, Chapter 15)

Ruth Raymond, head, public awareness unit, Bioversity International, Rome, Italy (co-author, Chapter 16)

Ehsan Dulloo, senior scientist, agricultural biodiversity conservation, understanding and managing biodiversity programme, Bioversity International, Rome, Italy (contributor to Chapter 8)

Many individuals have contributed boxes, tables and illustrations which, where possible, are clearly acknowledged in the text in the appropriate place. In particular, the book has benefited considerably from the dedication and contributions of the national project coordinators responsible for the implementation of project activities in the five participating countries:

Armen Danielyan, national project coordinator, Crop Wild Relatives Project, Yerevan, Armenia

Jeannot Ramelison, national project coordinator, Crop Wild Relatives Project, Centre National de la Recherche Appliquée au Développement Rural – FOFIFA, Antananarivo, Madagascar

Beatriz Zapata Ferrufino, national project coordinator, Crop Wild Relatives Project, Ministry for the Environment and Water, Vice-Ministry for Environment, Biodiversity and Climate Change, La Paz, Bolivia

Anura Wijesekara, national project coordinator, Crop Wild Relatives Project, Horticulture Crops Development and Research Institute, Peredeniya, Sri Lanka

Sativaldi Djataev, national project coordinator, Crop Wild Relatives Project, Institute of Genetics and Plant Experimental Biology, Academy of Sciences, Republic of Uzbekistan, Tashkent, Uzbekistan

Figure A.1 *The national project coordinators for the Crop Wild Relatives Project from left to right, Armen Danielyan, Jeannot Ramelison, Beatriz Zapata Ferrufino, Anura Wijesekara and Sativaldi Djataev*

Source: Sativaldi Djataev

Case study and other contributors

Teresa Borelli, Bioversity International, Italy; Vololoniaina Jeannoda, University of Antananarivo, Madagasacar; Tianjanahary Randriamboavonjy, RBG Kew, Antananarivo Madagascar, R.S.S. Ratnayake, Biodiversity Secretariat, Ministry of Environment and Natural Resources, Sri Lanka; Bhuwon Sthapit, Bioversity International, India; Feruza Mustafina and Gulayim Reimova, UNEP/GEF CWR Project, Uzbekistan; Naire Yeritsyan, UNEP/GEF CWR Project, Armenia; Janna Akopian, Institute of Botany of the National Academy of Sciences, Armenia; Siranush Muradyan, Bioresources Management Agency of the Ministry of Nature Protection, Armenia; Wendy Leslie Tejeda Perez, CWR Bolivia Project Assistant, Bolivia; Saul Cuellar from FAN-Bolivia.

The contributions of the following individuals, researchers, national executing agency and national partner organizations to the successful outcomes of the UNEP/GEF Crop Wild Relatives Project are gratefully acknowledged, some of which are captured in this manual.

Armenia

Siranush Muradyan, Ministry of Nature Protection of the Republic of Armenia (MoNP); Michael Oganesyan, Ministry of Agriculture of the Republic of Armenia (MoA); Margarita Harutunyan, Andreas Melikyan and Alvina Avagyan, Armenian State Agrarian University; Janna Akopian, Ivan Gabrielyan, Anush Nersesyan, Madlena Musaelyan and Estella Nazarova, Institute of Botany of National Academy of Sciences (IBoNAS), Siranush Nanagulyan and Nelli Hovhannisyan, Yerevan State University; Karen Poghosyan, Institute of Viticulture, Wine-making and Fruit-growing of the Ministry of Agriculture.

Bolivia

Omar Rocha, Aldo Claure and Rafael Murillo, Ministerio de Medio Ambiente y Agua (MMAyA), Viceministerio de Medio Ambiente, Biodiversidad y Cambios Climáticos (VMABCC), Direccíon General de Biodiversidad y Áreas Protegidas (DGBAP); Wendy Leslie Tejeda, Ministerio de Medio de Medio Ambiente y Agua (MMAyA), Viceministerio de Medio Ambiente, Biodiversidad y Cambios Climáticos (VMABCC) – Proyecto UNEP/GEF 'Conservación *in situ* de parientes silvestres de cultivos'; Stephan Beck, Renate Seidel and Prem Jai Vidaurre, Herbario Nacional de Bolivia (LPB-UMSA); Susana Arrazola, Milton Fernandez, Magaly Mercado, Nelly de la Barra and Saul Altamirano, Centro de Biodiversidad y Genética-Herbario Nacional Forestal Martín Cárdenas, Universidad Mayor de San Simón (CBG-BOLV, UMSS); Patricia Herrera, Moises Mendoza and José Maria Taquichiri, Museo de Historia Natural 'Noel Kempff Mercado', Herbario del Oriente, Universidad Autónoma Gabriel René Moreno (MHNNKM-USZ, UAGRM); Carlos Rivadeneira and Rolando Bustillos, Instituto de Investigaciones Agrícolas 'El Vallecito' – Universidad Autónoma Gabriel René Moreno (IIA 'El Vallecito' UAGRM); Ximena Cadima,

Wilfredo Rojas, Fernando Patiño, Milton Pinto, Andres Mamani and Eliseo Mamani, Fundación PROINPA; Gonzalo Avila, Lorena Guzman and Margoth Atahuachi, Centro de Investigaciones fitoecogenéticas de Pairumani – Fundación Simón Patiaño; Natalia Araujo, Humberto Gomez, Saul Cuellar and Yaqueline Bellot, Fundación Amigos de La Naturaleza (FAN-Bolivia); Aniceto Ayala and Lucas Zamora, Confederación de Pueblos Indígenas de Bolivia (CIDOB); Adrian Nogales, José Coello, Hector Cabrera and Jaime Galarza, Servicio Nacional de Áreas Protegidas (SERNAP).

Madagascar

Solo Hery Rapanarivo, Frank Rakotonasolo, Hélène Elisabeth Razanantsoa, Jacqueline Razanantsoa, Hanta Razafindraibe, Jacky Lucien Andriatianaina and Elysette Raherivololona, Parc Botanique et Zoologique de Tsimbazaza; Herivololona Mbola Rakotondratsimba, Domohina Noromalala Andrianasolo, Nivo Raharison, Mamy Tiana Rajaonah, Verohanitra Rafidison, Elisabeth Rabakonandrianina and Bakolimalala Rakouth, Université d'Antananarivo, Faculté des Sciences, Département de Biologie et Ecologie Végétales; Voahangy Raharimalala, Didier Andriamparany and Bakoly Andrianaivoravelona, National Office for the Environment – ONE; Jaotera, René Razafindrajery, Paul Ignace Rakotomavo, Jacqueline Razaiarimanana, Florent Razanakolona, Edouard Randriamanantsoa and Justin Rakotoarimanana, Madagascar National Parks – formerly ANGAP; Lolona Ramamonjisoa and Annick Razafintsalama, National Forest Seed Bank – SNGF; Naritiana Rakotoniaina, Support Service for the Management of Environment – SAGE; Yvonne Rabenantoandro, Raymond Rabevohitra, Nirina Rabemanantsoa, Hanitra Viviane Andriamampadry, Voahangy Andrianavalona and Harizoly Razafimandimby, Centre National de la Recherche Appliquée au Développement Rural – FOFIFA; Noasilalao Nomenjanahary and Hiarinirina Andrianizahana, MINENVEF – Ministry of the Environment and Forest Resources; Michelle Andriamahazo and Nirina Rajaonah, Ministry of Agriculture; Xavier Rakotonjanahary, Ministry of Research (actually Ministry of National Education); Razafinakanga and Rova Raharison Association du Réseau des Systèmes d'Information Environnementale – ARSIE; Hélène Ralimanana, Tianjanahary Randriamboavonjy and Landy Rajaovelona Royal Botanic Gardens, Kew; Sylvie Andriambololonera, Tantely Raminosoa, Richard Razakamalala, Faranirina Lantoarisoa and Brigitte Ramandimbisoa, Missouri Botanical Garden; Heritiana Ranarivelo and Letsara Rokiman, California Academy of Science; Andry Rakotomanjaka and Dimby Razafimpahana, REBIOMA (Network of Madagascar Biodiversity) – WCS (World Conservation Society)

Sri Lanka

M.A.R.D. Jayatillaka, Ministry of Environment and Natural Resources; L.K. Haturusinghe, Ministry of Agriculture and Agrarian Services; Gamini Gamage and Sujith Ratnayake, Biodiversity Secretariat; C. Kudagamage, Jinadari de

Zoysa, Rohan Wijekoon, K.N. Mankotte and Jarantha Illankoon, Department of Agriculture; D.H. Muthukudaarachchi, Srimathi Dissanayake, S.U. Liyanage, Gamini Samarasinghe, Plant Genetic Resources Center; P.V. Hemachandra, Rice Research and Development Institute; H.D. Ratnayake, Department of Wildlife Conservation; Sarath Fernando and Anura Saturushinghe, Forest Department; Dayangani Senasekara and Sudeepa Sugatadasa, Bandaranayake Memorial Ayurvedic Research Institute; Siril Wijesundara, Department of National Botanic Gardens; Subani Ranasinghe, National Herbarium; Lional Gunaratne, Department of Export Agriculture; Samantha Gunasekara, Department of Customs; Buddhi Marambe and D.K.N.G. Pushpakumara, Faculty of Agriculture, University of Peradeniya; Gamini Senanayake, Faculty of Agriculture, University of Ruhuna; Channa Bambaradeniya, IUCN Sri Lanka; and Jagath Gunawardane, Environmental Science Foundation.

Uzbekistan

Eugeniy A. Butkov, Galina M. Chernova and Yuvenaliy A. Karpenko, Republican Scientific Production Centre of Ornamental Gardening and Forestry of Ministry Agriculture and Water Resources Republic of Uzbekistan (MAWR RUz); Karim I. Baymetov and Fayzulla H. Abdullaev, Uzbek Scientific Research Institute of Plant Industry MAWR RUz; Abdusattor Abdukarimov, Mirakbar D. Yakubov and Svetlana I. Kim, Institute of Genetics and Plant Experimental Biology of the Academy of Sciences of Republic of Uzbekistan; Uktam P. Pratov and Akramjon S. Yuldashev, Scientific Plant Production Centre 'Botanika' of Academy of Sciences of Republic of Uzbekistan; Yuriy M. Djavakyants and Abduvakhid A. Abdurasulov, Uzbek Scientific Research Institute of Fruit Growing, Viticulture & Winemaking MAWR RUz; Muratbek Sh. Ganiev, Main Administration of Forestry Department of MAWR RUZ; Abdukarim A. Abdujamilov, Ergash Sarymsakov, Jasur T. Dustov and Alexandr V. Esypov, Ugam Chatkal National Natural Park and Chatkal Biosphere Reserve.

The Uzbek CWR Project team would also like to make particular mention of Ms Svetlana I. Kim, who made an invaluable contribution to the project from 2004 to 2007. Sadly, Ms Kim passed away at a young age, but the members of the CWR Project will continue to honour her memory.

International partners, steering committees and other international meetings

The contributions of the following individuals from the project's international partner organizations are gratefully acknowledged, especially: Suzanne Sharrock, Botanic Gardens Conservation International (BGCI), England; Diane Wyse Jackson, formerly BGCI, England; Juan Fajardo, formerly FAO, Italy; Kakoli Ghosh, FAO, Italy; Arturo Mora, IUCN, Ecuador; Harriet Gillett, UNEP-WCMC, England; Jane Smart, IUCN, Switzerland and Julie Griffin, IUCN,

Switzerland. The additional contributions and support of the following individuals from Bioversity International are duly acknowledged: Ehsan Dulloo, Toby Hodgkin, Laura Snook, Imke Thormann, Paul Quek, Muhabbat Turdieva, Ramanatha Rao, Marleni Ramirez, Michael Halewood and Annie Lane (former Global Project Coordinator). We would like to reserve a special mention and thanks for Teresa Borelli, Bioversity International, who spent many hours researching information for consideration in the manual and who had many ideas and suggestions to improve the final publication.

We especially acknowledge the implementation support of Marieta Sakalian, Senior Programme Management/Liaison Officer (CGIAR/FAO), Biodiversity UNEP Division of GEF Coordination, and her continuing support to the development of this manual.

The international steering committee (ISC) for the project met on six separate occasions and provided valuable guidance to the project. In particular, the last two meetings (Cochabamba, Bolivia 2008 and Rome, Italy 2009) were important in terms of providing inputs and advice to the development of this book. In addition to ISC members identified and acknowledged earlier, we would especially like to thank the following country members of the ISC: Ivan Gabrielyan, Armenia; Lala Aime Razafinjara, Madagascar; Aldo Claure and Omar Rocha, Bolivia; Abdusattar Abdukarimov, Uzbekistan; Siril Wijesundara, Sri Lanka; and the advisors involved in the project's Technical Advisory Committee: Vernon Heywood, England; Arthur Chapman, Australia; and Susan Bragdon, United States of America; who also participated in the ISC.

Figure A.2 *Participants, UNEP/GEF Crop Wild Relatives Project 2008 Steering Committee on a field visit to Morochata community of Cochabamba, Bolivia*

Source: Danny Hunter

CWR Manual Review Workshop

In October 2009, a writing workshop took place in the very beautiful setting of the Majella National Park, in the Abruzzo Region of Italy. The purpose of the workshop was to bring together those closest to the UNEP/GEF Crop Wild Relatives Project to review the first draft of this manual, as well as identify any major gaps and ways to address them. In addition, the workshop provided an opportunity to give the manual additional context and relevance by highlighting the main challenges and problems that partner countries faced in implementing *in situ* conservation actions, exchange lessons learned by comparing experiences from Armenia, Bolivia, Madagascar, Sri Lanka and Uzbekistan and come up with suggested good practices and brief case studies. Participants in the workshop are listed here in recognition of their support and constructive input to the content of the manual: Teresa Borelli, Italy; Armen Danielyan, Armenia; Sativaldi Djataev, Uzbekistan; Ehsan Dulloo, Mauritius; Beatriz Zapata Ferrufino, Bolivia; Danny Hunter, Ireland; Vololoniaina Jeannoda, Madagascar; Feruza Mustafina, Uzbekistan; Jeannot Ramelison, Madagascar; R.S.S. Ratnayake, Sri Lanka; Professor Vernon Heywood, England; and Anura Wijesekara, Sri Lanka.

Figure A.3 *Participants, UNEP/GEF Crop Wild Relatives Project Manual Review Workshop at Majella National Park headquarters, Abruzzo, Italy*

Source: Danny Hunter

Reviewers

A number of additional individuals generously contributed their expertise and time to the development of this manual. We are grateful for their constructive comments, feedback and efforts to ensure the usefulness and success of the publication and would like to thank the following:

Brian V. Ford-Lloyd, Director of the University Graduate School and Deputy Head of School of Biosciences, University of Birmingham, England; Suzanne Sharrock, Director Global Programmes, Botanic Gardens Conservation International, England; L. Jan Slikkerveer, Leiden University Branch of the National Herbarium of the Netherlands, Faculty of Sciences, Leiden Ethnosystems and Development Programme (LEAD); Toby Hodgkin, Principal Scientist, Bioversity International, Italy; José María Iriondo, Area de Biodiversidad y Conservación, Departamento de Biología y Geologia, ESCET, Universidad Rey Juan Carlos, Spain; Kate Gold, Training Manager, Millenium Seed Bank, Kew, England; Peter Taylor, Senior Program Specialist, Think Tank Initiative, International Development Research Centre, Canada; Jan Beniest, Training Unit Manager/Principal Training Scientist, World Agroforestry Centre (ICRAF), Kenya; Ramanatha Rao, Honorary Research Fellow, Biodiversity International and Adjunct Senior Fellow, ATREE, India; Kim Hamilton, Rainforest Seed Project Coordinator, Botanic Gardens Trust, Mount Annan Botanic Garden, Australia; Steve Waldren, Lecturer and Curator of Botanic Gardens, Trinity College Dublin, Ireland; Andy Jarvis, Programme Leader – Decision and Policy Analysis, International Centre for Tropical Agriculture (CIAT), Colombia; Nathan Russell, Senior Communications Officer, CGIAR Fund Office; Luigi Guarino, Senior Science Coordinator, Global Crop Diversity Trust, Italy; Bhuwon Sthapit, Regional Project Coordinator/*In Situ* Conservation Specialist, Bioversity International, India; Juan Carlos Moreno Saiz, Departamento de Biología (Botánica), Universidad Autónoma de Madrid, Spain.

Publishing and production

We would like to extend our sincere thanks to Tim Hardwick, Claire Lamont and other staff at Earthscan. We also sincerely thank Teresa Borelli and Nicole Hoagland at Bioversity International, both of whom have provided untiring support to the production of this manual.

Financial support

The development and publication of this manual was made possible first and foremost through support from the Global Environment Facility (GEF) and the United Nations Environment Programme (UNEP) within the framework of the global project '*In situ* conservation of crop wild relatives through enhanced information management and field application', as well as co-financing support made available from national and international partners. Without this support the implementation and outputs of the project would not have been realized and the preparation and publication of this manual would not have been possible.

We would also like to thank Jenessi Matturi, Co-publishing Programme, Technical Centre for Agricultural and Rural Cooperation (CTA) (ACP-EU) and Kakoli Ghosh, Agriculture Officer, Plant Production and Protection, FAO who were instrumental in facilitating additional support from CTA and FAO for the preparation and publication of this manual.

Finally, we would like to express our sincere thanks to the various host organizations that supported the authors and contributors.

Copyright permissions

To the best of our knowledge, materials presented in this manual have been accurately referenced and sourced. We have made every effort to ensure that the original source of copyright material has been provided within the text. If any errors or omissions are noted, we would be pleased to rectify them.

About Bioversity

Bioversity International is one of 15 centres supported by the Consultative Group on International Agricultural Research (CGIAR). Bioversity has worked for more than 35 years to support the improved use and conservation of agricultural biodiversity. Through international research, in collaboration with partners throughout the world, Bioversity strives to build the knowledge base needed to ensure effective use of diversity to increase sustainable agricultural production, improve livelihoods and meet the challenges of climate change. Financial support for Bioversity's research is provided by more than 150 donors, including governments, private foundations and international organizations.

Bioversity International
Via dei Tre Denari
00057 Maccarese,
Rome, Italy
www.bioversityinternational.org

About CTA

The Technical Centre for Agricultural and Rural Cooperation (CTA) was established in 1983 under the Lomé Convention between the ACP (African, Caribbean and Pacific) Group of States and the European Union Member States. Since 2000, it has operated within the framework of the ACP-EU Cotonou Agreement. CTA's tasks are to develop and provide products and services that improve access to information for agricultural and rural development, and to strengthen the capacity of ACP countries to acquire, process, produce and disseminate information in this area. CTA is financed by the European Union.

CTA
Postbus 380
6700 AJ Wageningen
The Netherlands
www.cta.int

Foreword

This manual, the first in a new series of *Issues in Agricultural Biodiversity* from Earthscan and Bioversity International, focuses on the *in situ*, or on the ground, conservation of crop wild relatives. These species represent a vital genetic resource for breeding the new and better varieties that will be needed to maintain and increase the productivity of our crops and to allow them to survive under the new conditions created by climate change.

Unfortunately, crop wild relatives are themselves at risk not only from climate change but also from other pressures such as overgrazing, fragmentation, habitat degradation and loss, invasive species and overexploitation. Until recently, the main conservation strategy adopted for crop wild relatives has been *ex situ*, through the maintenance of samples, including seeds or vegetative material, in various kinds of genebank or other facility.

However, many experts now recognize that conserving crop wild relatives in their natural surroundings can allow populations to continue to evolve and generate new genetic variation that is adapted to changing conditions. Until now, experience and knowledge in conserving the wild relatives of crops *in situ* has been very limited – an issue this manual clearly addresses by significantly enhancing the global body of knowledge on the subject. The research, highlighted in this manual and coordinated by Bioversity International in collaboration with country and international partners, has created a wealth of information on good practices and lessons learned.

The Global Environment Facility (GEF) and the United Nations Environment Programme (UNEP) have played a major role in supporting the pathbreaking project: In situ *conservation of crop wild relatives through enhanced information management and field application* that led to this manual as one of its outputs. Starting in 2004, GEF, in partnership with UNEP, invested US$5.8 million in the project, with another US$6.9 million contributed by other countries and partners.

Six years later, this project is now a success story, thanks in no small measure to the leadership and commitment of five partner countries: Armenia, Bolivia, Madagascar, Sri Lanka and Uzbekistan.

In the following pages, the authors detail the important practical experiences of our stakeholders that can be shared with the wider conservation community. This new publication also provides relevant information and guidance for scaling up actions targeting crop wild relative conservation around the world. It is our hope that through these pages we can conserve and promote crop wild relatives as an efficient way to build sustainable development and protect against famine and the effects of climate change worldwide.

Monique Barbut
Chief Executive Officer
Global Environment Facility

Achim Steiner
UN Under-Secretary General and Executive Director
United Nations Environment Programme

Preface

Considering the importance of biodiversity to human health and food security, the United Nations General Assembly has designated 2010 as the International Year of Biodiversity. According to the Food and Agricultural Organization of the United Nations, over one billion children, women and men are now going to bed hungry. The very first among the UN Millennium Development Goals, relating to reduction in hunger and poverty by half by the year 2015, is nowhere near achievement. It is clear that we should accelerate our efforts to improve the production and consumption of crop plants. It is in this context that the present manual on *in situ* conservation of crop wild relatives is a timely one. It will help to rekindle interest in the wild relatives of crop plants and help to initiate a climate-resilient food security system, based on the widening of the food basket.

Biodiversity provides the building blocks for sustainable food, health and livelihood security systems. It is the raw material for both the biotechnology industry and a climate-resilient farming system. Because of its importance for human well-being and survival, the Convention on Biological Diversity (CBD) was adopted at the UN Conference on Environment and Development held in Rio de Janeiro in 1992. The Convention's three goals are: conservation, sustainable use and equitable sharing of benefits of biodiversity. The Convention also recognizes that the biodiversity existing within a country is the sovereign property of its people.

In spite of the importance given to the conservation of biodiversity, genetic erosion is progressing in an unabated manner, both globally and nationally. For example, 12 per cent of birds, 21 per cent of mammals, 30 per cent of amphibians, 27 per cent of coral reefs and 35 per cent of conifers and cycads are currently facing extinction. According to the International Union for the Conservation of Nature (IUCN) over 47,677 species may soon disappear. A comprehensive study published in *Science* (29 April 2010) has revealed that there has been no notable decrease in the rate of biodiversity loss between 1970 and 2010. Even a very unique species like the orangutan, a close relative of man, is threatened with extinction in the island of Borneo. Leaders from 170 countries gathered at a UN Biodiversity Summit in Nagoya in Japan in October 2010 to adopt a road map for stopping biodiversity loss.

The challenge now is for every country to develop an implementation strategy for saving rare, endangered and threatened species through education, social mobilization and regulation. The Nagoya Summit will lead to meaningful results only if biodiversity conservation is considered in the context of sustainable development and poverty alleviation. The then Prime Minister of India, Indira Gandhi, pointed out at the UN Conference on the Human Environment held in Stockholm in 1972 that unless we attend concurrently to the needs of the poor and of the environment, the task of saving our environmental assets will not be easy. Biodiversity loss is predominantly related to habitat destruction largely for commercial exploitation as well as for alternative uses like roads, buildings, etc. Invasive alien species and unsustainable development are other important causes of genetic erosion. How can we reverse the paradigm and enlist development as an effective instrument for conserving biodiversity? Let me cite a few examples to illustrate how biodiversity conservation and development can become mutually reinforcing.

In 1990, I visited Maruthur Gopalan Ramachandran (MGR) Nagar Village near Pichavaram in Tamil Nadu, India to study the mangrove forests of that area. The families living in MGR Nagar were extremely poor and were not receiving government benefits since they had not been classified as either a Scheduled Caste (SC) or Scheduled Tribe (ST). The District Administrator collector mentioned that this matter is being reviewed. The village children had no opportunities for education and the fishermen were catching fish and shrimps by hand. When I asked the parents why they were not sending their children to school, the answer was that schools were far away and children were not being admitted due to the delay in their classification as a SC or ST. I then mentioned to my colleagues, '*saving mangrove forests without saving the children for whose well-being these forests are being saved makes no sense*'. With the help of a few donors, we started a primary school in the village and all children, irrespective of their age, joined the school. A few years later, the State Government took over the school and expanded its facilities. Following the tsunami, the huts were also replaced by brick buildings and the situation in MGR Nagar changed totally. Recently, the leader of the village met me and said that they would like the school developed into a higher secondary school with facilities for two additional classes. He also mentioned that they now know the value of mangroves since the root exudate from the mangrove trees enriches the water with nutrients and promotes sustainable fisheries. Further, during the 2004 tsunami, mangroves served as speed breakers and saved people from the fury of the tidal waves. He said that everyone in the village now understands the symbiotic relationship between mangroves and coastal communities. It is clear that hereafter mangroves in this region will be in safe hands.

Another example relates to the tribal families of Kolli Hills in Tamil Nadu. The local tribal population had been cultivating and conserving a wide range of millets and medicinal plants. However, due to the absence of a market for traditional foods, they had to shift to more remunerative crops like tapioca and pineapple. The millet crops cultivated and consumed by them for centuries were rich in protein and micronutrients. They were also much more climate resilient, since mixed cropping of millets and legumes minimizes risks arising from unfavourable rainfall. Such risk distribution agronomy is the saviour of food security in an era of climate change. How, then, can we revitalize the conservation traditions of tribal families, without compromising their economic well-being? Scientists at the M. S. Swaminathan Research Foundation (MSSRF) started a programme designed to create an economic stake in conservation, by both value-addition to primary products and by finding niche markets for their traditional food grains. Commercialization thus became the trigger for conservation. Today, many of the traditional millets are once again being grown and consumed. They now proudly sing, 'biodiversity is our life', which is also the key message of the International Year for Biodiversity.

A third example relates to the tribal areas of the Koraput region of Orissa, which is an important centre of rice diversity. Fifty years ago, there were over 3500 varieties of rice in this area. Now, this number has been reduced to about 300. To save these 300 varieties, it is essential that the tribal families derive some economic benefit from the preservation of such rich genetic variability in rice. Now, the local population, in partnership with scientists, has developed improved varieties like Kalinga Kalajeera, which fetch a premium price in the market. For too long, tribal and rural families have been conserving genetic resources for public good at personal cost. It is time that we recognize the importance of promoting a genetic conservation continuum, starting with *in situ* on-farm conservation of landraces by local communities, and extending with the preservation of a sample of genetic variability under permafrost conditions at locations like Svalbard near the North Pole, maintained by the government of Norway, or Chang La in Ladakh, India, where the Defence Institute of High Altitude Research (DIHAR), has created a germplasm storage facility under permafrost conditions at an altitude of 5360m.

How can we harness biodiversity for poverty alleviation? Obviously, this can be done only if we can convert biodiversity into jobs and income on a sustainable basis. Several institutional mechanisms have been developed at MSSRF for this purpose, such as biovillages and biovalleys. In biovillages, the conservation and enhancement of natural resources like land, water and biodiversity become priority tasks. At the same time, the biovillage community aims to increase the productivity and profitability of small farms and create new livelihood opportunities in the non-farm sector. Habitat conservation is vital for preventing genetic erosion. In a biovalley, the local communities try to link biodiversity, biotechnology and business in a mutually reinforcing manner. For example, the Herbal Biovalley, under development in Koraput, aims to conserve medicinal plants and local foods and convert them into value-added products based on assured and remunerative market linkages. Such sustainable and equitable use of biodiversity leads to an era of 'biohappiness'. Tribal families in Koraput have formed a 'Biohappiness Society'.

There is need for a Biodiversity Literacy Movement to be launched, so that from childhood onwards everyone is aware of the importance of diversity for the maintenance of food, water, health and livelihood security, as well as a climate-resilient food production system. The government of India has started programmes like DNA and Genome Clubs to sensitize school children about the importance of conserving biodiversity. Wherever there is strong interaction between biodiversity and cultural diversity, we see rich agrobiodiversity, i.e. diversity which is economically valuable and life sustaining. The government of India has also started recognizing and rewarding the contributions of rural and tribal families in the field of genetic resources conservation through Genome Saviour Awards. We need similar awards for those who are conserving breeds of animals, forests and fishes. National governments must ensure that all development programmes are subjected to a biodiversity impact analysis, so that economic advance is not linked to biodiversity loss. *Ex situ* preservation in cryogenic genebanks is no substitute for *in situ* conservation. This is why the present manual places emphasis on *in situ* conservation which will lead to both preservation and continuous evolution.

Our gratitude goes to Professor Vernon Heywood, Dr Danny Hunter and colleagues for their labour of love for biodiversity conservation and sustainable food security. I hope this book will be widely read and used for saving plants in order to save lives and livelihoods.

M.S. Swaminathan, FRS
Chairman, M.S. Swaminathan Research Foundation
World Food Prize, Mahatma Gandhi Gold Medal

List of Acronyms and Abbreviations

AFLP	amplified fragment length polymorphism
ANGAP	National Association for the Management of Protected Areas in Madagascar
AOO	area of occupancy
APBCI	Área de Patrimonio Biocultural Indígena
AVRDC	The World Vegetable Centre
BGCI	Botanic Gardens Conservation International
BRAHMS	Botanical Research and Herbarium Management System
CBD	Convention on Biological Diversity
CBNRM	community-based natural resource management
CCA	community conserved areas
CEM	climate envelope modelling
CENARGEN	Centro Nacional de Pesquisas de Recursos Genéticos e Biotecnologia, Brazil
CGIAR	Consultative Group on International Agricultural Research
CIAT	International Centre for Tropical Agriculture
CIFOR	Centre for International Forestry Research
CIP	International Potato Centre/Centro Internacional de la Papa
CITES	Convention on International Trade in Endangered Species of Wild Fauna and Flora
CLT	Conservation Land Trust
CMPA	collaboratively managed protected areas
CR	critically endangered
CSIRO	Commonwealth Scientific and Industrial Research Organisation
CWR	crop wild relatives
CWR-GRIS	CWR-Genetic Resources Information System
CWR SG	Crop Wild Relative Specialist Group
DD	data deficient
EC	European Commission
EN	endangered
ENSCONET	European Native Seed Conservation Network
EOO	extent of occurrence
ePIC	electronic Plant Information Centre
ESPC	European Strategy for Plant Conservation

EUFORGEN	European Forest Genetic Resources Programme
EW	extinct in the wild
EX	extinct
FAO	Food and Agriculture Organization of the United Nations
FFI	Fauna & Flora International
FOFIFA	Centre National de la Recherche Appliquée au Développement Rural, Madagascar
GAIN	Global Alliance for Improved Nutrition
GBIF	Global Biodiversity Information Facility
GCDT	Global Crop Diversity Trust
GCF	gene conservation forest
GCM	general circulation model
GEF	Global Environment Facility
GELOSE	Gestion Locale Sécurisée
GENRES	Information System Genetic Resources
GIS	geographic information system
GISIN	Global Invasive Species Information Network
GISP	Global Invasive Species Programme
GMZ	gene management zone
GNP	gross national product
GPA	Global Plan of Action for the Conservation and Sustainable Utilization of Plant Genetic Resources for Food and Agriculture
GRIN-USDA	Germplasm Resources Information Network-United States Department of Agriculture
GSPC	Global Strategy for Plant Conservation
HCP	habitat conservation planning
IABIN	Inter-American Biodiversity Information Network
IAEA	International Atomic Energy Agency
IAS	invasive alien species
IBPGR	International Board for Plant Genetic Resources (now Bioversity International)
ICCA	indigenous and community conserved area
IFPRI	International Food Policy Research Institute
IFS	International Foundation for Science
IGNARM	Network on Indigenous peoples, Gender and Natural Resource Management, Denmark
IIED	International Institute for Environment and Development
ILDIS	International Legume Database and Information System
ILPA	indigenous lands protected area
INGO	international non-governmental organization
INIBAP	International Network for the Improvement of Banana and Plantain
IPCC	Intergovernmental Panel on Climate Change
IPGRI	International Plant Genetic Resources Institute (now Bioversity International)

IPNI	International Plant Names Index
IRAP	inter-retrotransposon amplified polymorphism
IRRI	International Rice Research Institute
ITIS	Integrated Taxonomic Information System
ITPGRFA	International Treaty on Plant Genetic Resources for Food and Agriculture
IUCN	International Union for the Conservation of Nature
LC	least concern
MAB	Man and the Biosphere Programme
MASH	minimum available suitable habitat
MDG	Millennium Development Goals
MEP	minimum effective population
MOU	memoranda of understanding
MSB	Millennium Seed Bank, Royal Botanic Gardens, Kew
MVM	minimum viable metapopulation
MVP	minimum viable population
NBASP	National Biodiversity Strategy and Action Plan
NE	not evaluated
NGO	non-governmental organization
NISCWR	National Information System of Crop Wild Relatives
NPAS	National Protected Areas System
NPGS	National Plant Germplasm System
NT	near threatened
NTFP	non-timber forest product
OCR	optical character recognition
OECD	Organisation for Economic Co-operation and Development
PGR	plant genetic resources
PGRFA	plant genetic resources for food and agriculture
PLA	participatory learning and action
PMR	plant micro-reserve
PRA	participatory rural appraisal
PROINPA	Fundación para la Promoción e Investigación de Productos Andinos, Bolivia
PVA	population viability analysis
RECOFTC	Regional Community Forestry Training Centre for Asia and the Pacific
REDD	reducing emissions from deforestation and forest degradation
REMAP	retrotransposon-microsatellite amplified polymorphism
RRA	rapid rural appraisal
SGRP	System-wide Genetic Resources Programme
SNCO	state non-commercial organization
SSAP	sequence-specific amplified polymorphism
SSR	simple sequence repeat
TBA	transboundary protected area
TCP	The Climate Project

TDWG	Biodiversity Information Standards, formerly known as Taxonomic Database Working Group
TNC	The Nature Conservancy, USA
UNDP	United Nations Development Programme
UNEP	United Nations Environmental Programme
UNEP-WCMC	United Nations Environment Programme World Conservation Monitoring Centre
UNESCO	United Nations Educational, Scientific and Cultural Organization
UNFCCC	United Nations Framework Convention on Climate Change
UN-REDD	United Nations Collaborative Programme on Reducing Emissions from Deforestation and Forest Degradation in Developing Countries
UPOV	International Union for the Protection of New Varieties of Plants
USDA	United States Department of Agriculture
WCPA	World Commission on Protected Areas
WDR	World Development Report
WIEWS	World Information and Early Warning System of Plant Genetic Resources for Food and Agriculture, FAO
WRI	World Resources Institute
WWF	World Wide Fund for Nature

Part I

Introduction

This part sets the scene for Crop Wild Relative (CWR) *in situ* conservation. It outlines the different approaches to defining CWR, describes the importance of these species, makes a case for their conservation in the wild and illustrates the challenges involved in setting up actions to target their preservation.

Chapter 1

Introductory and Background Material

Without continued genetic enhancement using diverse germplasm from both wild and modified sources, the gains in crop yields obtained over the past seven decades are not sustainable, and yields might eventually grow more slowly or even decline. Agricultural production increasingly relies on 'temporal diversity,' changing varieties more frequently to maintain resistance to pests and diseases (Rubenstein et al, 2005).

Introduction: Crop wild relatives (CWR)

Crop wild relatives (CWR) collectively constitute an enormous reservoir of genetic variation that can be used in plant breeding and are a vital resource in meeting the challenge of providing food security, enhancing agricultural production and sustaining productivity in the context of a rapidly growing world population and accelerated climate change. They occur in a wide range of habitats but as numerous assessments testify, habitats continue to be lost or degraded across the world, putting many of these species at risk. It is therefore essential that urgent steps are taken to conserve them both in the wild (*in situ*) and in genebanks (*ex situ*) while the genetic diversity they contain is still available.

What are genetic resources?

Genetic resources were traditionally defined as genetic material (alleles) of known value used in plant or animal improvement, but the meaning has been widened by the Convention on Biological Diversity (CBD) to mean *any material of plant, animal, microbial or other origin containing functional units of heredity, of actual or potential value*. It thus covers both living (e.g. seeds) and preserved material (e.g. herbarium or museum specimens). The International Treaty on Plant Genetic

Resources for Food and Agriculture (ITPGRFA) adopts a similar definition. Crop Wild Relatives are a key component of plant genetic resources for food and agriculture.[1]

What is a crop wild relative?

In general terms, a crop wild relative (CWR) may be defined as a wild plant species that is more or less closely related to a particular crop and to which it may contribute genetic material, but unlike the crop species has not been domesticated (Heywood et al, 2007). It is difficult to give a more precise definition, yet we need one if we are to be able to assess how many CWR exist both nationally and globally. Being a CWR is a matter of degree – some are more closely related than others to the crop. Two ways of describing this relationship have been employed – genecological – based on the extent to which they can exchange genes with the crop – and taxonomic – based on their taxonomic relationship with the crop (see Table 1.1). The genecological approach often uses the Harlan and de Wet (1971) gene pool concept to define the degree of relatedness, based on the relative ease with which genes can be transferred from them to the crop. In the complete or partial absence of genetic data or information on crossability, use of the taxon group concept has been proposed by Maxted et al (2008), which relies on the likelihood of the existing taxonomic classification reflecting a degree of genetic relationship or crossability.

For the purposes of the United Nations Environment Programme (UNEP)/Global Environment Facility (GEF) CWR Project described in this manual (see p19), a CWR was defined as any species belonging to the same genus as the crop, based on the argument that species judged to be sufficiently

Table 1.1 *Taxonomic and genecological definitions of CWR*

Gene pool concept of CWR

Primary gene pool (GP1)
Contains close relatives that readily intercross with the crop

Secondary gene pool (GP2)
Contains all the biological species that can be crossed with the crop but where hybrids are usually sterile

Tertiary gene pool (GP3)
Comprises those species that can be crossed with the crop only with difficulty and where gene transfer is usually only possible with radical techniques

Taxon group concept of CWR

Taxon Group 1a – crop
Taxon Group 1b – same species as crop
Taxon Group 2 – same series or section as crop
Taxon Group 3 – same subgenus as crop
Taxon Group 4 – same genus as crop
Taxon Group 5 – different genus to the crop

similar to belong to the same genus are likely to be related genetically. A similar approach has been proposed by Meilleur and Hodgkin (2004) who suggest as a definition 'CWRs should include the wild congeners or closely related species of a domesticated crop or plant species, including relatives of species cultivated for medicinal, forestry, forage, or ornamental reasons'. A number of other recent major CWR projects follow this approach. Such a broad definition leads to large numbers of species being considered CWR. For example, Kell et al (2008) found that around 83 per cent of the Euro-Mediterranean flora comprises crop and CWR species. Faced with handling such large numbers of CWR, a priority determining mechanism needs to be used to select which species will be the subject of particular conservation actions (see Chapter 7). CWR are a very diverse group of plants and occur in a wide variety of habitats. They range from forest trees and shrubs to climbers, perennials, biennials and annuals. Some of them are widespread and may even occur as weeds while others have scattered or restricted distributions and some of them are rare and endangered.

Landmark events – a bit of history

Although genes from CWR have almost certainly been used in the development of crops from early times, recorded use of CWR in commercial plant breeding dates back to the end of the 19th century (Hodgkin and Hajjar, 2008) and the potential significance of CWR in plant breeding and crop improvement was recognized by Vavilov and other pioneers[2] of the genetic resources movement. Wider recognition of the value of genes from CWR in conferring desirable characteristics in crop cultivars developed in the 1940s and 1950s (see Hajjar and Hodgkin, 2007, for a summary of the early uses of CWR). It was not, however, until the 1960s that active steps were made to undertake coordinated conservation of the genetic diversity represented by landraces, local ecotypes and wild relatives of crops. The recommendations made by the Food and Agriculture Organization of the United Nations (FAO) Technical Meeting in Rome in 1961 represented a key development (Bennett, 1965). It recognized 'the great importance to this and future generations of preserving the gene pool of genetic variability which now occurs in the major gene-centres of the world, but which is threatened with destruction'. The FAO recommended the establishment of International Crop Centres within the gene-centres to be charged with the task of fully exploring the genetic potential of their respective regions on the basis of detailed local knowledge, of assessing and maintaining basic collections of crops and local races and of wild forms, and of setting up areas in genetic conservation to be managed in such a way as to preserve the evolutionary potential of local population–environment complexes (Bennett, 1965). The International Institute in Izmir (the Izmir Centre), Turkey, was established in 1964 with such terms of reference (Sencer, 1975).

In the 1970s and 1980s, there was increasing recognition of CWR as a significant component of plant genetic resources. In tune with the times, the main focus was on the collection and *ex situ* conservation of samples of genetic diversity, activities which accelerated in the mid-1980s, probably as a consequence of the introduction of ecogeographic surveying. It was only in the 1980s that a small number of agricultural and forestry scientists began to actively target CWR for *in situ* conservation, probably due to a growing awareness of habitat and species decline, followed by calls for the conservation of CWR by prominent international and conservation organizations. Although some time and resources began to be allocated to studying the possibilities of *in situ* CWR conservation, the necessary cross-sectoral approach was often lacking. A number of scientific meetings and publications followed, dealing with various aspects of *in situ* CWR conservation during the 1980s.

The entry into force of the Convention on Biological Diversity (CBD) in 1993, the endorsement of the Global Plan of Action for the Conservation and Sustainable Use of Plant Genetic Resources for Food and Agriculture (GPA) in 1996 and the International Treaty on Plant Genetic Resources (ITPGRFA) in 2001, whereby signatory countries adopted *in situ* CWR conservation as a national priority, and a series of books on *in situ* CWR conservation theory and methods, as well as some on-the-ground field projects, provided added impetus to our appreciation and understanding of the importance of CWR (Meilleur and Hodgkin, 2004).

Landmark publications on CWR

One of the first publications to draw attention to the importance of conserving CWR was the booklet *Conserving the Wild Relatives of Crops* by Erich Hoyt, published by the International Union for Conservation of Nature (IUCN), IBPGR [later to become IPGRI and today Bioversity International] and the World Wide Fund for Nature (WWF) in 1988.[3] Much of what it says is still valid and Hoyt's statement, 'The conservation of crop genetic resources – the plants that feed us and their wild relatives – is one of the most important issues for humankind today', remains true to this day. A major review of the use of CWR was published by Prescott-Allen and Prescott-Allen (1988).

A significant, although frequently overlooked, publication is the booklet *Plant Genetic Resources: Their Conservation* in situ *for Human Use* (FAO, 1989), which arose out of a decision taken during the first meeting of the ad hoc working group on *in situ* conservation of the Ecosystems Conservation Group in 1986, including members from FAO, the United Nations Educational, Scientific and Cultural Organization (UNESCO), UNEP, the IUCN and the International Board for Plant Genetic Resources (IBPGR). This included a series of cases studies from around the world, illustrating action planned or underway in *in situ* conservation of plant genetic resources.

Other important resources are the proceedings of the workshops initiated by the Council of Europe on 'Conservation of the Wild Relatives of European Cultivated Plants' (Valdés et al, 1997), which were held in Faro (Portugal), Neuchâtel (Switzerland) and Gibilmanna-Palermo (Sicily, Italy), and addressed a wide range of issues concerning the genetics, demography, ecology, conservation, management and protection of genetic variability through a series of case studies.

A further valuable resource is the global survey of *in situ* conservation of wild plant species (Heywood and Dulloo, 2005) that arose out of another UNEP/GEF-supported project 'Design, Testing and Evaluation of Best Practices for *In Situ* Conservation of Economically Important Wild Species'.

An additional landmark publication is *Crop Wild Relative Conservation and Use* (Maxted et al, 2008) which arose out of the first international conference on CWR, organized within the framework of the European Commission (EC)-funded Plant Genetic Resources (PGR) Forum project and held in Agrigento, Sicily, Italy in September 2005.[4]

The second report on the *State of the World's Plant Genetic Resources for Food and Agriculture*[5] was endorsed at the 12th Session of the Commission on Genetic Resources for Food and Agriculture (Rome, 18–23 October 2009). It updates the first report with the best data and information available, through a participatory process, and with a focus on changes that have occurred since 1996; the report provides a concise assessment of the status and trends of plant genetic resources for food and agriculture (PGRFA) and identifies the most significant gaps and needs in order to provide a basis to update the rolling Global Plan of Action. It contains several references to CWR, especially Section 1.2.3: Changes in the status of crop wild relatives; Section 2.2.1: Inventory and state of knowledge; and 2.2.2: *In situ* conservation of crop wild relatives in protected areas. Salient points are:

- while many new priority sites for conserving CWR have been identified around the world during the last decade, largely as a result of ecogeographic surveying, many species remain under threat as a result of land degradation, changes in land-use practices and other factors;
- since the publication of the first State of the World Report, most countries have carried out specific surveys and inventories of PGRFA, but the majority have been confined to single crops, small groups of species or limited areas;
- very little survey or inventory has been done on PGRFA in protected areas as compared with other components of biodiversity in these areas and *in situ* conservation of wild species continues to be an unplanned result of efforts to protect particular habitats or charismatic species; and
- relatively few countries have been active in conserving wild PGRFA in protected areas although some progress has been made.

The creation in 2003 of the Crop Wild Relative Specialist Group (CWR SG)[6] within the IUCN Species Survival Commission provided a network for those interested in the conservation and sustainable use of CWR. It publishes a regular newsletter, *Crop Wild Relative*.[7]

The value and use of CWR

The value of CWR is evident from the use that has been made of them in crop improvement, especially in the last few decades. In a recent review of their use,

Maxted and Kell (2009) cited 91 articles that reported the identification and transfer of useful traits from 185 CWR taxa into 29 crop species (see Figure 1.1). They found that the degree to which breeders had used CWR diversity varied markedly between crops, both in terms of CWR taxa usage and number of citations of CWR usage reported. The use of CWR has been particularly notable in barley, cassava, potato, rice, tomato and wheat. The crops in which CWR have been most widely used are rice and wheat, both in terms of the number of CWR taxa and number of successful attempts to introgress traits from the CWR to the crop.

The key to successful crop improvement is a continued supply of genetic variability and beneficial traits contained in this diversity (Dwivedi et al, 2008), and wild relatives of modern crops are the source of much of this novel diversity. It is not widely realized how high the turnover rate of cultivars is in many crops as a consequence of losing, for example, resistance or tolerance or because of the need for continual innovation. For example, in tomato (*Lycopersicum esculentum*) the average turnover time of commercial cultivars is approximately five years, largely because seed companies must continuously develop new cultivars with added value and hence commercial tomato breeding is very innovative (Bai and Lindhout, 2007).

The deployment of innovative biotechnology tools provides new opportunities to make greater and more effective use of wild species in crop improvement (Tanksley and McCouch, 1997; Dwivedi et al, 2007). The latter argue that, 'the tools of genome research may finally unleash the genetic potential of our wild and cultivated germplasm resources for the benefit of society.' Genes from wild plants have so far provided cultivars with resistance against pests (e.g. Malik et al, 2003) and diseases (e.g. Brar, 2005), improved tolerance to abiotic stresses (e.g. Farooq and Azam, 2001), tolerance of extreme temperatures and salinity; and resistance to drought and enhanced nutritional quality (e.g. Kovacs et al, 1998; Dillon et al, 2007). Indeed, modern cultivars of most crops now contain some genes that are derived from a wild relative. For example, genes from several wild species of *Aegilops*, which is closely related to *Triticum*, have been transferred to cultivated wheat, including those that confer resistance to leaf rust, stem rust, powdery mildew and nematodes (Schneider et al, 2008); many other valuable genetic resources in *Aegilops* species remain untapped. Likewise, wild rice species have proven to be important gene reservoirs that can be used to increase domesticated rice yield, quality and resistance to diseases and insects. They have furnished genes for the hybrid rice revolution, exhibit yield-enhancing traits and have shown tolerance to biotic and abiotic stress (Brar and Khush, 1997; Xiao et al, 1998). In Sri Lanka, wild *Oryza nivara* is being used to breed resistance to the pest brown plant hopper into cultivated rice varieties (see Box 1.2). In cotton (*Gossypium*), the narrow genetic base of the primary cotton breeding gene pool is one of the major constraints in cotton breeding programmes worldwide. This underlies the necessity to enrich the gene pool with genetic diversity from landraces and CWR (Abdurakhmonov et al, 2007). The use of CWR in breeding stress- and disease-resistant cotton in Uzbekistan is summarized in Box 1.4.

Box 1.1 Examples of the use of CWR

In tomato, extensive use has been made of the genetic variation present in wild species (Rick and Chetelat, 1995; Bai and Lindhout, 2007; Robertson and Labate, 2007) in developing today's commercial varieties. Over 130 genes associated with drought responsiveness have been identified at AVRDC (The World Vegetable Center) and those from its wild relatives in the Chilean deserts are being introgressed into commercial lines. However, compared with the rich reservoir in wild species, the cultivated tomato is genetically poor and it is estimated that the genomes of tomato cultivars contain only 5 per cent of the genetic variation of their wild relatives (Miller and Tanksley, 1990). It is expected that the potential of tomato breeding using only cultivated germplasm will reach a ceiling, necessitating that future plant breeding initiatives explore the diversity available in related wild species (see review by Bai and Lindhout, 2007). With techniques like EcoTILLING,[8] allele mining will greatly facilitate the identification of useful genes in wild tomato germplasm (Comai et al, 2004).

It is clear that CWR represent a vast unexplored potential for future crop improvement. For example, in wild emmer wheat (*Triticum turgidum* subsp. *dicoccoides*) accessions, Chatzav et al (2010) found wide genetic diversity for all grain nutrients, with the concentrations of grain zinc, iron and protein being twice as much in wild accessions as in domesticated genotypes. They consider that wild

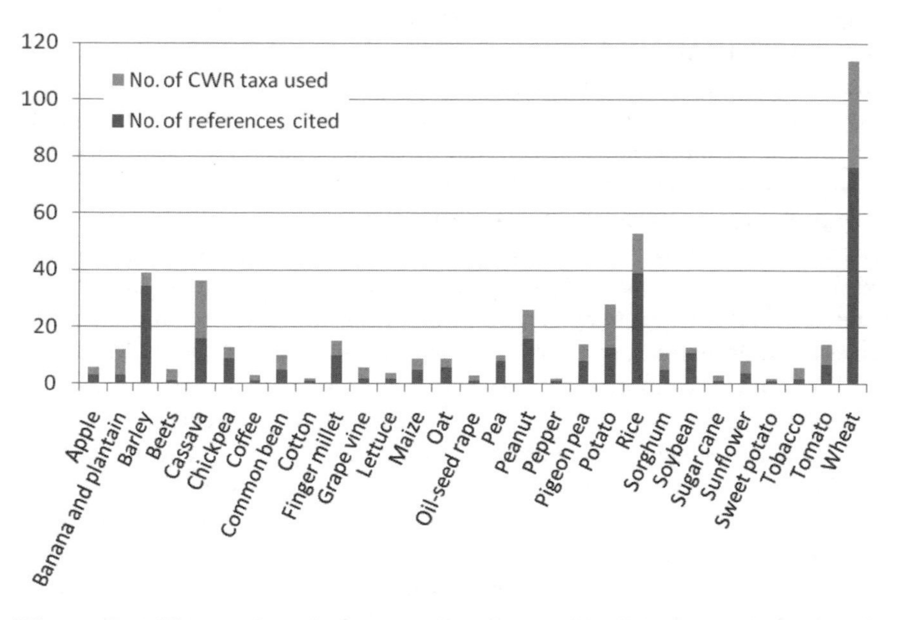

Figure 1.1 *The number of references reporting the identification and transfer of useful traits from 185 CWR taxa to 29 crop species, showing the number of CWR taxa used in each crop*

Source: Maxted and Kell, 2009

Figure 1.2 *Crossing cultivated rice with wild* Oryza nivara *at the Rice Research and Development Institute, Batalagoda, Sri Lanka*

emmer germplasm offers unique opportunities to exploit favourable alleles for grain nutrient properties excluded from the domesticated wheat gene pool. In maize (*Zea mays*), Ortiz et al (2009) found that only a small portion of the wide array of genetic diversity found in wild relatives of the crop is represented in current elite breeding pools. Given that growing demands for food production, feed and bio-energy are estimated to require a 2 per cent annual increase in global maize production, it can be expected that the diversity found in CWR will be tapped by breeders to meet these needs. On the other hand, as Hajjar and Hodgkin (2007) point out, CWR have contributed less than might be expected to the development of new cultivars, despite improved procedures for intercrossing species from different gene pools, advances in molecular methods for managing backcrossing programmes, increased numbers of wild species accessions in genebanks and the substantial literature available on beneficial traits associated with wild relatives. Heywood et al (2007) suggest the main reasons for the neglect of CWR conservation have to do with practicality, priorities and economics. There is, in fact, widespread uncertainty as to the benefits to be obtained from CWR *ex situ* and, especially, *in situ* conservation.

It is exceedingly difficult to quantify the monetary or commercial benefits to be obtained from the conservation and use of plant genetic resources and of CWR in particular (see NRC, 1991a, 1993; Rubenstein et al, 2005). It has been suggested that, on average, genetic contributions from wild species increase crop productivity by about 1 per cent each year, and this increase in productivity has

Box 1.2 Rice breeding programme with wild Oryza nivara in Sri Lanka

Brown plant hopper (BPH) is one of the major pests of rice in Sri Lanka. Annually, it affects an average of 5–10 per cent of the extent of total paddy cultivation. Presently, BPH resistance is incorporated into all new rice varieties; the source of the resistance was found decades ago in rice variety PTB 33. Due to continued use of the single resistance source, new biotypes of BPH have developed and the crop's resistance has been compromised. Rice breeders in Sri Lanka have been looking for a new source of resistance and have investigated wild rice as a possible genetic resource. There are five wild *Oryza* species in Sri Lanka, namely *O. nivara*, *O. rufipogon*, *O. eichingeri*, *O. rhizomatis* and *O. granulata*. Of these five species, *O. nivara* and *O. rufipogon* are in the same genome group as cultivated rice, *Oryza sativa*. Hence, both species are relatively easy to hybridize with cultivated rice.

With assistance from the UNEP/GEF Crop Wild Relatives project, plant breeders at the Central Rice Research and Development Institute in Sri Lanka collected 40 different accessions of *O. nivara* during 2006–2008. These accessions were tested for BPH resistance using standard screening procedures, and it was found that 3 accessions were highly resistant to BPH while 15 accessions were within the moderately resistant category. It was found that these three accessions survived even after the death of the resistant variety PTB 33 from the intensity of BPH attack, indicating the resistance in the three *O. nivara* accessions was different from that of PTB 33. Ten crosses were made between *O. nivara* and cultivated rice and eight were successful. Forty-two F_1 seeds were obtained from the successful crosses. All F_1 seeds were germinated and produced seeds, but only 10 per cent of the seeds were filled. Screening of the F_2 generation for resistance showed 30 per cent of the seedlings were resistant to BPH. F_3 seed formation from resistant lines resulted in 60 per cent filled seeds and F_3 screening results revealed that 50 per cent of seedlings were resistant to BPH. In the F_4 generation, empty seeds were reduced to 10 per cent and 92 per cent of seedlings were resistant to BPH. Currently, seeds of the F_6 generation have been harvested and are being used as parental material in the National Rice Breeding Programme. Yield observations of the new lines are expected to be conducted shortly. Rice Breeder: P.V. Hemachandra.

been valued at US$1 billion (NRC, 1991b). Some idea of the scale of benefits may, however, be obtained from published estimates referring to a selected number of crops. For example, the desirable traits of wild sunflowers (*Helianthus* spp.) are worth an estimated US$267 to US$384 million annually to the sunflower industry in the United States; one wild tomato variety has contributed to a 2.4 per cent increase in solids content worth US$250 million; and three wild peanuts have provided resistance to the root knot nematode, which costs peanut growers around the world US$100 million each year. Of course, the commercial contribution of the majority of CWR is likely to be on a much smaller scale.

Examples of CWR from the UNEP/GEF project countries and their desirable traits are given in Table 1.2.

Table 1.2 *Wild species being evaluated for their potential to improve the tolerance of their crop relatives to biotic and abiotic stresses as part of the UNEP/GEF project*

Country	*Wild relative of*	*Desirable traits*
Armenia	Wheat, pear	Resistance to adverse environmental conditions
Bolivia	Potato, quinoa, cañahua (*Chenopodium pallidicaule*)	Pest and diseases resistance of selected species from three genera Nutritious properties of quinoa and cañahua
Madagascar	Coffee, rice, yam	No or low caffeine, high content of chlorogenic acid Resistance to rice yellow mottle virus (RYMV) Potential for domestication
Sri Lanka	Rice	Resistance to biotic and abiotic stresses
Uzbekistan	Apple, pistachio	Resistance to adverse environmental conditions

Source: http://www.underutilized-species.org/Documents/PUBLICATIONS/sbstta_cwr_final.pdf

Box 1.3 Breeding potential of CWR in Madagascar

Rice breeders from the Centre National de la Recherche Appliquée au Développement Rural (FOFIFA) managed to obtain approximately 100 lines derived from inter-specific crosses with the wild species *Oryza longistaminata* and the cultivated species *Oryza sativa*, as well as multiple back crosses from the hybrid plant. They are different phenotypes, consistent and stable, and are believed to possess the genes of *Oryza longistaminata* in their gene pool. These lines are selected primarily for their trait of resistance to the rice yellow mottle virus (RYMV), which makes the panicles sterile, causing a drop in grain yield. It is transmitted mechanically by contact and by insects, mainly *Trichispa sericea* or *Hispa gestroy*. The disease occurs in the rice producing regions of the north Andapa Basin, northwest and west of the island. It has not been identified in the highlands, but it may be occasionally observed in the region of Lake Alaotra, especially during high rainfall periods, and more rarely in the southwest. It was observed that the wild species *Oryza longistaminata* is never attacked by the disease. However, many defects are observed, since it has rhizomes like a weed. Its seeds have a very low percentage of fertility and shatter easily, even when immature. In addition, its panicles are very loose, and the stigma is extruded. Recently, the prospect of improvement through inter-specific crossing between the wild species and the cultivated species *Oryza sativa* has become feasible. The goal is to introgress resistance to RYMV from the wild relative to the cultivated lines, while avoiding the inclusion of disadvantageous traits. Several attempts with 100 different crosses with cultivated lines have already been made, but they were not successful as there was no fertilization, the embryo being aborted before maturity. Although hybridization between the two species was a very laborious process, it was possible to fertilize a spikelet using a cultivated line 'Miandry Bararata' as a female parent and the wild species as pollinator. The resulting embryo was immature and needed a suitable culture medium to result in an adult plant with intermediate phenotype. The F_1 plant obtained possessed rhizomes and further backcrossing followed using multiple crosses with other lines to eliminate or reduce this disadvantageous feature.

Source: Rakotonjanahay Xavier pers.comm. to J. Ramelison (April 2008)

Box 1.4 Use and potential of cotton CWR in Uzbekistan

The Institute of Genetics and Experimental Plant Biology in Uzbekistan holds a collection of 45 wild cotton species and forms of *Gossypium*. The genetic potential of wild cotton relatives was used in inter-species hybridization whereby valuable features of wild species were successfully transmitted into cultivated species. Complex synthetic hybrids were created on the basis of trigenetic hybrids of *G. hirsutum* × (*G. harknessii* × *G. thurberi*) and prospective hybrid lines were obtained as the result of *G. hirsutum* × (*G. thurberi* × *G. raimondii*) crosses. These hybrids possess valuable features such as high fertility and fibre quality. Wild relatives of cultivated cotton species represent very valuable material with potential for adaptation, through resistance to environmental stress factors and agricultural pests. Wilt-resistant forms of *G. hirsutum* subsp. *mexicanum* and ruderal forms of *G. hirsutum* 'El Salvador' were used in breeding programmes as the basis for the creation of a series of new forms. Wild accessions of *G. herbaceum* L. and *G. arboreum* L. are characterized by hygroscopic fibres of high quality. They were used as donors in genetic breeding programmes to create intra- and inter-specific forms. *G. hirsutum* L. was used in obtaining wilt-, heat- and drought-resistant varieties (subsp. *mexicanum* var. *nervosum*, subsp. *punctatum*) and *G. barbadense* L. was used as the basis for the salt-resistant variety *G. barbadense* subsp. *darwinii*. Wild cotton relatives which were used to produce synthetic hybrids with valuable features are shown in Figure 1.3.

Source: Sativaldi Djataev

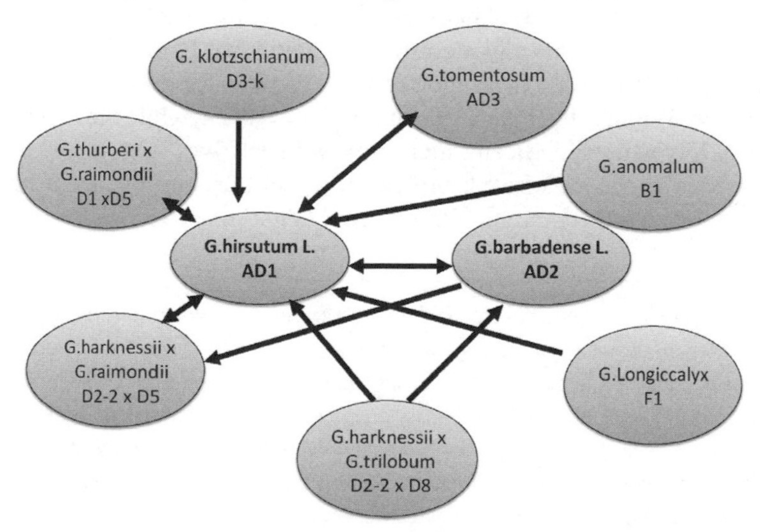

Figure 1.3 *Relationships of synthetic hybrids of cotton produced in Uzbekistan*

Source: Sativaldi Djataev

Why is *in situ* conservation of CWR important?

Despite the fact that the importance of *in situ* conservation for CWR has been widely recognized, until recently the main conservation strategy of the plant genetic resource sector has been to collect material of cultivars, landraces and, to a lesser extent, CWR and to store these material *ex situ* in genebanks for use or potential use in plant breeding (see Chapter 12). Little attention was paid to *in situ* approaches. Although a handful of reserves for the *in situ* conservation of CWR were established in the 1980s – the Sierra de Manantlán Biosphere Reserve for the maize wild relative, *Zea diploperennis*, in Mexico; the Erebuni Reserve in Armenia and the Ammiad Project Reserve in Israel for wheat wild relatives; and the National Citrus Gene Sanctuary-cum-Biosphere Reserve in the West Garo Hills, India, for citrus wild relatives – only in the last 10–15 years have serious efforts been made to conserve CWR in their natural wild habitats (*in situ*). In a major GEF/World Bank project on conservation of genetic diversity in Turkey (Tan and Tan, 2002), a wide range of crop wild relatives (*Triticum*, *Lens*, *Pisum*, *Castanea*, *Abies* and *Pinus*) were selected as target species for *in situ* conservation in 'gene management zones' (GMZs) – natural and semi-natural areas set aside for maintaining genetic diversity in a natural setting for the species of interest.

Practical experience is therefore very limited and there are no generally agreed procedures to follow. The reason the genetic resources sector is now paying attention to the conservation of CWR *in situ* is due to the recognition that such initiatives allow CWR to remain in their natural surroundings with associated species where populations can not only be maintained as a source of potentially useful variation for crop improvement, but also to continue to evolve and generate new variation, some of which might be valuable for use in future breeding efforts. There may also be additional economic benefits of *in situ* conservation, as will be discussed later (see Chapter 3). The importance of conserving CWR and other wild plants *in situ* was specifically identified in the Global Plan of Action for the Conservation and Sustainable Utilization of Plant Genetic Resources for Food and Agriculture (1996) under the Plan's Priority Activity Area 4, while the Convention on Biological Diversity specifically mentions 'wild relatives of domesticated or cultivated species' in the indicative list of categories of the components of biological diversity to be identified and monitored given in its Annex 1.

> In situ *conservation is the only practical method presently available to conserve a great variety of ecosystems, species and genes which are today vulnerable, threatened or endangered. In addition to allowing conservation of a range of different species and co-evolution of biological systems,* in situ *conservation of genetic resources can be compatible with their management for the sustained production of goods to meet day-to-day requirements of local populations, such as food, fodder and medicines; and for the harvesting of timber, wood and fuel.*
>
> *Source:* FAO, 1989

Populations of many CWR species occur in existing protected areas, although the absence of proper inventories means detailed information on such species is not available. It may be assumed that because they are found in protected areas, CWR may be afforded some degree of protection, provided the area is well managed. However, as will be elaborated later, this alone does not, in many cases, represent effective *in situ* conservation as some degree of management or intervention targeted at CWR populations is necessary, particularly if the species is threatened. Moreover, reliance on the continued existence of protected areas in their current location is a risky strategy in the face of global change, especially climate change (see Chapter 14). What is more, the majority of CWR occur outside protected areas and there has been little experience thus far of how to safeguard CWR in such a context. It should also be emphasized that *in situ* conservation is not a short-term approach: on the contrary the timescale of concern is effectively open-ended. This presents major logistical, scientific, technical, economic, political and financial challenges for long-term sustainability.

Threats to the maintenance of CWR

As discussed in detail in Chapter 10, like many other wild species, CWR are increasingly threatened, primarily from habitat loss, fragmentation and degradation, changes in disturbance regimes and invasive alien species. An additional threat that must be addressed is the impact of accelerated global change. The loss of genetic material from CWR has profound implications for agriculture. It reduces the potential for continuous improvement in crop productivity and quality and in the ability of crops to adapt to changing environmental conditions. These assets are critical to reduce hunger and poverty across the developing world. Such loss in diversity could be especially serious in areas containing a wide range of wild progenitors and related wild species and may be exacerbated in some regions by the effects of global change such as demographic growth, population movements, changes in disturbance regimes and climate change.

Few studies have yet been made focusing on the impacts of climate change on the survival rates of CWR, but the evidence published to date, based on the use of bioclimatic modelling, suggests many will be at risk (see Box 1.5). There is an urgent need, therefore, to identify priority species and areas for conservation and, as elaborated in Chapter 12, to develop integrated *in situ* and *ex situ* conservation strategies to ensure that the rich genetic diversity of CWR is protected for the benefit of future generations.

The adaptation of crops to gradual change in climatic conditions will require screening of existing cultivars and breeding of new ones for adaptation to drought, temperature stresses, sustained productivity, disease resistance and other factors, highlighting the importance of maintaining the pools of genetic variation in CWR.

Box 1.5 Evaluating the impact of climate change on CWR

The survival of crop wild relatives is now threatened by the impacts of climate change. An evaluation was conducted by Andy Jarvis and colleagues at the International Centre for Tropical Agriculture (CIAT), the Global Biodiversity Information Facility (GBIF) and Bioversity International, using data accessible through the GBIF, of the possible threats posed by climate change on 11 wild gene pools of major crops worldwide, comprising a total of some 343 species.

For each species, data from both herbarium specimens and germplasm accessions were used to determine the potential distribution of each species and, based on 18 global climate models for the year 2050 under gas emissions scenario A2a[9] and assuming unlimited migration, their future geographical distribution was also mapped.

A map was then generated to illustrate the current richness of crop wild relatives, future predicted richness and the predicted change in richness. The map reveals the hotspots of change where significant loss of diversity is expected to occur. These sites, mostly in sub-Saharan Africa, eastern Turkey, the Mediterranean region and parts of Mexico, are priority areas for collection and conservation of genetic resources.

Another study by Lira and colleagues in Mexico used bioclimatic modelling and two possible scenarios of climatic change to analyse the distribution patterns of eight wild cucurbits closely related to cultivated species. The results showed that all eight taxa displayed a marked contraction in area under both climate scenarios and, that under a drastic climatic change scenario, the eight taxa would only be maintained in 29 of the 69 protected areas in which they currently occur.

Source: Jarvis et al, 2008 and Lira et al, 2009

The challenge of *in situ* conservation of CWR

As is evident in later chapters of this manual, the *in situ* conservation of crop wild relatives is a complex and multidisciplinary process and one that creates many challenges and difficulties. Not only are there complex issues to be addressed, such as the location and selection of populations for conservation, demography and size of populations, the nature of threats to both habitats and the CWR populations and how to manage them, the design of genetic reserves and the need for detailed management protocols, but the multiplicity and complexities of national political and administrative structures also render it extremely difficult to implement a common strategy or framework, assuming one could be agreed.

The limited practical experience in conserving CWR *in situ* to date means that there are no generally agreed protocols or recommendations, and good practice is limited by the shortage of successful examples for reference. On the other hand, there is much to be learned from the experience of *in situ* conservation of endangered wild species through recovery programmes in many European countries, the US, Australia and South Africa, supported by extensive conservation biology literature. Also, the forestry sector has been engaged in *in situ* conservation of forest

Box 1.6 Sierra de Manantlán and maize and its wild relatives

The discovery in the mid-1970s of the wild maize – the endemic perennial *Zea diploperennis* – in its natural habitat in Jalisco in western Mexico, led to the establishment of the Sierra de Manantlán Biosphere Reserve in 1987. Populations of the wild annual relative, *Z. mays* subsp. *parviglumis*, and the Tabloncilo and Reventador races of maize traditional in this area, are further targets for conservation. Although limits on external inputs (such as exotic improved germplasm and chemicals) may need to be set so as not to endanger the wild relative, plant geneticists are optimistic that *Z. diploperennis* and the three other taxa can be conserved *in situ*, as long as ways to provide opportunities for the cultivators involved in managing the system continue to be identified. Indeed, research has shown that populations of *Z. diploperennis* virtually require cultivation and grazing in adjacent fields to prosper.

Source: http://www.unesco.org/mab/sustainable/chap2/2sites.htm

genetic resources for several decades with support from FAO, which has reviewed this topic on a regular basis. Unfortunately, there are practically no examples of *in situ* conservation of CWR in the tropics, apart from the establishment of some genetic reserves for various species of fruit trees such as the Gene Sanctuary-cum-Biosphere for citrus in the Garo Hills of Meghalaya in northeast India. This reserve is located within the Nokrek National Park and was created in 1981; it is the first reserve specifically established for the conservation *in situ* of a tropical shrub (Singh, 1981; Smith et al, 1992). Further, in Mexico an *in situ* reserve was created in 1987 within the Biosphere Reserve of the Sierra de Manantlán for *Zea diploperennis*, a wild relative of maize (*Zea mays*) (Box 1.6).

Given the heterogeneity of species, environments, threats and needs, there is certainly no blueprint or 'one size fits all' approach to *in situ* conservation of CWR. While many of the challenges are of a technical nature, there are an equal number of political, institutional, cultural, legal and social issues that must be addressed and resolved. The sectors that must work together, i.e. the agricultural, forestry and environmental agencies, often have no linkages or tradition of collaboration. Frequently, there is no collaborative framework to guide the activities required to support conservation decision-making. The current disconnect existing among such agencies presents considerable challenges for partnership and coordination, as well as for establishing a suitable policy/legal enabling environment for CWR conservation. In addition, there may well be other complex political and social issues related to land ownership/tenure, access to resources and benefit-sharing. Such complexity usually guarantees that obstacles will need to be addressed to integrate CWR conservation into national programmes.

The situation is made more difficult by the fact that CWR are not usually considered to be flagship or iconic species; therefore, attracting interest and resources is a further challenge. As a result, there is often a lack of funding for

CWR research and conservation, as well as for capacity building and training. This, combined with a general lack of information about CWR, results in a limited understanding and awareness of the importance of CWR and the threats posed to their very existence by global change. The term *crop wild relative* is not readily comprehensible to most people and it might be preferable to replace it with another term such as 'gene donor species for crops'.

The way in which CWR are defined and the application of priority-determining mechanisms to focus resources are important issues that have a bearing on the number of candidate species a programme will need to consider, as well as financial and resource implications. The prioritization or selection of areas for CWR conservation also presents its own challenges.

A major limitation most countries and agencies will face when implementing a CWR conservation programme is the capacity and tools to bring together and use existing information. A substantial amount of relevant and useful information is often available within different institutions at both the national and international levels; however, it is typically highly dispersed and difficult to compile. Such information can include: data on species distribution and biology, held in national herbaria and botanic gardens, and in key international collections in other countries (such as the Royal Botanic Gardens, Kew, UK; Missouri Botanical Garden, USA; and Muséum National d'Histoire Naturelle, Paris, France); information on distribution and scope of existing protected areas held nationally and by organizations such as United Nations Environment Programme World Conservation Monitoring Centre (UNEP-WCMC); and information on species status and existing *ex situ* collections, conserved in genebanks. Mapped national survey data from different sources (geography, town planning, soil survey, etc.) provide further information to aid in the conservation planning process through the increasing power of GIS analysis. It should be noted that GBIF is a major repository of georeferenced data used in bioclimatic modelling.

Further, conservation activities often are sponsored by grants from agencies or fall within traditional project implementation and funding cycles, which adds to existing challenges. By their nature, grants and projects are time-bound, presenting obstacles for long-term conservation planning. Project-driven conservation also faces important issues in relation to sustainability and institutionalization of processes and activities, which means when the project finishes so do the activities. This problem may be mitigated to some extent if projects are more locally driven, with close involvement of the stakeholders most directly concerned, so that long-term conservation actions are not mainly dependent on externally funded sources. Some of these issues are dealt with in more detail in Chapters 4 and 5.

Many of the above issues have been addressed in a European context by the EC-funded project 'European Crop Wild Relative Diversity Assessment and Conservation Forum (PGR Forum)' for the assessment of taxonomic and genetic diversity of European CWR and the development of appropriate conservation methodologies (http://www.pgrforum.org/Publications.htm) and by the GEF/World Bank project on conservation of genetic diversity in Turkey (Tan and Tan, 2002).

Box 1.7 Goals of the UNEP/GEF CWR Project

1. To develop international and national information systems on CWR that include data on species biology, ecology, conservation status, distribution, actual and potential uses, conservation actions and information sources.
2. To build the capacity of national partners to use this information for developing and implementing rational and cost-effective approaches to conserving CWR *in situ*.
3. To raise awareness among policy-makers, conservation managers, plant breeders, educators and local users of the potential of CWR for improving agricultural sustainability.

The UNEP/GEF Crop Wild Relatives Project

The Global Environment Facility (GEF) is the financial mechanism for the Convention on Biological Diversity (CBD) and helps countries fulfil their obligations under the CBD. Biodiversity conservation constitutes one of the GEF's major priorities; since 1991, the GEF has invested nearly US$4.2 billion in grants and co-financing for biodiversity conservation in developing countries. Over the last ten years the GEF has supported a number of projects at the national, regional and global levels that seek to enhance the conservation and use of CWR, in line with its goal and objectives (see Box 1.8). Many developing countries, located within centres of plant diversity and centres of crop diversity, contain large numbers of important crop relatives. Although most of these countries have listed the conservation of CWR within their national biodiversity strategies and their agricultural development strategies, they generally possess such limited resources that they have not yet been able to invest in programmes to support the effective conservation and optimum use of CWR. The UNEP/GEF-supported project, '*In situ* conservation of crop wild relatives through enhanced information management and field application' (CWR Project) was specifically designed to address these issues and aims to seek ways of satisfying national and global needs to improve global food security through effective conservation and use of CWR (see Box 1.7). Five countries are involved in the project though their national governments – Armenia, Bolivia, Madagascar, Sri Lanka and Uzbekistan. Each country has significant numbers of CWR, many of which are at risk and in need of conservation. Details of the institutions involved in the partner countries are provided in the acknowledgements section at the beginning of the manual.

To bring the necessary expertise and multidisciplinary skills to bear on a project of this complexity, international partners were identified and invited to collaborate and provide resources and technical support. The international partners are Botanic Gardens Conservation International (BGCI), the FAO, the IUCN and the United Nations Environment Programme World Conservation Monitoring Centre (UNEP-WCMC). The executing agency of the project is Bioversity International (formerly IPGRI).

Box 1.8 Major GEF projects in support of CWR conservation

Kibale Forest Wild Coffee Project (Uganda) – This project assisted Uganda's implementation of its national biodiversity strategy and action plan by helping maintain biodiversity in the landscape mosaics beyond the boundaries of protected areas of global importance.

http://www.gefonline.org/projectDetailsSQL.cfm?projID=490

***In Situ*/On-Farm Conservation and Use of Agricultural Biodiversity (Horticultural Crops and Wild Fruit Species) in Central Asia** (multi-country) – The project provides farmers, institutes and local communities with knowledge, methodology and policies to conserve globally significant *in situ*/on-farm horticultural crops and wild fruit species in Central Asia.

http://www.gefonline.org/projectDetailsSQL.cfm?projID=1025

***In-Situ* Conservation of Andean Crops and their Wild Relatives in the Humahuaca Valley, the Southernmost Extension of the Central Andes** (Argentina) – The project aimed at ensuring that indigenous farmers in the Humahuaca Valley of Argentina adopted improved on-farm conservation and management practices, based on traditional production practices that contribute to *in situ* conservation of selected globally significant Andean crop varieties and their wild relatives

http://www.gefonline.org/projectDetailsSQL.cfm?projID=1732

Conservation and Sustainable Utilization of Wild Relatives of Crops (China) – The project aims at supporting plans to establish protected areas with an integrated and landscape approach and with participation from local communities, so as to secure the wild relatives of soybean, wheat and rice, including their natural habitats.

http://www.gefonline.org/projectDetailsSQL.cfm?projID=1319

***In Situ* Conservation of Native Cultivars and Their Wild Relatives** (Peru) – The project aimed at conserving the agrobiodiversity in one of the world's most important centres of origin of crop and plant genetic diversity. This project targeted 11 important crop species, including several local varieties and wild relatives, for conservation of their genetic diversity within functioning agroecosystems.

http://www.gefonline.org/projectDetailsSQL.cfm?projID=500

***In situ* Conservation of Native Landraces and their Wild Relatives** (Vietnam) – The project targeted the conservation of six important crop groups (rice, taro, tea, litchi-longan, citrus and rice bean) including native landraces and wild relatives in three local ecogeographical areas rich in biodiversity of native landraces and their wild relatives.

http://www.gefonline.org/projectDetailsSQL.cfm?projID=1307

Conservation and Sustainable Use of Cultivated and Wild Tropical Fruit Diversity (Asia) – The aim of the project is to improve the conservation and use of tropical fruit genetic diversity by strengthening the capacity of farmers, local communities and institutions to sustainably manage and utilize tropical fruit trees.

http://www.gefonline.org/projectDetailsSQL.cfm?projID=2430

Box 1.9 National Information System of Crop Wild Relatives of Bolivia

The Bolivian National Information System of Crop Wild Relatives was designed and developed in the framework of the UNEP/GEF project: '*In situ* conservation of crop wild relatives through enhanced information management and field application'. Now operative, the system comprises eight institutional databases, each located at one of the national institutions that participated in the project: three herbaria, three genebanks, one agricultural research institution, and one Organization of the Indigenous Peoples of Bolivia. In addition, the National Portal and GisWeb are part of the system. The databases can be visited online through the National Portal website: http://www.cwrbolivia.gob.bo. The Google Maps application has been customized to function as an integrated GisWeb and is integrated into the National Portal.

The information system contains data on species from 15 genera (*Anacardium, Ananas, Annona, Arachis, Bactris, Capsicum, Chenopodium, Cyphomandra, Ipomoea, Manihot, Phaseolus, Rubus, Solanum, Theobroma, Vasconcellea*), regarding taxonomy, accessions, population and ecology. The database of the system has approximately 3223 records of 190 species, of which 33 species are endemic to Bolivia. It also incorporates a map gallery containing roughly 150 different types of maps, e.g. maps of current and potential distribution of CWR species, collection and other sites, and an image gallery with approximately 152 photos of different CWR species. The National Portal also contains an Atlas of Bolivian CWR.

The information contained in the database is released through the national and international portals, based on a data-sharing agreement between Bioversity International and the government of Bolivia. The system has tools for the identification and prioritization of species, implementation and monitoring of conservation actions and use of CWR. It is also a support tool for decision-makers regarding strategies and policies on CWR in the context of genetic resource management in Bolivia. This information is important to support the improvement of food security in Bolivia and the world.

The immediate objective of the UNEP/GEF CWR Project was to enhance conservation of CWR in each of the project countries. It aimed to achieve this through a series of coordinated components, including the development of a national information system in each country (see Box 1.9 for a description of the Bolivian system), a global information system, enhanced national capacity and conservation actions and public awareness. A major focus of the project was the systematic compilation, enhanced access to and use of information related to CWR. Analysis of this information is a first step towards developing and implementing national-level *in situ* conservation and monitoring strategies. The recently launched Crop Wild Relatives Global Portal (www.cropwildrelatives.org) (see Box 1.10) serves as a gateway through which CWR information can be made widely available. Users can search through databases maintained by national and international partners to obtain information for better decision-making, which leads to more effective conservation and sustainable use of crop wild relatives.

Box 1.10 Information included in the CWR Global Portal

The global UNEP/GEF CWR Project includes a component on information management, an important aspect for enhanced decision-making and conservation. Earlier studies, as well as baseline studies for the project showed that, although information on CWR was available, it was often scattered and hard to access, since it was not in digital format. The five partner countries – Armenia, Bolivia, Sri Lanka, Madagascar and Uzbekistan – set up national inventory databases on CWR, storing previously existing data from various sources, which in most cases were digitized during the life of the project, as well as many additional records gathered during field surveys. Given the different national and institutional contexts and varying levels of expertise and use of software programs, all five national inventories were designed according to appropriate national preferences and settings. Armenia developed a web-based system with PHP and MySQL, which is used in the institutions that have CWR data. Data is sent through modem connection from the institutions to the central database, which now contains more than 30,000 records for 104 species. The Uzbek national database was developed in Access, while in Madagascar and Sri Lanka the newly digitized data was first entered into Excel worksheets. Bolivia compiled at least 3010 records for over 160 CWR species. The development of the national systems allowed countries to map distribution of wild relatives in their countries, identify areas for CWR conservation and prioritize protected areas where CWR should be included in the protected areas management plans. In addition to the national information systems, a global portal was developed to provide access to CWR information at the global level. The national CWR inventories are all searchable through the global portal and are linked to it using TapirLink as the providing software. Further information and resources on CWR provided by the portal include publications, projects and experts, news and images. The choice of freely available and easy-to-use tools, as well as approved and widely used standards, make it easy to link additional national CWR inventories to the portal in the future and to provide a CWR-viewpoint on plant genetic resources data and distribution. Ideally, the global portal will be further developed by Bioversity International to link to all relevant information sources on CWR so as to provide a convenient information gateway.

The portal provides information on the following:

- species-level data on CWR;
- *ex situ* conservation;
- taxonomy;
- conservation status;
- distribution;
- the presence of CWR in protected areas;
- relevant contacts, literature sources, latest news and photos.

Information sources include: country partners (Armenia, Bolivia, Madagascar, Sri Lanka and Uzbekistan); international partners (BGCI, FAO, IUCN and UNEP-WCMC); other countries' data accessible via the Global Biodiversity Information Facility (GBIF).

Source: www.cropwildrelatives.org

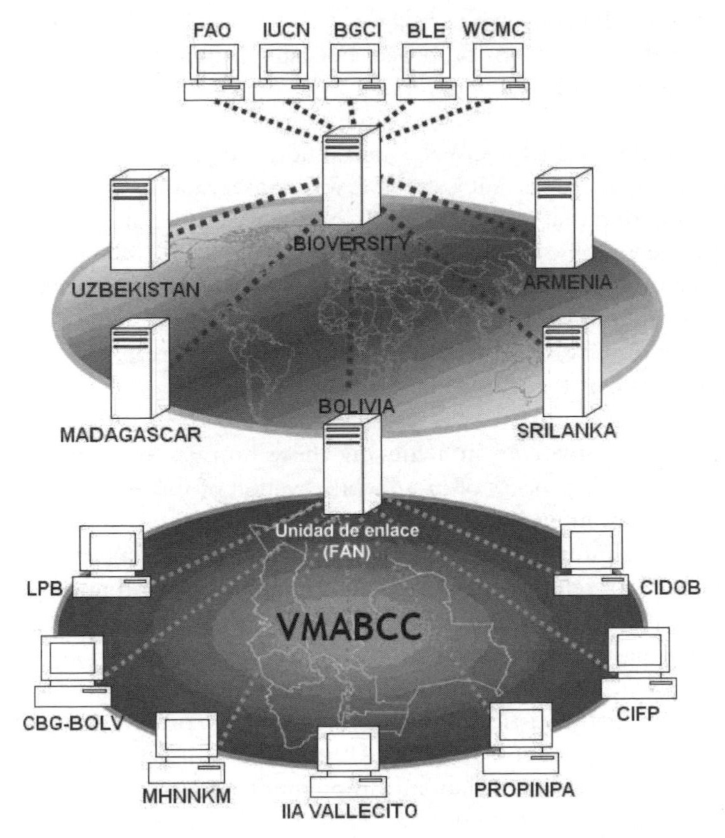

Figure 1.4 *Bolivian National Information System linked to the CWR International Portal*

In addition to addressing *in situ* conservation needs of target species, the project was also concerned with use of selected taxa for crop improvement. Hence, economic value for breeding, actual and potential, has been an important consideration in selecting target species for conservation action. They may possess characteristics, for example, which could provide resistance against disease or pests or difficult growing conditions such as a shortage or an excess of water, extreme heat or cold, or soil salinity.

About this manual

As already noted, *in situ* conservation of CWR has gained a certain momentum in the past 5–10 years but is still a poorly understood process and only a limited amount of practical experience can be drawn upon. The aim of this manual, therefore, is to share the experience obtained during this UNEP/GEF CWR Project of planning and implementing the *in situ* conservation and sustainable use of CWR, both on the part of the individual partner countries and institutional partners, and by the consortium as a whole. These include the difficulties faced, lessons learned and solutions proposed. Focusing primarily on the *in situ* conservation aspects of the project, it covers:

- national action plans for CWR conservation and use;
- identification of important areas of CWR conservation;
- assessment of threat status using IUCN Red List criteria;
- maps of geographic distribution of CWR species;
- adapting protected area management plans for CWR conservation;
- development of management plans for target CWR;
- guidelines for CWR conservation outside protected areas;
- monitoring plans for crop wild relative species.

The various steps involved in achieving these outputs are summarized in an overall scheme, 'The process of *in situ* conservation of CWR', presented in Table 1.3. The manual is intended to provide practical guidance on all the operations involved, such as information gathering, field assessment, taxon and area selection, and on the development, organization, implementation and monitoring of management plans and interventions to conserve CWR *in situ*. The manual will thus provide national and international conservation practitioners (including agrobiodiversity and conservation researchers, educators and students, NGO staff, genetic resource institutions, funding agencies, protected area managers, policy-makers and project managers) with practical information as well as tried and tested tools needed to plan and implement effective *in situ* conservation actions targeting the conservation of CWR. In this way, it goes well beyond the titles and literature already available.

Case studies from the five project countries are used to illustrate practical applications and real outcomes. While the valuable and complementary role of *ex situ* conservation is acknowledged, its detailed coverage is beyond the scope of both the project and this manual. The reader is referred to a number of key references on *ex situ* conservation listed in the references section.

This manual deals with the essential steps needed to achieve effective *in situ* conservation of CWR. After an introduction, it summarizes the importance of CWR in the five project partner countries, followed by an introduction to *in situ* conservation, looks at the planning issues involved and then details the major areas of work involved in CWR conservation, with illustrations and examples from the five countries.

It should also be pointed out that the materials in this manual are complemented by information and resources available through the CWR Global Portal described in Box 1.10. A page on the CWR Global Portal is, in fact, dedicated to the *In Situ* Manual at: http://www.cropwildrelatives.org/training/in_situ_conservation_manual.html. Chapter summaries, as well as other resources, including a glossary, additional annexes, examples of national action plans and management plans, and PowerPoint presentations are available for download at: http://www.cropwildrelatives.org/capacity_building/elearning/elearning.html. As

Table 1.3 *The process of* in situ *conservation of CWR*

The conservation of CWR *in situ* involves a series of procedures and actions which ideally should be undertaken in a logical sequence, for example:

1 Selection of priority/target species
2 Verification of taxonomic identity
3 Assessment of their geographical distribution, ecology, soil preferences
4 Assessment of their demography and population structure
5 Assessment of their phenology, reproductive biology and breeding systems
6 Assessment of their conservation status; and threat analysis
7 Assessment of their genetic variation and distribution of key alleles
8 Selection of the target populations to be conserved
9 Selection of the area(s) in which the target species are to be conserved: existing protected natural or semi-natural areas; or non-protected natural or semi-natural areas
10 Determination of the spatial scale of conservation needed – location, number and size of populations to be conserved; decision on whether to adopt a single-species or multi-species approach
11 Identification of aims of conservation and the appropriate conservation measures
12 Preparation of a conservation management plan for the target populations, if threatened, or monitoring plan if not currently threatened
13 Organization and planning of specific conservation activities
14 Identification and involvement of stakeholders
15 If the target area is already protected, assessment of the management status of the protected areas in which the target populations occur; and proposals for modification of management guidelines as appropriate
16 Consultation with protected area managers, local communities and other stakeholders
17 If the area or reserve/genetic reserve/gene management zone has to be created *de novo*, design of the reserve including boundaries, zoning and protection, and development of a management plan and guidelines
18 Determine statutory and legal requirements involved and arrange for necessary legislative approval (e.g. publication of management plan, gazetting new protected area/reserve) or legislative changes (e.g. modification of management plan of protected area) to be submitted to competent authorities
19 Development of a monitoring strategy for the area(s)
20 Development of a monitoring plan for assessing the effectiveness of the management interventions on the target populations and their conditions, genetic variability and needs
21 Development of a monitoring plan for assessing the impacts of human activities
22 Consideration of the possibilities of developing conservation strategies for species/populations occurring off-reserve/outside protected areas, such as easements, covenants, trusts, partnerships
23 Submit the management and monitoring plans and the whole conservation strategy to review
24 Prepare outreach and publicity materials
25 Preparation of a budget
26 Development of a timeline
27 Build a project team
28 Field implementation

In practice, as the circumstances and context of each *in situ* conservation project are unique, the actual sequence and emphasis given to each component will vary considerably.

additional relevant information and resources become available, they will be added to the online version of the *In Situ* Manual.

Further sources of information

A selection of useful sources of further information on CWR:

Bennett, A. (1965) 'Plant introduction and genetic conservation: genecological aspects of an urgent world problem', *Scottish Plant Breeding Station Record*, pp17–113.

Hamilton, A. and Hamilton, P. (2006) *Plant Conservation: An Ecosystems Approach*, Earthscan, London.

Heywood, V.H. and Dulloo, M.E. (2005) In Situ *Conservation of Wild Plant Species – A Critical Global Review of Good Practices*, IPGRI Technical Bulletin, no 11, FAO and IPGRI, IPGRI, Rome, Italy.

Hodgkin, T. and Hajjar, R. (2008) 'Using crop wild relatives for crop improvement: trends and perspectives', pp535–548, in N. Maxted, B.V. Ford-Lloyd, S.P. Kell, J.M. Iriondo, M.E. Dulloo and J. Turok (eds) *Crop Wild Relative Conservation and Use*, CAB International, Wallingford, UK.

Iriondo, J., Maxted, N. and Dulloo, M.E. (eds) (2008) *Conserving Plant Genetic Diversity in Protected Areas*, CAB International, Wallingford.

Maxted, N., Ford-Lloyd, B.V. and Hawkes, J.G. (eds) (1997) *Plant Genetic Conservation: The* In Situ *Approach*, Chapman and Hall, London.

Maxted, N., Ford-Lloyd, B.V., Kell, S.P., Iriondo, J.M., Dulloo, M.E. and Turok, J. (eds) (2008) *Crop Wild Relative Conservation and Use*, CABI, Wallingford.

Meilleur, B.A. and Hodgkin, T. (2004) '*In situ* conservation of crop wild relatives: status and trends', *Biodiversity and Conservation*, vol 13, pp663–684.

Stolton, S., Maxted, N., Ford-Lloyd, B., Kell, S.P. and Dudley, N. (2006) *Food Stores: Using Protected Areas to Secure Crop Genetic Diversity*, World Wide Fund for Nature (WWF) Arguments for protection series, WWF, Gland, Switzerland.

Thormann, I., Jarvis, D., Dearing, J. and Hodgkin, T. (1999) 'International available information sources for the development of *in situ* conservation strategies for wild species useful for food and agriculture', *Plant Genetic Resources Newsletter*, 118, pp38–50.

Tuxill, J. and Nabhan, G.P. (2001) *People, Plants and Protected Areas: A Guide to In Situ Management*, Earthscan, London.

Valdés, B., Heywood, V.H., Raimondo, F. and Zohary, D. (eds) (1997) *Conservation of the Wild Relatives of European Cultivated Plants*, *Bocconea* 7, Palermo, Italy.

A selection of important websites follows:

FAO home page; www.fao.org/

CGIAR home page; www.cgiar.org/

CWR Global Portal; www.cropwildrelatives.org/

Bioversity International home page; www.bioversityinternational.org/

IUCN Species Survival Commission Crop Wild Relative Specialist Group (CWRSG); www.cwrsg.org/

European Crop Wild Relative Diversity Assessment and Conservation Forum (PGR-Forum); www.pgrforum.org/

UNEP/GEF CWR project website http://www.bioversityinternational.org/research/conservation/crop_wild_relatives.html (accessed 23 November 2010)

Notes

1. As explained later, CWR also include those of fibre, oil, ornamental and medicinal species, not just agricultural (food) crops.
2. Although not specifically aimed at CWR, proposals for genetic resource centres were made as far back as 1890 by Emmanuel Ritter von Proskowetz and Frans Schindler at the International Agricultural and Forestry Congress, Vienna, and in 1914 Bauer warned of the dangers of the loss of local landraces through replacement by uniform bred varieties that could lead to a serious reduction in the genetic resource base, i.e. genetic erosion (see Flitner, 1995), both long before Vavilov.
3. French and Spanish versions were also published.
4. http://www.pgrforum.org/Conference.htm
5. http://typo3.fao.org/fileadmin/templates/agphome/documents/PGR/SoW2/Second_Report_SOWPGR-2.pdf (last accessed 27 October 2010)
6. CWR SG http://www.cwrsg.org/index.asp
7. http://www.cwrsg.org/Publications/Newsletters/crop%20wild%20relative%20Issue%207.pdf
8. EcoTILLING is a variation of TILLING (Targeting Induced Local Lesions IN Genomes) – a technique that can identify polymorphisms in a target gene by heteroduplex analysis – that aims to determine the extent of natural variation in selected genes in crops.
9. One of the emission scenarios reported in the Special Report on Emissions Scenarios (SRES) by the Intergovernmental Panel on Climate Change, IPCC (http://www.grida.no/climate/ipcc/emission/).

References

Abdurakhmonov, I.Y., Buriev, Z.T., Saha, S., Pepper, A.E., Musaev, J.A., Almatov, A., Shermatov, S.E., Kushanov, F.N., Mavlonov, G.T., Reddy, U.K., Yu, J.Z., Jenkins, J.N., Kohel, R.J. and Abdukarimov, A. (2007) 'Microsatellite markers associated with lint percentage trait in cotton, *Gossypium hirsutum*', *Euphytica*, vol 156, pp141–156

Bai, Y. and Lindhout, P. (2007) 'Domestication and breeding of tomatoes: What have we gained and what can we gain in the future?', *Annals of Botany*, vol 100, no 5, pp1085–1094

Bennett, A. (1965) 'Plant introduction and genetic conservation: genecological aspects of an urgent world problem', *Scottish Plant Breeding Station Record*, pp17–113

Brar, D.S. (2005) 'Broadening the gene pool of rice through introgression from wild species', in K. Toriyama, K.L. Heong, and B. Hardy (eds) *Rice is Life: Scientific Perspectives for the 21st Century*. Proceedings of the World Rice Research Conference held in Tokyo and Tsukuba, Japan, 4–7 November 2004, pp157–160, International Rice Research Institute, Los Baños, the Philippines, and Japan International Research Center for Agricultural Sciences Tsukuba (Japan)

Brar, D. and Kush, G. (1997) 'Alien introgression in rice', *Plant Molecular Biology*, vol 35, pp35–47

Chatzav, M., Peleg, Z., Ozturk, L., Yazici, A., Fahima, T., Cakmak, I. and Saranga, Y. (2010) 'Genetic diversity for grain nutrients in wild emmer wheat: potential for wheat improvement', *Annals of Botany Preview*, published on 3 March 2010, doi:10.1093/aob/mcq024

Comai, L., Till, B.J., Reynolds, S.H., Greene, E.A., Codom, C., Enns, L.C., Johnson, J.E., Burtner, C., Odden, A.R. and Henikoff, S. (2004) 'Efficient discovery of DNA polymorphisms in natural populations by EcoTILLING', *The Plant Journal*, vol 37, no 5, pp778–786

Dillon, S.L., Shapter, F.M., Henry, R.J., Cordeiro, G., Izquierdo, L. and Lee, L.S. (2007) 'Domestication to crop improvement: genetic resources for sorghum and saccharum (Andropogoneae)', *Annals of Botany*, vol 100, pp975–989

Dwivedi, S.L., Crouch, J.H., Mackill, D.J., Xu, Y., Blair, M.W., Ragot, M., Upadhyaya, H.D. and Ortiz, R. (2007), 'The molecularization of public sector crop breeding: progress, problems, and prospects', *Advances in Agronomy,* Chapter 3, pp163–319, doi:10.1016/S0065-2113(07)95003-8

Dwivedi S.L., Upadhyaya, H.D., Thomas Stalker, H., Blair, M.W., Bertioli, D.J., Nielen, S. and Ortiz, R. (2008), 'Enhancing crop gene pools with beneficial traits using wild relatives', *Plant Breeding Reviews*, vol 30, pp180–230

FAO (1989) *Plant Genetic Resources: Their Conservation* in situ *for Human Use*, Food and Agriculture Organization of the United Nations (FAO), Rome, Italy

Farooq, S. and Azam, F. (2001) 'Production of low input and stress tolerant wheat germplasm through the use of biodiversity residing in the wild relatives', *Hereditas,* vol 135, pp211–215

Flitner, M. (1995) *Sammler, Räuber und Gelehrte: die Politische Interessen an Pflanzengenetischen Resourcen 1895–1995,* Campus Verlag, Frankfurt/Main, New York

Flynn, J. (2006) 'Reflections on two ecosystem services: "The Production of Ecosystem Goods" and the "Generation and Maintenance of Biodiversity"', http://www.google.co.uk/search?hl=en-GB&q=Flynn%2C+J.+%282006%29+Reflections+on+two+Ecosystem+services%3A+%E2%80%9CThe+Production+of+Ecosystem+Goods%E2%80%9D+and+the+%E2%80%9CGeneration+and+Maintenance+of+Biodiversity%E2%80%9D.&sourceid=navclient-ff&rlz=1B3GGGL_enGB269GB269&ie=UTF-8, accessed 26 April 2010

Hajjar, R. and Hodgkin, T. (2007) 'The use of wild relatives in crop improvement: a survey of developments over the last 20 years', *Euphytica,* vol 156, pp1–13

Harlan, J.R. and de Wet, J.M.J. (1971) 'Towards a rational classification of cultivated plants', *Taxon,* vol 20, no 4, pp509–517

Heywood, V.H. and Dulloo, M.E. (2005) In Situ *Conservation of Wild Plant Species – A Critical Global Review of Good Practices,* IPGRI Technical Bulletin, no 11, FAO and IPGRI, IPGRI, Rome, Italy

Heywood, V., Casas, A., Ford-Lloyd, B., Kell, S. and Maxted, N. (2007) 'Conservation and sustainable use of crop wild relatives', *Agriculture, Ecosystems and Environment,* vol 121, pp245–255

Hodgkin, T. and Hajjar, R. (2008) 'Using crop wild relatives for crop improvement: trends and perspectives', pp535–548, in N. Maxted, B.V. Ford-Lloyd, S.P. Kell, J.M. Iriondo, M.E. Dulloo and J. Turok (eds) *Crop Wild Relative Conservation and Use*, CAB International, Wallingford, UK

Jarvis, A., Lane, A. and Hijmans, R. (2008) 'The effect of climate change on crop wild relatives' *Agriculture, Ecosystems and Environment,* vol 126, pp13–23

Kell, S.P., Laguna, L., Iriondo, J. and Dulloo, M.E. (2008) 'Population and habitat recovery techniques for the *in situ* conservation of genetic diversity', in J. Iriondo, N. Maxted and M.E. Dulloo (eds) *Conserving Plant Genetic Diversity in Protected Areas*, Chapter 5, pp124–168, CAB International, Wallingford, UK

Kovacs, M.I.P., Howes, N.K., Clarke, J.M. and Leisle, D. (1998) 'Quality characteristics of

durum wheat lines deriving high protein from *Triticum dicoccoides* (6b) substitution', *Journal of Cereal Science,* vol 27, pp47–51
Malik, R., Brown-Guedira, G.L. Smith, C.M., Harvey, T.L. and Gill, B.S. (2003) 'Genetic mapping of wheat curl mite resistance genes *Cmc3* and *Cmc4* in common wheat', *Crop Science,* vol 43, pp644–650
Lira, R., Tellez, O. and Davila, P. (2009) 'The effects of climate change on geographic distribution of Mexican wild relatives of domesticated cucurbitaceae', *Genetic Resources and Crop Evolution,* vol 56, pp 691–703.
Maxted, N. and Kell, S.P. (2009) *Establishment of a Global Network for the* In Situ *Conservation of Crop Wild Relatives: Status and Needs,* FAO Commission on Genetic Resources for Food and Agriculture, Rome, Italy
Maxted, N., Dulloo, M.E., Ford-Lloyd, B.V., Iriondo, J. and Jarvis, A. (2008) 'Gap analysis: a tool for complementary genetic conservation assessment', *Diversity and Distributions,* vol 14, no 6, pp1018–1030
Meilleur, B.A. and Hodgkin, T (2004) '*In situ* conservation of crop wild relatives: status and trends', *Biodiversity and Conservation,* vol 13, pp663–684
Miller, J.C. and Tanksley, S.D. (1990) 'RFLP analysis of phylogenetic relationships and genetic variation in the genus *Lycopersicon', Theoretical and Applied Genetics,* vol 80, pp437–448
NRC (National Research Council) (1991a) *Managing Global Genetic Resources: The US National Plant Germplasm System,* National Academy Press, Washington, DC
NRC (1991b) *Managing Global Genetic Resources: The US National Plant Germplasm System,* National Academy Press, Washington, DC
NRC (1993) *Managing Global Genetic Resources: The US National Plant Germplasm System: Agricultural Crop Issues and Policies,* National Academy Press, Washington, DC
Ortiz, R., Taba, S., Tovar, V.H.C., Mezzalama, M., Xu, Y., Yan, J. and Crouch, J.H. (2009) 'Conserving and enhancing maize genetic resources as global public goods – a perspective from CIMMYT', *Crop Science,* vol 50, pp13–28
Prescott-Allen, R. and Prescott-Allen, C. (1988) *Genes from the Wild: Using Wild Genetic Resources for Food and Raw Materials,* Earthscan Publications Limited, London, UK
Rick, C.M. and Chetelat, R.T. (1995) 'Utilization of related wild species for tomato improvement', *Acta Horticulturae,* vol 412, pp21–38
Robertson, L. and Labate, J. (2007) 'Genetic resources of tomato', in M.K. Razdan and A.K. Mattoo (eds) *Genetic Improvement of Solanaceous Crops,* vol 2, *Tomato,* Science Publishers, Enfield, NH
Rubenstein, K.D., Heisey, J.P., Shoemaker, R., Sullivan, J. and Frisvold, G. (2005) *Crop Genetic Resources: An Economic Appraisal,* Economic Information Bulletin Number 2, United States Department of Agriculture (USDA), Washington, DC
Schneider, A., Molnár, I. and Molnár-Láng, M. (2008) 'Utilisation of *Aegilops* (goatgrass) species to widen the genetic diversity of cultivated wheat', *Euphytica,* vol 163, pp1–19
Sencer, H.A. (1975) 'Recent and proposed activities of the Izmir Centre', in O.H. Frankel and J.G. Hawkes (eds) *Crop Genetic Resources for Today and Tomorrow,* International Biological Programme 2, Cambridge University Press, Cambridge.
Singh, B. (1981) *Establishment of first gene sanctuary in India for Citrus in Garo Hills,* Concept Publishing Co., New Delhi, India
Smith, N.J.H., Williams, J.T., Plucknett, D.L. and Talbot, P. (1992) *Tropical Forests and their Crops,* Comstock Publishing Associates, Cornell University Press, Ithaca, NY and London, UK

Tan, A. and Tan, A.S. (2002) 'In situ conservation of wild species related to crop plants: the case of Turkey', in J.M.M. Engels, V. Ramantha Rao, A.H.D. Brown and M.T. Jackson (eds) *Managing Plant Genetic Diversity*, pp195–204, CAB International, Wallingford, UK

Tanksley, S.D. and McCouch, S.R. (1997) 'Seed banks and molecular maps: unlocking genetic potential from the wild', *Science*, vol 277, pp1063–1066

Valdés, B., Heywood, V.H., Raimondo, F. and Zohary, D (eds) (1997) *Conservation of the Wild Relatives of European Cultivated Plants*, *Bocconea* 7, Palermo, Italy

Xiao, J., Li, J., Grandillo, S., Ahn, S.N., Yuan, L., Tanksley, S.D. and McCouch, S.R. (1998) 'Identification of trait-improving quantitative trait loci alleles from a wild rice relative, *Oryza rufipogon*', *Genetics*, vol 150, pp899–909

Chapter 2

Crop Wild Relatives in the Project Countries

Increasing our knowledge of the biodiversity of a country with great riches of natural and cultural resources like ours, and contributing to the sustainable development of the natural resources that lead to a reduction in poverty, is not only a major need but a great challenge (René Orellana Halkyer and Juan Pablo Ramos Morales, 2009).

This chapter provides background information on the five countries involved in the UNEP/GEF CWR Project and reviews their experience and policies regarding CWR conservation.

The background for *in situ* conservation in the project countries

Although the five participating countries of the UNEP/GEF CWR Project include significant numbers of globally important taxa of crop wild relatives (CWR), by 2004 little progress in CWR conservation had been made. Armenia and Uzbekistan executed limited CWR surveys decades before, and a small number of reserves were created in each country with some consideration given to CWR; however, neither country established CWR management plans for these reserves and no conservation projects or CWR monitoring actions were initiated. In Bolivia and Madagascar, governments were aware of the importance of CWR and some plant genetic resources (PGR) materials were conserved *ex situ*. Nonetheless, national inventories had not yet been undertaken and information management focusing on CWR was non-existent. Both countries had established protected areas, but none included management plans concerned with CWR use and conservation. In Sri Lanka, several CWR conservation and awareness-raising projects had been conducted for selected taxa.

Reasons for the relative weakness of CWR conservation efforts include limited technical capacity to develop conservation plans for such a diverse range of species; absence of coordination and partnership between disciplines (agricultural and conservation sectors and social and economic sciences); and political, administrative and infrastructural obstacles.

At the time the project was initiated, none of the CWR Project countries had developed clear, coherent national strategies or action plans to conserve and use CWR, although all countries recognized the need to improve national agrobiodiversity conservation programme planning, decision-making and implementation frameworks to support effective *in situ* CWR conservation. Collaborative agreements, necessary for coordinating and implementing conservation actions, were largely absent or occurred only on an *ad hoc* basis in these countries. Notable limitations also existed in identifying priority actions and developing necessary management plans for the conservation of target taxa and priority areas.

While the CWR Project countries were aware that relevant information was available to assist in the planning process, they noted that such data was usually dispersed and not readily accessible. Useful information necessary to determine the likely location of CWR populations existed in herbaria and *ex situ* genebanks for each country. Further, information on the extent and distribution of protected areas was available from responsible agencies in the Ministries of environment, forestry, planning and so forth. Institutions linked to the Ministry of Agriculture, universities and colleges also possessed data on CWR utilization. However, in Armenia and Uzbekistan, little information was actually available in computerized form; in all countries, most location data had not yet been digitized. Where information was available in an electronic format (e.g. Bolivia, Madagascar and Sri Lanka), different agencies had developed independent information management systems with unique data structures and formats. Combining information from different sources for the necessary integrated analyses was, therefore, difficult and complex.

In common with most other countries, the absence of a supportive legal framework for the conservation and utilization of CWR proved to be a further impediment. The CWR Project countries did not have any legislation in place consistent with new international agreements such as the International Treaty on Plant Genetic Resources for Food and Agriculture (ITPGRFA) and the Convention on Biological Diversity (CBD), resulting in little commitment by governments to apply constitutional provisions and recognize international norms as part of their national legislative framework. Further, none of the countries had developed legislation and procedures adequately addressing benefit-sharing issues for CWR.

Generally, the limited development of CWR conservation efforts in each country reflected low levels of awareness among decision-makers and the general public of the importance of these resources and the need to maintain and use them wisely. This was evident in the low priority given to CWR in national budgets and research agendas, as well as the general lack of enabling policies and actions.

Armenia

The mountainous nature of Armenia, and the Caucasus Mountains in particular, determines much of the character of the country's landscapes, climate, vegetation, soils and biodiversity.

Armenia is home to around 3600 species of vascular plants, including more than half of the flora of the Caucasus (about 7200 species), even though the country only occupies 6.7 per cent of the Caucasus region. Over 125 species are endemic to Armenia. As one of the centres of origin of cultivated plants, the country is known for its diversity of native species of cereals; vegetables, in particular cucurbits; oil-bearing plants and fruit crops.

Forests cover some 20 per cent of the country and are generally found at mid-elevations on mountains, at altitudes between 500m and 2100m in the north (up to 2500m in the south). In central Armenia, forests occur in small areas rather than as a continuous zone and can also be found on steep slopes and in other areas with limited human access.

Protected areas

A network of specially protected areas was first established in Armenia in 1958, to protect ecosystems, habitats and rare, endemic and threatened species. There are currently five state reserves, 22 state reservations and one national park registered, which together cover around 311,000ha, or 10 per cent of the surface of the country.

The Erebuni Reserve, located in close proximity to Yerevan city, was established in 1981 specifically to protect wild relatives of grain crops. It covers roughly 89ha on either side of the road from Yerevan to Garni and harbours populations of *Triticum araraticum*, *T. boeoticum*, *T. urartu*, *Secale vavilovii* and *Hordeum spontaneum* (Damania 1994, 1998; Damania et al, 1998; Harutyunyan et al, 2008).

Crop wild relatives

Armenia has many species of wild relatives of domestic crops, including three of the four known wild species of wheat (*Triticum boeticum*, *T. urartu* and *T. araraticum*), many species belonging to the genus *Aegilops* (i.e. *Ae. tauschii*, *Ae. cylindrica*, *Ae. triuncialis*, etc.), and wild relatives of rye and barley. Wild apple and pear species grow in most of Armenia's forests, together with wild forms of other fruits and nuts (e.g. quince, apricot, sweet and sour cherry, walnut, pistachio and fig). A survey of the wild relatives of food crops of Armenia was made by Gabrielian and Zohary (2004). During the course of the CWR Project, 2518 species out of about 3600 vascular plants reported for Armenia's flora (about 70 per cent), were identified as CWR. They represented 431 genera and 119 families.

Bolivia

Bolivia possesses great biological richness in terms of plant and animal species and is home to a diversity of environments and ecosystems. It houses approximately 20,000 species of higher plants and more than 2600 species of vertebrates. Bolivia is a country of deserts and tropical rainforests, deciduous forests, savannas, lakes and rivers, with elevations ranging from 150m to 6500m and an annual rainfall of between 0mm and 6000mm (MDS-VRFMA-DGBAP, 2004). The country's location within the Andean region, where several important biomes are represented within a limited geographical area, and where mountain ecosystems form one of the major components, means that it is rich in natural biodiversity.

In this natural environment domestication took place of some of the most important crop species feeding much of world population, including potatoes, squash, peanuts, chilli peppers and other crops, some of which are only now beginning to receive attention, such as quinoa and cañahua (*Chenopodium pallidicaule*), grown in the Andean region of Bolivia (MDS-VRFMA-DGBAP, 2004). In the lowlands of Bolivia, more than 100 species of wild fruits occur (Vasquez and Coimbra, 1996) and nearly 3000 medicinal plant species with potential as genetic resources for industrial, pharmaceutical and cosmetic uses are found (Ibisch and Mérida, 2003).

Threats to biodiversity

The genetic diversity found in the production systems of rural communities and indigenous peoples, as well as in the wild ecosystems of Bolivia, is now facing various threats. The genetic diversity of cultivated plants is increasingly threatened by:

- increased substitution of crops and native varieties by introduced crops and varieties of more value or appreciation in the market;
- insufficient land, leading farmers to prioritize which crops and cultivated varieties to grow;
- weakness of traditional knowledge regarding the marketing of genetic quality of seeds;
- effects of climate change on rural economies, leading to abandonment of fields and farmer migration to cities, destroying the systems of traditional production;
- climatic change – drought, hail and frost.

Crop wild relatives

Bolivia lies within one of the world's centres of crop domestication and within the centres of diversity of important crops such as potato (*Solanum* spp.), sweet potato (*Ipomoea batatas*), maize (*Zea mays*), peanut (*Arachis hypogaea*), cassava (*Manihot esculenta*), cotton (*Gossypium barbadense*), tobacco (*Nicotiana tabacum*), cocoa (*Theobroma cacao*), beans (*Phaseolus* spp.) and peppers (*Capsicum* spp.), as well as several local Andean tubers (e.g. *Ullucus tuberosus*, *Oxalis* spp.), quinoa (*Chenopodium quinoa*), tarwi (*Lupinus mutabilis*), and others. Most of the CWR of these and other Bolivian species are characterized by environmental and soil stress tolerance, disease resistance and other adaptive traits useful for crop improvement programmes.

Bolivia has published the 'Red Book of the Crop Wild Relatives of Bolivia' (*Libro Rojo* de *Parientes Silvestres de Cultivos de Bolivia*) (VMABCC-Bioversity, 2009 in hard copy and as an interactive CD ROM). In addition, an atlas of CWR (http://www.cwrbolivia.gob.bo/atlaspsc/) was prepared by the Fundación Amigos de la Naturaleza (FAN-Bolivia) in 2001–2002, under the scope of a Letter of Agreement signed between the International Plant Genetic Resources Institute (IPGRI, now Bioversity International) Colombia, the United States Department of Agriculture (USDA) and FAN-Bolivia on 25 July 2001, to support the elaboration of the atlas of Bolivian CWR.

The database for the atlas includes records of 2486 samples from herbaria and accessions in genebanks, representing 14 families, 18 genera and 161 species of CWR. The atlas also includes a series of maps of the country (political divisions, roads, populated towns, hydrology, climate and ecoregions), maps of the current distribution of 161 species of CWR, their distribution in protected areas and communal lands of indigenous peoples, potential distribution maps for 57 of the most abundant species (using FloraMap and DIVA-GIS), and maps of diversity and richness for gene pools, as well as for CWR species. The atlas provided key information for the National Report of Bolivia on CWR, elaborated in the preparatory (PDF-B) phase of the UNEP/GEF CWR Project.

National legal framework on genetic resources

The legislation of Bolivia pertaining to access to genetic resources was approved by the Supreme Decree No. 24676 on 21 June 1997. It stipulates that in order to access genetic resources of which Bolivia is a country of origin, users must sign an access agreement or contract with the national competent authority. This legislation considered the elements agreed by the CBD and by the Decision 391: Common Regime on Access to Genetic Resources of the Andean Community Countries, adopted on 2 July 1996.

Bolivia adopted a National Strategy for Conservation and Sustainable Use of Biodiversity on 19 March 2002, for a period of ten years. The strategy recognizes the importance of CWR, which are useful for genetic improvement of crops, but does not establish a set of specific actions for the conservation of these; emphasis is instead on the *ex situ* conservation of plant genetic resources. In February 2009, Bolivia approved a new state policy which, unlike previous versions, includes articles related to genetic resources and established the following responsibilities for the state:

- The native species of plants and animals are a natural heritage and the state shall establish the necessary measures for their conservation, utilization and development.

- The state shall protect all genetic resources and micro-organisms that are found in the ecosystems of the territory, as well as the knowledge associated with their use and exploitation. For their protection, a register system will be established to help safeguard their existence, and the intellectual property of the state or local social subjects that claim it. For those resources that are not registered, the state shall establish procedures for their protection by law.
- The entry and exit of genetic resources of the country shall be controlled and mechanisms established for the repatriation of genetic material obtained by other countries or international research centres and to ensure their preservation in *ex situ* conservation centres within the country.

The management of natural resources located in the territories of indigenous people will be shared, subject to the particular rules and procedures of indigenous nations and farmers. Where overlaps exist in protected areas and indigenous territories, management of the areas must be shared and will be carried out subject to the particular rules and procedures of the indigenous peoples and farmers, while respecting the objective for the creation of these areas.

Protected areas

The National Protected Areas System (NPAS) for the management of protected areas in Bolivia was established in 1997 through Supreme Decree No. 24781. Its objective is to 'maintain representative samples of biogeographic provinces, through the implementation of policies, strategies, plans, programmes and rules to generate sustainable processes within the protected areas to achieve the objectives of biodiversity conservation by incorporating the participation of the local population and benefits for actual and future generations'. The NPAS comprises more than 66 protected areas of national, departmental, municipal or private interest. They account for more than 15 per cent of the national territory. There are five categories of management that define the kind and extent of use of natural resources within the protected areas. The categories of 'Park', 'Shrine' and 'Natural Monument' are aimed at strict protection and preservation of the richness in biodiversity of the protected areas, while the categories of 'Wildlife Reserve' and 'Integrated Management Natural Area' allow the sustainable management of the natural resources under legal and technical conditions. Finally, there exists a transitional legal regime that defines the category of 'Natural Reserve of Immobilization', which corresponds to areas that are deemed protected after a preliminary assessment, but for which further studies are required for their definitive characterization and zoning (MDS-SERNAP, 2001).

Some protected areas have a dual category, such as the National Parks and Integrated Management of Natural Areas and National Parks and Indigenous Territories (MDS-SERNAP, 2001).

Madagascar

Madagascar is one of the most important biodiversity hotspots in the world and is characterized by the richness of its flora (12,000 spp. of vascular plants) and the great diversity of its ecosystems.

Vegetation and ecosystems

The variety of the ecosystems can be explained by (1) the existence of many types of soils and rocky substrates; (2) an altitudinal gradient ranging from 0m to more than 2500m; (3) the contrasting climate among the eastern, western and southern regions; and (4) the fact that the country extends over about 13° of latitude, from 12.2°S to 25°S.

The last vegetation classification was established in 2007 as a result of collaboration among the Royal Botanic Gardens, Kew; Missouri Botanical Garden; and Conservation International; with the contribution of national expertise from research centres and universities (see www.vegmad.org). The vegetation of Madagascar comprises various ecosystems that belong to five main domains: the wet eastern domain; the wet Sambirano domain in the northern part of Madagascar; the centre domain, wet on its eastern part and dry on its western part; the western domain, which is dry; and the south-western domain, which is arid.

The different types of ecosystems encountered in Madagascar are divided into several categories: humid forest, littoral forest (east); western humid forest; western sub-humid forest; western dry forest; south-western dry spiny forest-thickets; south-western coastal bush land; mangroves, tapia or *Uapaca* sclerophyllous forest; wetlands; degraded humid forest; degraded south-western dry spiny forest; wooded grassland–bush land mosaic; and plateau grassland–wooded grassland mosaic (Moat and Smith, 2007).

Flora

The flora of Madagascar is characterized by an 85 per cent endemism level. Schatz (2000) has shown that endemism is as high as 90 per cent for the tree flora. The endemism at the generic level is also high (30 per cent). In addition, there are seven families that are only found in Madagascar, the largest of which is the Sarcolaenaceae. Regarding particular groups, the Pteridophytes of Madagascar comprise 586 species and 106 genera, representing 6 per cent of the Pteridophyte flora of the world (Rakotondrainibe, 2003). A monograph of the Madagascar palms (Dransfield and Beentjee, 1995) has shown that there are 175 species in the country, while the whole flora of the nearby African continent only contains 110 species. A recent taxonomic study of the family of Balsaminaceae, which is not yet treated in the flora of Madagascar and Comoros, ended with the description of 50 new species (Gautier and Goodman, 2003). At the genus level, the case of the baobabs should be cited: six out of the eight species of *Adansonia* are endemic to Madagascar. Considering the *Coffea* genus, Madagascar possesses about 50 wild species that are largely caffeine free, belonging to the *Mascarocoffea* section. The genus *Dioscorea* has at least 40 endemic species in Madagascar, representing 10 per cent of world diversity in the genus. The same applies to the genus *Helichrysum*, which numbers about 180 species endemic to Madagascar. Even the traveller's tree, *Ravenala madagascariensis*, previously thought to be a single species, has been shown to contain at least six different variants that can be considered as subspecies (Blanc et al, 2003; Hladik et al, 2000). It should be noted, however, that knowledge of Madagascar plant diversity is incomplete and much taxonomic and inventory work is still required.

Useful plants

Madagascar flora contains a multitude of useful plants, including more than 5300 species of medicinal plants, which corresponds to about 50 per cent of the Malagasy flora. Many woody species are used for timber, some of them providing prized cabinet wood such as *Santalina*, *Diospyros*, *Dalbergia* (rosewood and palissander), *Ocotea* and *Canarium*. Timber species are heavily exploited and many are now endangered. Ornamental plants are also very well represented in the country's flora, including the flagship species, *Ravenala madagascariensis* and *Delonix regia*, which are now grown throughout the tropics. Other species are listed under the Convention on International Trade in Endangered Species of Wild Fauna and Flora (CITES) Appendices, among them orchids and species of *Pachypodium*, *Aloe* and *Euphorbia*.

Although Madagascar is not a centre of origin for food plants, parts of several wild species are used as food, among them: fruits (*Eugenia*, *Sygygium*, *Adansonia* or *Uapaca*), tubers (*Dioscorea*, *Tacca*), leaves (*Moringa*) or apices (hearts) (*Dracaena*, *Ravenala*, various palms). Many species, both herbaceous (*Lepironia*, *Heleocharis*, *Cyperus*) and woody, are utilized by local populations for crafts.

Malagasy farmers grow many different cereals (mainly rice and maize), tubers (potato, cassava, taro and sweet potatoes), legumes (beans, peas, voandzou) and leafy greens which are important in the diet of the Malagasy people. Fruit plants, both tropical and temperate, are also grown.

Protected areas

The greatest conservation effort in Madagascar has been the creation of a system of protected areas. Before 2003, the protected areas network covered 2 million hectares and was managed by Madagascar National Parks. The areas included natural integral reserves, national parks and special reserves. In 2003, Madagascar pledged to triple the area under protection by 2010, bringing the total area to 6 million hectares, corresponding to 10 per cent of the country's surface area. These 6 million hectares are now part of the System of Protected Areas of Madagascar (SAPM) and correspond to IUCN Categories 4, 5 and 6. They will be managed by Madagascar National Parks, NGOs or a consortium of different managers, including local communities. To date (2009) all potential protected areas have been identified and about 2 million hectares are now under temporary protection status.

Crop wild relatives

Madagascar is home to more than 150 CWR distributed among approximately 30 genera. Some are relatives of food plants such as *Ficus*, *Ipomoea*, *Oryza*, *Prunus*, *Rubus*, *Asparagus*, *Vanilla*, *Poupartia*, *Ensete*, *Solanum*, *Eugenia* or *Sygygium*. They include two wild relatives of rice (*Oryza staminata* and *O. punctata*), which possess virus and pest resistance, one wild relative of sorghum (*Sorghum verticiflorum*), two wild relatives of *Vigna* (*V. vexillata* and *V. angivensis*) and a wild relative of banana (*Musa perrieri*). The two most important genera are *Coffea*, which contains more than 50 caffeine-free or low-caffeine species (Sect. *Mascarocoffea*), and *Dioscorea* with 40 species, most of which are consumed by local populations, even when they are known to be toxic. Some other genera are relatives to ornamental plants such as *Delonix*, *Bauhinia*, *Mimosa*, *Gardenia*, *Hibiscus* and *Caesalpinia*. Finally, there are some wild species belonging to *Gossypium* or *Linum*, which contain textile species of global economic importance. Mention should be made of a wild species of *Jatropha* related to *Jatropha multifida*, which is now cultivated in Madagascar as a source of biofuel.

These different CWR are distributed throughout the country, but the majority are found in the forest ecosystems of the island. They are subject to a range of threats, mainly habitat loss because of forest exploitation leading to deforestation, slash and burn practices, soil impoverishment due to bush fires and mining.

Sri Lanka

Sri Lanka is a biodiversity hub of worldwide significance; the country possesses globally significant agricultural ecosystems and agrobiodiversity central to the livelihood strategies of small-scale farmers, rural communities and indigenous peoples. It is currently estimated that about 1.8 million families and 75 per cent of the country's labour force depend on agriculture and on the diversity in agricultural ecosystems, which includes some 237 fruit species, 82 vegetable species, 16 cereal and legume species, 20 species of spices and 1550 medicinal plant species.

Sri Lanka's ecosystems include forest, inland wetland, coastal, marine and agricultural ecosystems.

Agricultural ecosystems are represented by paddy lands, horticultural farms, small crop holdings, crop plantations, home gardens, *chena* lands, village small tank systems and *owita* agroecosystems. Sri Lanka has been an agrarian-based society for over 2000 years. Agriculture currently contributes around 20 per cent of the country's gross national product (GNP), second only to the manufacturing sector. The agricultural landscape is dominated by paddy cultivation and rice is the major staple crop. Sri Lanka's traditional agricultural systems, such as forest gardens, represent diverse landscapes and play a vital role in the *in situ* conservation of agrobiodiversity selected by farmers over generations, but today they are threatened and efforts are needed to encourage and sustain the multi-cropping practices and high agrobiodiversity inherent in these systems. Although Sri Lanka is an important centre for CWR diversity, many populations are under threat due to habitat destruction and other human activities.

Crop wild relatives

Prior to 2004, little attention was given to conserving and utilizing CWR and few had been comprehensively studied or researched. An inventory of food CWR in Sri Lanka was compiled using already published material on the Sri Lankan flora (Hasanuzzaman et al, 2003) and the records of the national herbarium. The list includes 410 species of food CWR, belonging to 47 families and 122 genera. Of these, 366 are native species and 77 are endemic relatives of food crops, while 44 species are naturalized exotics. This is only a preliminary list, which needs to be further refined. To recognize the true genetic relationships of these species, detailed studies must be carried out.

These CWR species of agricultural importance generally occur as members of disturbed communities within the major vegetation types of the country. Open canopy forest areas, secondary forests, disturbed grasslands and shrub jungles are rich in these plants. However, the relatives of fruit plants are largely associated with semi-evergreen, intermediate and wet evergreen forests. There are a large number of wild species of agricultural importance in different crop groups.

Protected areas

The total land area of Sri Lanka is 65,000km^2, a quarter of which is reserved for forests and administered by the Department of Forests and the Department of Wildlife Conservation. Currently, the country's 501 protected areas occupy around 26.5 per cent of the total land area of the country (see Table 2.1). A major part of the protected area system is under the control of the Department of Wildlife Conservation. However, within the 1 million hectares of state forests under the control of the Forest Department, there are a number of important protected areas, most notably: Hurulu and Sinharaja Biosphere Reserves, and Knuckles and Kanneliya-Dediyagala-Nakiyadeniya (KDN) Forest Reserves.

The Kanneliya Forest Reserve is notable for having the highest percentage of endemic woody species of any single wet zone forest in the country. Detailed studies of the floristic composition of the forest demonstrate that no single part of

Table 2.1 *Protected areas in Sri Lanka*

Extent of protected areas by IUCN Category (000ha), 2003:	
Nature reserves, wilderness areas, and national parks (categories I and II)	419
Natural monuments, species management areas, and protected landscapes and seascapes (categories III, IV, and V)	218
Areas managed for sustainable use and unclassified areas (category VI and 'other')	1129
Total area protected (all categories)	1767

it is representative of the whole, due to microclimatic differences (Ministry of Environment and Natural Resources, 1999). Kanneliya is also notable for having important wild relative species of *Cinnamomum.*

Threats to agrobiodiversity and crop wild relatives

Sri Lanka's natural forests contain a wide range of useful plant species. At the beginning of the last century, 70 per cent of the land area is said to have been covered by natural forests. The latest figures, however, show that natural forest cover has decreased to about 22 per cent of the land area. There are two factors that have posed serious threats to the preservation of natural floristic diversity in Sri Lanka. One is the heavy rate of deforestation due to various development projects, village expansion and settlement schemes. The second is selective felling of trees for timber and removal of plant species, particularly those with medicinal value. Thus many species, once plentiful, are now considered to be seriously threatened. In addition, unplanned land use, pollution and fragmentation have contributed to the loss of CWR.

Uzbekistan

Uzbekistan was identified by Vavilov as one of the centres of origin of many modern crop plants. It has some of the closest wild relatives of cultivated onion (*Allium oschaninii, A. vavilovii, A. praemixtum, A. pskemense*), as well as many wild fruit and nut species (*Vitis vinifera, Pistacia vera, Malus sieversii, Pyrus turkomanica*, and *Rubus caesius*). The flora of Uzbekistan contains some 4800 species. According to Professor U.P. Pratov (personal communication), more than 2500 useful wild species grow in the territory of Uzbekistan. Seventy species belonging to 48 genera of CWR are present, including nutritional, medical and ornamental plants of various life forms – trees, bushes and grasses.

Uzbekistan is a landlocked country of some 447,000km^2, bordered by Afghanistan to the south, Kazakhstan to the north and northeast, Kyrgyzstan and Tajikistan to the east and southeast, and Turkmenistan to the west and southwest. Most of the territory is steppe, desert (the Karakum and Kyzyl Kum deserts), semi-desert and mountains, while about 10 per cent comprises broad, flat intensely irrigated fertile valleys along the course of the rivers Amu Darya, Syr Darya (Sirdaryo) and Zarafshon. The Fergana Valley in the east is surrounded by the mountains of Tajikistan and Kyrgyzstan. Uzbekistan is one of the world's biggest producers of cotton and is rich in natural resources, including oil, gas and gold.

Main biogeographic zones

The main part of Uzbekistan's territory is occupied by valleys (almost 80 per cent); mountains are common only in the eastern part of the country. The valleys are occupied by desert vegetation; the low foothills by mountain, semi-desert vegetation; the high foothills by different grass and wheat steppe vegetation; the mountains by wood and bush vegetation; and the high mountains by subalpine and alpine meadows.

Among priority species, only barley is widespread in the low and high foothills. The rest of the prioritized species – apple, walnut, pistachio, almond and onion – grow in the mountain zone and pistachio and almond in the high foothills.

In the eastern regions, desert valleys are bordered by a strip of loess mountain valleys and foothills. They account for 18 per cent of the land area and they are occupied by ephemerals, with a small number of perennial grass species. Mountains are characterized by unusual diversity of climate and nature. The richest vegetation of grass and wood species grows well on the northern slopes of mountains. Vegetation on the southern slopes is less developed but includes grass species as well as wood and bush species. In the low belts of the mountains, vegetation is represented by xerophytes, in the middle belts by mesophytic deciduous plant species and in the high mountains vegetation is represented by only coniferous plants, tree-like juniper with rare populations of deciduous plants. All five prioritized wild relatives identified during the UNEP/GEF CWR Project grow in the mountain belt.

Protected areas

Currently, the protected areas system consists of nine state reserves (Zapovedniks), with an area of 2164km^2; two national parks, with a total area of 6061km^2; one biosphere reserve (452km^2); nine special state reserves (Zakazniks), with an area totalling 12,186.5km^2; and one captive breeding centre for rare animals. The total protected area in Uzbekistan is 20,520km^2, which represents 4.6 per cent of the Republic's territory. However, in terms of strict/long-term protection (i.e. IUCN Category I and II, including the national parks, biosphere reserve and state reserves) only 8171km^2 or 1.8 per cent of the Republic's territory is covered (see Table 2.2).

Table 2.2 *Strictly protected areas in Uzbekistan*

State strict reserves (Zapovedniks) (IUCN category I)	*Area km^2*
Chatkal Mountain Forestry Biosphere Reserve 1947	356.8
S Gissar Mountain Archa (Juniper) Reserve 1983	814.3
Zaamin Mountain Archa (Juniper) Reserve 1926, 1960	268.4
Badai-Tugai Steppe-tugai 1971	64.6
Kyzylkum Tugai-sand Reserve 1971	101.4
Zerafshan Lowland Tugai Reserve 1975	23.5
Nuratin Mountain Walnut-tree Reserve 1975	177.5
Kitab Geological Reserve 1979	53.7
Surkhan Mountain Forestry Reserve 1987	276.7
State national parks (IUCN category II)	*Area km^2*
Zaamin People's Park 1976	241.1
Ugam-Chatkal Natural National Park 1990	5745.9

Further sources of information

The CWR Global Portal

A more detailed version of this background chapter, including illustrative maps and tables, can be found at the CWR Global Portal at: http://www.cropwildrelatives.org/index.php?id=2916.

National project websites

Project websites have been set up in each partner country to increase national knowledge and awareness of the importance and value of conserving CWR, to document the progress made by the project activities, and to disseminate results obtained to policy-makers and the wider public. Links to national project websites are provided below:

Armenia – www.cwr.am/
Bolivia – www.cwrbolivia.gob.bo/inicio.php
Madagascar – www.pnae.mg/cwr/index.php
Sri Lanka – www.agridept.gov.lk/other_pages.php?heading=CWR
Uzbekistan – www.cwr.uz/en

National information system websites

To gather as much data as possible on CWR and enable informed decision-making, the project also included a component on information management, which required countries to pool together existing information on these species. This led to the creation of five national databases, where detailed information for hundreds of CWR species was collected and is now available for others to use. National inventories can be accessed through the CWR Global Portal (http://www.cropwildrelatives.org/national_inventories.html).

State of the World PGR Country and Regional Reports

The Second Report on the *State of the World's Plant Genetic Resources for Food and Agriculture* was published in 2010. This report updates the first report with the best data and information available and focuses on changes occurring since 1996. The report provides a concise assessment of the status and trends of plant genetic resources and identifies the most significant gaps and needs. Country reports for Armenia, Bolivia, Madagascar, Sri Lanka and Uzbekistan can be viewed by visiting the website below: http://www.fao.org/agriculture/crops/core-themes/theme/seeds-pgr/sow/sow2/country-reports/en/.

National biodiversity strategies and action plans

Further information regarding plans and actions to support the conservation and sustainable use of biodiversity in Armenia, Bolivia, Madagascar, Sri Lanka and

Uzbekistan can be found by searching the document database of the Convention on Biological diversity at: https://www.cbd.int/reports/search/.

References

Blanc, P., Hladik, A., Rabenandrianina, N., Robert, J.S. and Hladik, C.M. (2003) 'Plants: Strelitziaceae: The variants of *Ravenala* in natural and anthropogenic habitats', in *The Natural History of Madagascar,* S.M. Goodman and J.P. Benstead (eds), pp472–476, University of Chicago Press, Chicago, USA

Damania, A.B. (1994) '*In situ* conservation of biodiversity of wild progenitors of cereal crops in the near East', *Biodiversity Letters,* vol 2, pp56–60

Damania, A.B. (1998) 'Domestication of cereal crop plants and *in situ* conservation of their genetic resources in the Fertile Crescent', in A.B. Damania, J. Valkoun, G. Willcox and C.O. Qualset (eds) *The Origins of Agriculture and Crop Domestication,* International Center for Agricultural Research in the Dry Areas (ICARDA), Aleppo

Damania, A.B., Valkoun, J., Willcox, G. and Qualset, C.O. (eds) (1998) *The Origins of Agriculture and Crop Domestication*, International Centre for Agricultural Research in the Dry Areas (ICARDA), Aleppo

Dransfield, J. and Beentje, H. (1995) *The Palms of Madagascar,* Royal Botanic Gardens, Kew and International Palm Society

Gabrielian, E. and Zohary, D. (2004) 'Wild relatives of food crops native to Armenia and Nakhichevan', *Flora Mediterranea,* vol 14, pp5–80

Gautier, L. and Goodman, S.M. (2003) 'Introduction to the flora of Madagascar', in S.M. Goodman and J.P. Benstead (eds) *The Natural History of Madagascar,* p229, University of Chicago Press, Chicago, USA

Harutyunyan, M., Avabyan, A. and Hovhannisyan, M. (2008) 'Impoverishment of the gene pool of the genus *Aegilops* in Armenia', in N. Maxted, B.V. Ford-Lloyd, S.P. Kell, J.M. Iriondo, M.E. Dulloo and J. Turok (eds) *Crop Wild Relative Conservation and Use,* pp309–315, CAB International, Wallingford, UK

Hasanuzzaman, S.M., Dhillon, B.S., Saxena, S., Upadhyaya, M.P., Joshi, B.K., Ahmad, Z., Qayyum, A., Ghafoor, A., Jayasuriya, A.H.M. and Rajapakse, R.M.T. (2003) *Plant Genetic Resources in SAARC Countries: Their Conservation and Management*, SAARC Agricultural Information Centre, Dhaka, Bangladesh

Hladik, A., Blanc, P., Dumetz, N., Jeannoda, V., Rabenandrianina, N. and Hladik, C.M. (2000) 'Données sur la répartition géographique du genre *Ravenala* et sur son rôle dans la dynamique forestière à Madagascar', in W.R. Lourenco and S.M. Goodman (eds) *Diversity and Endemism in Madagascar,* pp93–104, Mémoires de la Société de Biogéographie de Paris

Ibisch P. and Mérida, G. (2003) *Biodiversidad: la riqueza de Bolivia. Estado de conocimiento y conservación.* Editorial FAN, Santa Cruz, Bolivia

MDS-SERNAP (2001) *Sistema Nacional de Áreas Protegidas de Bolivia,* Ministerio de Desarrollo Sostenible y Planificación, Servicio Nacional de Áreas Protegidas (MDS-SERNAP), La Paz, Bolivia

MDS-VRFMA-DGBAP (2004) *Diagnosticos sobre el Biocomercio en Bolivia y Recomendaciones para la puesta en marcha del Programa Nacional de Biocomercio Sostenible,* Ministerio de Desarrollo Sostenible, Viceministerio de Recursos Naturales y Medio Ambiente, Direccion General de Biodiversidad. Fundación Bolivia Exporta – Programa Nacional de Biocomercio Sostenible (MDS-VRFMA-DGBAP), La Paz, Bolivia

Ministry of Environment and Natural Resources (1999) *Conservation of Biological Diversity in Sri Lanka: A Framework for Action*

Moat, J. and Smith, P. (2007) *Atlas of the Vegetation of Madagascar*, Kew Publishing, Royal Botanic Gardens, Kew, UK

Orellana Halkyer, R. and Ramos Morales, J.P. (2009) 'Presentation' in VMABCC-Bioversity, *Libro Rojo de Parientes Silvestres de Cultivos de Bolivia*, PLURAL Editores, La Paz, Bolivia

Rakotondrainibe, F. (2003) 'Checklist of the pteridophytes of Madagascar', in S.M. Goodman and J.P. Benstead (eds) *Natural History of Madagascar*, pp295–313, University of Chicago Press, Chicago, US

Schatz, G.E. (2000) 'Endemism in the Malagasy tree flora', in W.R. Lourenço and S.M. Goodman (eds) *Diversité et endémisme á Madagascar*, pp1–9, Mémoires de la Société de Biogéographie, Paris

Vasquez, R. and Coimbra, G. (1996) *Frutas Silvestres Comestibles de Santa Cruz*, Santa Cruz, Bolivia

VMABCC-Bioversity (2009) *Libro Rojo de Parientes Silvestres de Cultivos de Bolivia*, PLURAL Editores, La Paz, Bolivia

Chapter 3

What Do We Mean By *in situ* Conservation of CWR?

There is a need for more effective policies, legislation and regulations governing the in situ and on farm management of PGRFA, both inside and outside of protected areas (Second Report on the State of the World's Plant Genetic Resources for Food and Agriculture, 2010).

General and specific aims of *in situ* species conservation

It might appear to be a simple matter to explain what is meant by *in situ* conservation, but it has proved extremely hard to provide a clear and generally agreed definition of this key component of biodiversity conservation. As noted in the introductory chapter, most countries have not attempted to conserve CWR *in situ*. The reasons for this are various and complex, but there are two basic explanations for such neglect: the first lies in the difference in perceptions by the conservation and genetic resources sectors as to what *in situ* conservation means, how it is practised and why it is undertaken; the second is simply the complexity of the process and the wide degree of interdisciplinary cooperation it requires.

In situ conservation is a term that is applied to a variety of situations (see Box 3.1). It deals principally with (a) the conservation of natural habitats, notably in protected areas and other kinds of reserves; and (b) the conservation, maintenance or recovery of viable population of species in their natural habitats. In the case of CWR, the conservation of the widest range of genetic traits of potential use in plant breeding is of great concern and the term *genetic conservation* is often applied (see below).

Box 3.1 The various forms of *in situ* conservation

- conservation of natural or semi-natural ecosystems in various types of reserves or protected areas;
- conservation of agricultural biodiversity, including entire agroecosystems and the maintenance of domesticates (on-farm);
- conservation and maintenance of target species in their natural or semi-natural habitats;
- genetic conservation;
- species recovery programmes; and
- habitat restoration.

Long-term aims of *in situ* conservation of CWR

The main general aim and long-term goal of *in situ* conservation of target species is *to ensure their survival, evolution and adaptation to changing environmental conditions such as global warming, changed rainfall patterns, acid rain and habitat loss, through taking steps to protect, manage and monitor selected populations in their natural habitats so that the natural evolutionary processes can be maintained, thus allowing new variation to be generated in the gene pool.*

Most importantly, according to Frankel et al (1995), '*in situ* conservation is the method that preserves biological information on genetic diversity in context. Not only does it conserve the genetic diversity relevant to intra-specific and inter-specific interactions among organisms and their associated pests and beneficial species, it is also present in populations that are or have been host to the relevant biotypes of the pathogen or symbiont'.

In addition, various additional specific goals may be recognized (see Box 3.2):

In situ conservation of exploited species

Many of the species that may be targeted for *in situ* conservation because of their economic use are subject to exploitation, among them wild fruit trees, and medicinal and aromatic plants. It should not be assumed that the conservation objective is simply to maintain the species in such a way that they will continue to evolve as natural viable populations; it may be that the emphasis will be more on sustaining the use of the species itself for the benefit of various stakeholders, and this will affect the management objectives. As a recent review of sustainable use and incentive-driven conservation points out, these management objectives may include the conservation of the species (or its populations), the ecosystem in which they occur, or the livelihoods that depend on the species' exploitation (Hutton and Leader-Williams, 2003).

On-farm conservation

In the case of domesticates or cultivated species, *in situ* conservation refers to the maintenance of landraces or cultivars, not of wild species, in the surroundings

Box 3.2 Specific goals for *in situ* conservation of CWR

- Ensuring continuing access to these populations for research and availability of germplasm; for example, native tree species may be important plantation species within the country or elsewhere and thus *in situ* conservation will allow access to these forest genetic resources in the future, if needed.
- Ensuring continuing access to or availability of material of target populations maintained and used by local people, as in the case of medicinal plants, extracted products (e.g. rubber, palm hearts), and fuelwood.
- Selection for yield potential, i.e. genetic potential that confers desirable phenotypic traits (Hattemer, 1997), for example in forest trees, fruit- or nut-producing trees (Reid, 1990).
- Conserving species that cannot be established or regenerated outside their natural habitats, such as: species that are members of complex ecosystems (e.g. tropical forests, where there is a high degree of interdependency between species); species with recalcitrant seeds or with fugacious germination; or species with highly specialized breeding systems (e.g. those dependent on specific pollinators, which in turn depend on other ecosystem components) (FAO, 1989).
- Enabling some degree of conservation of other species occurring in the same habitats as the CWR, some of which may be of known economic value or of importance in maintaining a healthy ecosystem. This may provide additional justification for single-species conservation programmes.
- Minimizing human threats to genetic diversity and supporting actions that promote genetic diversity in target populations (Iriondo and De Hond, 2008).
- Minimizing the risk of genetic erosion from demographic fluctuations, environmental variation and catastrophes (Iriondo and De Hond, 2008).

where they have developed their distinctive properties, along with their pollinators, soil biota and other associated biodiversity; this is commonly referred to as '**on-farm conservation**'[1] (see Box 3.3). On-farm conservation has been defined as 'the sustainable management of genetic diversity of locally developed traditional crop varieties, with associated wild and weedy species or forms, by farmers within traditional agricultural, horticultural or agri-silvicultural cultivation systems' (Maxted et al, 1997). It is a form of conservation of agricultural biodiversity but is quite distinct from the conservation of CWR and is not considered further in this manual.

National and international mandates for *in situ* species conservation

The conservation of species and their populations *in situ* is mandated by the Convention on Biological Diversity (CBD), which includes, in Article 8, '…*the conservation of ecosystems and natural habitats and the maintenance and recovery of*

Box 3.3 *In situ* conservation on-farm

In situ conservation on-farm, sometimes referred to as 'on-farm conservation', has been defined as 'the continuous cultivation and management of a diverse set of populations by farmers in the agroecosystems where a crop has evolved' (Bellon et al, 1997). On-farm conservation concerns entire agroecosystems, including immediately useful species (such as cultivated crops, forages and agroforestry species), as well as their wild and weedy relatives that may be growing in nearby areas. Within this definition, it is possible to identify a wide range of objectives that may shape an on-farm conservation programme. These include:

- to conserve the processes of evolution and adaptation of crops to their environments;
- to conserve diversity at different levels – ecosystem, species and within species;
- to integrate farmers into a national plant genetic resources system;
- to conserve ecosystem services critical to the functioning of the earth's life-support system;
- to improve the livelihood of resource-poor farmers through economic and social development;
- to maintain or increase farmers' control over and access to crop genetic resources.

Source: Jarvis et al, 2000

viable populations of species in their natural surroundings and, in the case of domesticated or cultivated species, in the surroundings where they have developed their distinctive properties'. Specifically, *in situ* conservation is also addressed by the CBD's Global Strategy for Plant Conservation (GSPC) by both target vii, '60 per cent of the world's threatened species conserved *in situ*' and target viii, '10 per cent of threatened plant species included in recovery and restoration plans'. However, as Heywood and Dulloo (2005) note, none of the CBD's decisions or work programmes have specifically focused on how the *in situ* conservation or maintenance of viable populations of species is to be achieved, even though it is recognized in the Preamble to the Convention as a fundamental requirement for the conservation of biological diversity. Likewise, efforts to address this subject through the GSPC under targets vii and viii have not made much progress and are currently (September 2010) under review.

The Global Plan of Action (GPA) on Plant Genetic Resources for Food and Agriculture (FAO, 1996), together with the first report on the *State of the World's Plant Genetic Resources for Food and Agriculture*, was adopted by representatives of 150 countries during the Fourth International Technical Conference on Plant Genetic Resources, held in Leipzig, Germany from 17 to 23 July 1996. The report presents a global strategy for the conservation and sustainable use of plant genetic resources and, to some extent, complements the provisions of the CBD. The GPA specifically recognizes the need to promote *in situ* conservation of wild crop relatives and wild plants for food production (Priority Activity Area 4: Promoting

Box 3.4 Promoting *in situ* conservation of wild crop relatives and wild plants for food production

The long-term objective of this activity is to promote the conservation of genetic resources of crop wild relatives and wild plants for food production, in protected areas and on other lands not explicitly listed as protected areas. The Plan calls for some recognition of the valuable role crop wild relatives and wild plants play in food production, which should be taken into account in planning management practices. In addition, the importance of women in terms of their knowledge of the uses of wild plants for food production and as sources of income is acknowledged. Another important objective is to create a better understanding of the contribution of plant genetic resources for food and agriculture to local economies, food security and environmental health, and to promote complementarity between conservation and sustainable use in parks and protected areas by broadening the participation of local communities as well as other institutions and organizations engaged in *in situ* conservation. The importance of conserving genetic diversity for these species in order to complement other conservation approaches is also highlighted.

The activities of the International Treaty (ITPGRFA) relevant to *in situ* conservation are (see Article 5 – *Conservation, exploration, collection, characterization, evaluation and documentation of plant genetic resources for food and agriculture*):

- **Survey** and inventory plant genetic resources for food and agriculture, taking into account the status and degree of variation in existing populations, including those that are of potential use and, as feasible, assess any threats to them;
- **Promote** *in situ* conservation of crop wild relatives and wild plants for food production, including in protected areas, by supporting, *inter alia*, the efforts of indigenous and local communities;
- **Monitor** the maintenance of the viability, degree of variation and the genetic integrity of collections of plant genetic resources for food and agriculture.

Source: FAO, 1996

in situ conservation of wild crop relatives and wild plants for food production – see Box 3.4). The GPA notes that:

- Natural ecosystems hold important plant genetic resources for food and agriculture, including endemic and threatened wild crop relatives and wild plants for food production.
- Many such ecosystems and resources are not managed sustainably.
- This genetic diversity, because of interactions that generate new biodiversity, is potentially an economically important component of natural ecosystems and cannot be maintained *ex situ*.
- Unique and particularly diverse populations of these genetic resources must be protected *in situ* when they are under threat.

- Most of the world's 8500 national parks and other protected areas, however, were established with little specific concern for the conservation of crop wild relatives and wild plants for food production.
- Management plans for protected and other areas are not usually broad enough to conserve genetic diversity for these species to complement other conservation approaches.

While both the GPA and ITPGRFA recognize the importance of conserving CWR, the former has no dedicated funding mechanism for any of its activities and the latter does not have a specific funding arrangement for *in situ* conservation, as opposed to *ex situ* conservation, of plant genetic resources, including CWR. In view of the major contribution that CWR make to enhanced food production through the provision of genetic materials for breeding improved crops, as recognized by the Consultative Group on International Agricultural Research (CGIAR) in its latest draft strategy (CGIAR, 2009),[2] it would be appropriate to create a new fund to finance a major global initiative in this area, comparable to the Global Crop Diversity Trust. Without such a fund, it is highly unlikely that significant progress will be made in conserving CWR.

At a country level, there is considerable variation in national mandates for *in situ* conservation of target species. In some countries (e.g. several European countries, the US, Australia) considerable attention is paid to this topic and management or recovery plans are in place for some species, while in others there is an avowed interest but little action; in yet others, the subject is not even recognized in national conservation/biodiversity strategies. The GSPC should serve to focus attention on this issue through target vii.

Strategic planning for *in situ* species conservation

Until the recent interest displayed by the time-limited targets of the European Union, Millennium Commission and CBD, little attention has been paid to the strategic needs for species conservation. An exception is the very perceptive essay by Woodruff (1989) on the problems of conserving genes and species in the volume *Conservation for the Twenty-First Century* (Western and Pearl, 1989). He writes:

> *If we are really serious about species conservation, we might launch a Species Defence Initiative (SDI). The goals of the programme would include conserving selected species to prevent further environmental degradation. ... The SDI would require a planning policy shift toward maintaining the evolutionary potential of species. This will, in turn, shift the emphasis from simple censuses to determining the genetic quality of the managed populations.*

He then goes on to say that 'far more population-level intervention will be required to conserve most species'. This contrasts with the widely expressed view that, for most wild species, little if any specific conservation action is needed unless the species are seriously threatened. Such a hands-off approach, which is discussed in more detail below, was predicated on the premise that plant and animal diversity (biodiversity as we now call it) is safely protected in the world's ecosystems and that when a particular habitat or species became threatened, appropriate protective action could be taken. While this may have been true 50 years ago, we now face a situation in which it is estimated that about a quarter of the world's plant species are threatened and the proportion will only worsen, largely as a result of the widespread and continuing degradation, fragmentation, simplification and loss of terrestrial and aquatic habitats, caused by population movements and growth, changes in disturbance regimes, spread of invasive species, urbanization, industrialization, expanding agriculture and over-consumption and, of particular concern today, climate change. As discussed in Chapter 14, the problems of relying on a static system of protected areas in a period of accelerated climate change are causing us to reconsider traditional conservation strategies.

In such a situation, a static approach to species conservation is no longer justified. With a 100,000, or possibly more, threatened plant species today, many of these being CWR, action must be taken to ensure that threats are contained, if not removed; this represents a major global challenge. Also, we cannot take comfort in the likelihood that the remaining 300,000 species will continue to be safe in their natural habitats. For one thing, in many cases we simply do not know what their status is or the threats they now face and or will face in the coming decades.

On the other hand, when one considers that most biodiversity probably occurs outside existing protected areas – although precise data are not available – it follows that reliance on protected areas alone is not a viable approach. The *in situ* management of species outside protected areas represents a major challenge and demands considerable innovation and thinking. This is discussed in detail in Chapter 11.

In situ conservation in context

The underpinning of the conservation strategies of most countries is a protected areas system; this is reflected in the CBD where the main thrust of *in situ* biodiversity conservation is through the development of a system of protected areas. This has been criticized by some as being a somewhat restricted or protectionist approach to conservation with little regard for the interests of local communities (Mathews, 2005). As Adams and Mulligan (2003) comment, 'international conventions like the Convention on Biological Diversity (CBD) have come to drive a protectionist programme, including reinforcing the protected area strategy based largely upon a U.S. model of national parks and wilderness reserves ...'. The adoption by the CBD of the so-called 'ecosystem approach', discussed below, addresses these concerns to some extent.

In situ conservation of target species covers a broad spectrum of activities including the preparation and implementation of detailed single-species recovery

Box 3.5 Key distinguishing features of the ecosystem approach

- It is designed to balance the three CBD objectives of conservation, sustainable use and equitable sharing of benefits.
- It places people at the centre of biodiversity management.
- It extends biodiversity management beyond protected areas while recognizing that they are also vital for delivery of the objectives of the CBD.
- It engages the widest range of sectoral interests.

Source: Smith and Maltby, 2003, http://data.iucn.org/dbtw-wpd/edocs/CEM-002.pdf

plans, in the case of those species that are critically endangered; single-species management plans; monitoring for those species that are rare, not threatened or only vulnerable; multi-species recovery plans; and management plans and habitat protection. It should be viewed in the context of a mosaic of land-use options, each of which requires its own range of management approaches: it may be undertaken in nature reserves and other protected areas; in private and publicly owned natural forests, plantations and other types of habitat; as trees, shrubs and herbs in agroforestry systems of various types, including home gardens; in homesteads; and along rivers and roads.

Moreover, as we shall see (in Chapter 12), various forms of *ex situ* conservation may be needed to supplement *in situ* actions, such as conservation collections in arboreta and botanic gardens, properly sampled accessions in seed banks, clone banks, field trials and seed production areas (Palmberg-Lerche, 2002).

In recent years, it has been increasingly recognized by conservation practitioners that because of the limitations of both species-based and ecosystem-based approaches, integrative (sometimes called **holistic** or **complementary**) methods for deciding conservation strategies should be adopted. Essentially, this recognizes that one should adopt whatever scientific and social techniques or approaches (such as *in situ*, *ex situ*, *inter situs*, reintroduction or population reinforcement) are judged to be appropriate to a particular case and circumstances. A similar, but less unambiguous, strategy has been endorsed by the CBD in its promotion of the 'ecosystem approach', in which what is essentially a holistic approach is adopted. The ecosystem approach is defined by the CBD as 'a strategy for the integrated management of land, water and living resources that promotes conservation and sustainable use in an equitable way. Application of the ecosystem approach will help to reach a balance of the three objectives of the Convention' (Box 3.5). It aims to put people and their natural resource-use practices at the centre of decision-making and can be used to seek an appropriate balance between the conservation and use of biological diversity in areas where there are both multiple resource users and important natural values (Masundire, 2004). The core concept of the approach has been described as 'integrating and managing the range of demands we place on the environment, such that it can

Box 3.6 Differences between an ecosystem approach and *in situ* conservation

- There may be more human interventions in *in situ* approaches.
- Ecosystem approaches are more process- or function-oriented.
- *In situ* conservation may be more species-specific and species-centred than ecosystem approaches.
- *In situ* approaches are geographically more restricted.
- Ecosystem approaches primarily conserve habitats, often with little or no knowledge of the genetic resources present in those habitats, whereas *in situ* approaches often target specific genetic resources.

Source: Poulsen, 2001

indefinitely support essential services and provide benefits for all without deterioration to the natural environment' (UK Clearing House Mechanism for Biodiversity).[3]

An annotated bibliography of the ecosystem approach is available at: http://www.icsu-asia-pacific.org/resource_centre/Ecosystem%20Approach%20Annoted%20Bibliography2004.pdf (accessed 23 November 2010).

In situ conservation differs from an ecosystem approach in a number of ways (Box 3.6). In the case of CWR it is much more species-oriented than a purely ecosystem approach.

Complementary conservation strategies, combining *in situ* and *ex situ* approaches, may be necessary in cases where species are highly threatened and/or very valuable. *Ex situ* conservation involves the conservation of the components of biological diversity outside their natural habitats (see Chapter 12) and can act as an insurance policy in case *in situ* measures are unsuccessful and the target species becomes unviable or extinct. Complementary approaches are becoming increasingly important in light of climate change: populations of many species are unlikely to be able to keep evolutionary pace with the rate of change or to migrate to climatically suitable areas.

Interplay between species and habitats

The conservation of species *in situ* logically requires that the sites in which they occur are themselves effectively protected, a condition that does not often apply. Likewise, if threatened species are to be effectively conserved within the boundaries of protected areas, it requires that they be adequately managed and monitored. Unfortunately, as a World Wide Fund for Nature (WWF) survey notes (WWF, 2004), very few protected areas report having comprehensive monitoring and management programmes.

In practice, the conservation of species *in situ* is critically dependent on identifying the habitats in which they occur and then ensuring the protection of both the habitat and the species through various kinds of management and/or monitoring. In the case of threatened species, conservation *in situ* also requires that threats are removed or at least contained. Thus, although *in situ* species conservation is essentially a species-driven process, it also necessarily involves habitat protection. In terms of *in situ* conservation of target species, there is a very close relationship between taking action at the area/habitat level and action at the species population level (Heywood, 2005).

Coarse and fine filter approaches

The targets of conservation range from genes, populations and species to communities, habitats, ecosystems, landscapes and bioregions. In establishing biodiversity conservation goals, either a coarse or fine filter approach may be adopted. The conservation of genes, populations and species is sometimes known as the '**fine filter**' approach whereas the conservation of communities and habitats is known as the '**coarse filter**' approach. The original coarse filter concept of conserving entire plant and animal communities in reserves was viewed as an efficient approach to conserving biodiversity that would protect 85–90 per cent of all species, without requiring inventories or the planning of reserves for those species, individually.

In effect, setting aside entire ecosystems in reserves is considered an efficient way to maintain biodiversity because large numbers of species are protected. The idea behind using a coarse filter for ecosystems management is that if intact functioning ecological communities are maintained, the species living in those communities will thrive. To this extent, the coarse filter approach relates to the ecosystem approach but with a much more restricted focus. While it has been suggested that the coarse filter approach protects a large majority of species, this seems highly unlikely today, given the pressures on habitats from various components of global change. In addition, a coarse filter approach neglects a proportion of species and does not address the conservation needs of target species which require a specific and tailored conservation strategy. A complementary fine filter must then be applied to those species that slip through the coarse filter, to ensure their protection. Examples of species needing a fine filter approach are those exploited by humans, such as medicinal plants, CWR or rare species that have a specialized ecology that the coarse filter approach may well not capture.

The dilemma is that most conservationists would argue the number of species requiring some form of targeted conservation action is so great that entire communities rather than single species need to be the focus of conservation efforts. This is almost certainly true for CWR, where a single country may house scores to hundreds of CWR. In Bolivia, for example, nearly 200 CWR have been identified while in Armenia, 2518 CWR species have been inventoried (http://cwr.am/index.php?menu=list).

There is no obvious solution to this dilemma and each country must determine its own CWR conservation strategy. As we discuss later (in Chapter 7), some form of triage is usually employed, giving priority to those wild relatives that are closely related to crops, those that are endangered and therefore in need of urgent action if they are to survive, and so on. Even so, some countries will find themselves in a situation whereby there are still too many priority species to manage. If appropriate conservation action cannot be organized locally, and given that CWR in any country may be relevant to the crops of other countries, the problem assumes an international dimension. In other words, if it is decided that particular CWR are of such importance that their conservation is a global imperative, then international agencies must step in. At present, there is no provision made for such action even though it should logically fall under the mandate of the ITPGRFA.

Active and passive conservation

The assumption is often made that if a species is found to occur within a protected area then, provided the area is adequately managed, the continued survival of the species is likely without further intervention or management action. This is referred to as *passive* conservation, or the 'hands-off' approach, in that the existence of a particular species is coincidental and passive, and not the result of active conservation management. It contrasts with active conservation, which requires positive action to promote the sustainability of the target taxa and the maintenance of the natural, semi-natural or artificial (e.g. agricultural) ecosystems that contain them, thereby implying the need for associated habitat monitoring. Certainly, this assumption is likely to be valid in areas (whether protected or not) that are not subjected to unusual or exceptional pressure and provided the target species is not threatened by other factors. As Simberloff (1998) puts it, 'keep the ecosystem healthy ... and component species will all thrive'. This was regarded as the norm until recently. Unfortunately, it is now increasingly unlikely due to accelerating human-induced environmental pressures characterized collectively as global change (see Box 3.7); much more management intervention is necessary to ensure the survival of viable populations of target species. The implications of global change for CWR are discussed in detail in Chapter 14.

Without effective management, the populations of target species in existing protected areas are at risk of change in size and genetic composition because of the dynamics involved, and the habitats themselves are being put at risk through population pressure or movements, deforestation, the increasing demand for land for growing crops and other forms of anthropogenic change, or by the effects of climate change (see Chapter 14). As a consequence of these changes, the number of threatened species, although not known with any precision, is likely to increase substantially over the coming decades.

Referring specifically to the conservation *in situ* of wild species that are actual or potential genetic resources, Frankel et al (1995) comment that conservation in their natural habitats, within the communities of which they form a part, is the best option and that only when such communities, or individual species within them, are threatened, may some form of protection be necessary – in forestry

Box 3.7 CWR and protected areas

... presence in a protected area, provided the area is adequately managed, will afford some degree of protection to the species housed within it, and by definition it obviates the need to seek and place an area under reserve for the target species concerned. Obviously, if the target species is dominant in its ecosystem, such as forests of Cedrus *or* Abies *in Lebanon and Turkey, then the conservation of the habitat will effectively safeguard it and it will logically be included in the area's management plan. For species that are threatened or endangered, the removal or containment of the factors causing the threat means that some form of intervention is necessary so that a hands-off approach is not appropriate. But even if the wild populations of target CWR taxa selected for* in situ *conservation need little management, the processes involved in the assessment of their distribution, ecology, demography, reproductive biology and genetic variation, and in the selection of number and size of populations and sites to be conserved, are still onerous.*

Source: Heywood, 2008

reserves, genetic reserves or *ex situ*. They consider, however, 'that the genetic resources of the majority of species used by humans can be regarded as reasonably safe in at least a proportion of their natural habitats, although in some instances there is a need for protection, in others for continuing watchfulness'. Such an optimistic perspective can no longer be justified today for the reasons mentioned above. Many CWR are already threatened to some degree and the numbers are almost certain to increase considerably under conditions of global change, notably accelerated climate change. Monitoring of the status of CWR ('continuing watchfulness') will need to be undertaken on a much more extensive and substantial scale than has been customary hitherto. If the target species is threatened, the absence of any management intervention to counter the threats (i.e. passive conservation) will compromise its longer-term survival. Consequently, for such species, habitat protection will need to be supplemented by action at the species/population level.

Moreover, it should be noted that the ways in which protected areas and their component ecosystems are managed varies widely and may not favour the maintenance of populations of the target species. For example, if management is focused on processes or on ecosystem health, it would appear that losses of species would be permitted so long as they did not greatly affect processes like nutrient-cycling.

Genetic conservation/genetic reserve conservation

As noted above, the term '**genetic conservation**' (Frankel, 1974)[4] is often used for the conservation of CWR,[5] and a commonly used approach is known as 'genetic reserve conservation'. It may be defined as '*the location, management and monitoring of genetic diversity in natural wild populations within defined areas designated for long-term conservation*' (Maxted et al, 1997). The focus is on the conservation and utilization of genetic diversity. A *genetic reserve* is essentially a protected area managed in such a way as to maintain suitable ecological conditions for the conservation needs of one or more target species. The goal is to make available as much of the gene pool of the target species as possible for actual or potential use, with a specific focus on conserving genetic traits of potential use in plant breeding, rather than on maintaining as wide a range as possible of the biodiversity of the target species/populations.

Traditionally, in the sampling and conservation of plant genetic resources, the focus has been on maximizing the conservation of genes and alleles of potential value in plant breeding. As Maxted et al (1997) and Iriondo and De Hond (2008) state, the purpose of CWR conservation is to maintain the potential of existing genetic diversity in CWR populations for crop breeding to obtain cultivars that better suit the needs of humankind at each moment. In conservation biology and species recovery programmes, the emphasis has been on the maintenance of the genetic diversity of the population(s) so as to ensure its survival and continued evolution. In light of global change, there are many uncertainties as to what parts of the genetic variation of a species will be of potential value, and this distinction is probably no longer valid. Nonetheless, in the case of both CWR and threatened species, the following actions apply:

- minimize the risk of extinction from demographic fluctuation, environmental variation and catastrophes;
- maintain genetic diversity and potential for evolutionary adaptation;
- minimize human threats to target populations;
- support actions that promote a positive balance between births and deaths in target populations.

Additional actions that apply to CWR (Iriondo and De Hond, 2008) are:

- support actions that promote genetic diversity in target populations;
- ensure access to populations for research and plant breeding;
- ensure availability of material of target populations that are exploited and/or cultivated by local people.

Genetic reserve conservation, as practised so far,[6] has tended to focus more on groups of species occurring together in selected areas rather than on single target species, largely on the grounds of cost-effectiveness, given that the number of target species is likely to exceed available resources for a species-by-species approach. This parallels the multi-species approach recently adopted for recovery programmes by Australia, Canada, the United States and some European Union countries (through the Habitats Directive), although previously the single-species approach has been the norm. The scientific rationale behind the use of

Box 3.8 Examples of genetic reserves and gene management zones

Costa Rica – Corcovado National Park; genetic reserve for avocado (*Persea americana*), nance (*Byrsonima crassifolia*) and sonzapote (*Licania platypus*).

India – National Citrus Gene Sanctuary, Nokrek Biosphere Reserve, Garo, Meghalayas; known for preserving a rich diversity in indigenous citrus varieties including Indian wild oranges (*Citrus indica*, *C. macroptera*).

Palestine – Wadi Sair Genetic Reserve, Hebron; for legumes, fruit trees.

Syria – Sale-Rsheida Reserve; for *Triticum dicoccoides*, *Hordeum* spp.

Turkey – Ceylanpinar State Farm; includes seven genetic reserves for wild wheat relatives *Aegilops* spp., *Triticum* spp.

Kasdagi National Park; includes ten genetic reserves for wild plum (*Prunus divaricata*), chestnut (*Castanea sativa*), *Pinus brutia*, *P. nigra* and *Abies equi-trojani*.

Bolkar Mountains; includes five genetic reserves for *Pinus brutia*, *Pinus nigra* subsp. *pallasiana*, *Cedrus libani*, *Abies equi-trojani*, *Juniperus excelsa* and *Castanea sativa*.

Vietnam – Gene Management Zone in Huu Lien Nature Reserve, Lang Son Province; for *Colocasia* (Taro), litchi, longan, rice, *Citrus* spp. and rice bean.

Uzbekistan – Nurata State Reserve for walnut (*Juglans regia*) stands.

multi-species plans is mainly the assumption that the target species share the same or similar threats. On the other hand, the effectiveness of multi-species recovery conservation programmes for CWR has yet to be sufficiently assessed, but there is evidence from surveys of multi-species plans for wild species undertaken in Australia, Canada and the United States, that insufficient attention/detail is given to individual species within multi-species plans and that to be effective, as much effort would need to be placed on each species as in a series of single-species plans. One report found that nearly half of the multi-species plans failed to display threat similarity greater than that for randomly selected groups of species and concluded that, as currently practised, multi-species recovery plans are less effective management tools than single-species plans (Clark and Harvey, 2002). Another report (Sheppard et al, 2005) concluded that the effectiveness of multi-species recovery planning has yet to be sufficiently assessed and that the primary criticism is the lack of adequate attention to detail being paid to individual species within multi-species plans. In the case of CWR, the limited experience of multi-species genetic reserves means that their longer-term effectiveness has yet to be demonstrated and they should therefore be employed with caution.[7]

Genetic reserves, also referred to as gene management zones (Tan and Tan, 2002) or gene sanctuaries, are usually located in existing protected areas or may be established *de novo* on state-owned or privately owned land that is not currently protected. For examples see Box 3.8.

Special requirements for forestry species

Forests are estimated to cover over a quarter of the land surface of the globe (Kanowski, 2001); however, even though timber trees play a major role in the world economy, in practice, only a limited number are used commercially on an extensive scale. The situation may be summarized as follows (Heywood and Dulloo, 2005):

- Commercial timber is increasingly obtained from intensively managed plantations of a small number of species.
- A relatively small forest area is devoted to enterprises such as agroforestry and urban forestry, which play a small role commercially in global terms but are important nationally in poverty alleviation, in the provision of fuelwood, fruit trees, medicinal plants and other useful products.
- The vast bulk of forest is wild, natural or semi-natural, and not managed.

The conservation of forest genetic resources is often considered a special case and has tended to follow a different and wider set of approaches than those used for CWR and other exploited wild species (Hattemer, 1997). It includes not only the setting aside of areas of natural forest habitat as reserves, but also the regeneration or rehabilitation of forests that have been affected by logging or depleted through other causes, both stochastic and human-induced (see Box 3.9). However, as highlighted by Thomson et al (2001), 'artificial regeneration and establishment of plantations can expose trees to conditions that are very different from those under which they develop in natural forest'. The conservation of forest genetic resources has been described as being at the interface between the conservation of the genetic resources of cultivated species and the conservation of sites (Lefèvre et al, 2001).

The different approaches to forest genetic resource conservation reflect both the nature and special characteristics of trees and their economic role. For example, trees often contain greater genetic diversity than other species (Müller-Starck, 1995; 1997); there may be poor differentiation between and within populations with respect to nuclear markers; there is generally high differentiation among populations for adaptive traits; and the individuals often have long lifespans. It should also be noted that the tree crop and the wild relative are often

Box 3.9 *In situ* conservation of forestry species

In situ *conservation means the conservation of the genetic resources of a target species 'on site', within the natural or original ecosystem in which they occur, or the site formerly occupied by that ecosystem. Although frequently applied to populations regenerated naturally,* in situ *conservation may include artificial regeneration whenever planting or sowing is done without conscious selection and in the same area where the seed or other reproductive materials were randomly collected.*

Source: Palmberg-Lerche, 1993

the same species. In other words, many of the cultivated forms of tree species are usually particular provenances or ecotypes that have been selected from within the natural stands of the species.[8]

There is a need, of course, to distinguish between the conservation of forests as such and their wide range of economic, social, productive and protective values and the genetic management of targeted forestry species. The prospects for *in situ* conservation of forestry species has been reviewed by Namkoong (1986) who concludes that even for the relatively small number of forestry species that have a currently recognized commercial value, the amount of genetic management is limited and 'only very meagre funding is available for any but the most important commercial species in industrialized forestry'. Given that the vast majority of forest plant species have little known or potential commercial value or function that is not served by other species, he believes it is simply not feasible or desirable to consider conserving these on a species-by-species basis; in practice, the management objective most often followed is likely to be that of ensuring the continued existence of a sample of these populations or species in protected areas such as reserves or parks. Even this may be difficult to achieve in view of the lack of information available on the precise distribution and ecology of the species concerned, not to mention their demography, reproductive biology and other key attributes. Based on this view, it follows that the widespread *in situ* conservation of target species is not seen to be practicable, and therefore unlikely to be attempted, by forest authorities.

Despite the somewhat pessimistic assessment by Namkoong cited above, if we adopt a wider conservation perspective (Kanowski, 2001), many tree species play an important part in local economies, either for their wood or for a variety of non-timber forest products (NTFPs) (Ruiz Pérez and Arnold, 1996; Emery and McLain, 2001), although their potential is not always realized. To what extent these lesser-used species should be the subject of targeted *in situ* conservation action is a matter that has to be decided at national or local level.

Protected areas and forest conservation

Setting aside specific areas of forest to protect the features for which they are valued, including particular species, is an ancient and widespread practice. Many forestry species are found in various kinds of protected areas which serve, to some extent, as genetic reserves for these species, even though they are seldom sufficient or adequate for this purpose. It is widely agreed that conservation of forest species requires not only a series of protected areas or genetic reserves, but a comprehensive multi-scale approach that includes both reserves and non-reserve areas, as well as management of the wider matrix in which forestry species occur, from the landscape to the individual stand (Lindenmayer and Franklin, 2002).

Kanowski (2001) summarizes the advantages and limitations of protected areas for effective forest conservation:

> *It is clear that existing protected areas make important contributions to forest conservation, that they do protect many forest values, and that they represent very considerable effort and achievement on the part of all concerned in their establishment and management. It is also clear, however, that existing protected areas are not, in themselves, sufficient to achieve or sustain forest conservation goals. Many are in the wrong place, of inadequate size or inappropriate configuration, too disconnected from their surrounding environment, and inadequately protected from pressures that impact adversely on their conservation values. They seldom comprise more than 10% of any forest ecosystem, seldom protect forests on tenures other than public lands, and are often culturally inappropriate. They are subject to a range of social and economic pressures which may not be compatible with the protection of their conservation values, and which many cannot sustain.*

A considerable number of commercially important forest tree species have been the subject of *in situ* conservation/management action (FAO/DFSC/IPGRI 2001; FAO/FLD/IPGRI, 2004). In fact, some of the most detailed *in situ* genetic conservation studies have been made on forestry species such as the Monterey pine (*Pinus radiata* D. Don) and have been published by the University of California Genetic Resources Conservation Program (Rogers, 2002). In addition to a detailed account of the biology and genetics of this species, the publication contains a series of principles and recommendations for species' *in situ* conservation. The European Forest Genetic Resources Programme (EUFORGEN) network (see http://www.euforgen.org) also deals with a range of species for which management guidelines have been produced. For further information on such guidelines see Heywood and Dulloo (2005, Annex 3).

The term gene conservation forest is sometimes applied to forested areas reserved with the objective to protect the genetic resources of local tree species. An example is the Khong Chiam *In Situ* Gene Conservation Forest (GCF) in the Ubon Ratchathani Province of northeast Thailand. The GCF was set aside specifically to conserve the lowland form of *Pinus merkusii*, one of only six known lowland populations in Thailand, all of which are highly threatened (Granhof, 1998).

Economic and social considerations

Although strong arguments can be made for the conservation of CWR (see Chapter 1), these are often not obvious to either the general public or to local stakeholders. Setting aside large areas of land for the conservation of species whose economic potential is uncertain or cannot be easily perceived is difficult to justify and can be a serious constraint when selecting target species. This is discussed by Rubenstein et al (2005) who note that, 'because the full economic values of wild relatives can rarely be captured by landowners, the use of land to preserve habitats for wild relatives remains undervalued compared with alternative uses such as clearing for agricultural or urban use'. In most cases, the involvement and acquiescence of local inhabitants, farmers, officials and other interested parties is crucial for the successful implementation of *in situ* conservation projects (Damania, 1996); examples of participatory approaches to conservation of CWR are given in Chapter 5.

Further sources of information

Frankel, O.H., Brown, A.H.D. and Burdon, J.J. (1995) *The Conservation of Plant Biodiversity*, Cambridge University Press, Cambridge (see Chapter 6).

Heywood, V.H. and Dulloo, M.E. (2005) In Situ *Conservation of Wild Plant Species – A Critical Global Review of Good Practices*, IPGRI Technical Bulletin, no 11, FAO and IPGRI, International Plant Genetic Resources Institute (IPGRI), Rome, Italy

IPGRI/FAO/DFSC (2002, 2004a, 2004b) *Forest Genetic Resources Conservation and Management* vol 1: *Overview, Concepts and Some Systematic Approaches* (2004a); vol 2: *In Managed Natural Forests and Protected Areas* (In Situ) (2002); vol 3: *In Plantations and Genebanks* (Ex Situ) (2004b), IPGRI, Rome. Volume 2 of the series is a guide to *in situ* conservation of forest genetic resources in managed natural forests and protected areas (*in situ*). It contains guidance and a checklist for developing a programme of *in situ* conservation of target species or a group of species, based on local conditions and specific objectives, and includes a step-by-step approach to enhancing the conservation role of protected areas for forest genetic resources. Further information and examples can be found in volumes 1 and 3 of the series.

Maxted, N., Ford-Lloyd, B.V. and Hawkes, J.G. (eds) (1997) *Plant Genetic Conservation: The* In Situ *Approach*, Chapman and Hall, London.

Meilleur, B.A. and Hodgkin, T. (2004) '*In situ* conservation of crop wild relatives: status and trends', *Biodiversity and Conservation*, vol 13, pp 663–684.

Kanowski, P. (2001) '*In situ* forest conservation: a broader vision for the 21st century', in B.A. Thielges, S.D. Sastrapradja and A. Rimbawanto (eds) In Situ *and* Ex Situ *Conservation of Commercial Tropical Trees*, Faculty of Forestry, Gadjah Mada University and International Tropical Timber Organization, Yogyakarta, pp11–36.

Kanowski, P. and Boshier, D. (1997) 'Conserving the genetic resources of trees *in situ*', in N. Maxted, B.V. Ford-Lloyd and J.G. Hawkes (eds) *Plant Genetic Conservation: The* In Situ *Approach*, Chapman and Hall, London.

Palmberg-Lereche, C. (2002) 'Thoughts on genetic conservation in forestry', *Unasylva*, vol 53, pp57–61.

Notes

1. Jarvis and Hodgkin, 1998; Jarvis et al, 2000.
2. In *Progress Report No. 4: Toward a Strategy and Results Framework for the CGIAR* (CGIAR, 2009), which identifies as one of the proposed mega-programmes – Crop Germplasm Conservation, Enhancement, and Use.
3. http://uk.chm-cbd.net/Default.aspx?page=7707

4. The term genetic conservation was apparently introduced by Erna Bennett (Fowler and Mooney, 1990).
5. It also covers the conservation of traditional crop varieties (on-farm) as well as wild species (Frankel, 1974).
6. Most genetic reserve conservation has been undertaken in Turkey and other countries in the Middle East/SW Asia. For example, see Al-Atawneh et al (2008), Tan and Tan (2002).
7. For a detailed summary of strengths and weaknesses of multi-species and ecosystem-based approaches see Table 1 in Sheppard et al (2005) and Table 3.14 in Moore and Wooller (2004).
8. The same is also true of many medicinal, aromatic and ornamental species.

References

Adams, W.M. and Mulligan, M. (2003) 'Introduction', in W.M. Adams and M. Mulligan (eds) *Decolonizing Nature: Strategies for Conservation in a Post-Colonial Era*, Earthscan

Al-Atawneh, N., Amri, A., Assi, R. and Maxted, N. (2008) 'Management plans for promoting *in situ* conservation of local agrobiodiversity in the West Asia centre of plant diversity', in N. Maxted, B.V. Ford-Lloyd, S.P. Kell, J. Iriondo, E. Dulloo And J. Turok (eds) *Crop Wild Relative Conservation and Use*, pp340–361, CABI Publishing, Wallingford, UK

Bellon, M.R., Pham, J.L. and Jackson, M.T. (1997) 'Genetic conservation: a role for rice farmers', in N. Maxted, B.V. Ford-Lloyd, and J.G. Hawkes (eds) *Plant Genetic Conservation: The* In Situ *Approach*, pp263–289, Chapman and Hall, London, UK

CGIAR (2009) *Progress Report No. 4: Toward a Strategy and Results Framework from the Strategy Team*, Joachim von Braun (chair), Derek Byerlee, Colin Chartres, Tom Lumpkin, Norah Olembo, Jeff Waage, 17 September 2009, http://8270334765023298965-a-cgxchange-org-s-sites.googlegroups.com/a/cgxchange.org/alliance/strategy-and-results-framework-team-reports/StrategyProgressNo4_18909.pdf?, accessed 10 May 2010

Clark, J.A. and Harvey, E. (2002) 'Assessing multi-species recovery plans under the Endangered Species Act', *Ecological Applications*, vol 12, no 3, pp655–662

Damania, A.B. (1996) 'Biodiversity conservation: A review of options complementary to standard *ex situ* methods', *Plant Genetic Resources Newsletter*, no 107, pp1–18

Emery, M. and McLain, R.J. (eds) (2001) *Non-Timber Forest Products: Medicinal Herbs, Fungi, Edible Fruits and Nuts, and Other Natural Products from the Forest*, Food Products Press, Binghamton, NY, USA

FAO (1989) *Plant Genetic Resources: Their Conservation* In Situ *for Human Use*, Food and Agriculture Organization of the United Nations (FAO), Rome, Italy

FAO (1996) 'Global Plan of Action for the Conservation and Sustainable Utilization of Plant Genetic Resources for Food and Agriculture and the Leipzig Declaration', adopted by the International Technical Conference on Plant Genetic Resources, Leipzig, Germany, 17–23 June 1996, Food and Agriculture Organization of the United Nations, http://www.fao.org/ag/AGP/agps/GpaEN/leipzig.htm, accessed 10 May 2010.

FAO/DFSC/IPGRI (2001) *Forest Genetic Resources Conservation and Management, Vol 2: In Managed Natural Forests and Protected Areas* (In Situ), International Plant Genetic Resources Institute (IPGRI), Rome, Italy

FAO/FLD/IPGRI (2004) *Forest Genetic Resources Conservation and Management, Vol 1: Overview, Concepts and Some Systematic Approaches,* International Plant Genetic Resources Institute (IPGRI), Rome, Italy

Fowler, C. and Mooney, P.R. (1990) *Shattering: Food, Politics, and the Loss of Genetic Diversity,* University of Arizona Press, Tuscon, AZ, USA

Frankel, O.H. (1974) 'Genetic conservation: our evolutionary responsibility', *Genetics* vol 78, pp53–65

Frankel, O.H., Brown, A.H.D. and Burdon, J.J. (1995) *The Conservation of Plant Biodiversity,* Cambridge University Press, Cambridge

Granhof, J. (1998) *Conservation of Forest Genetic Resources without People's Participation: An Experience from Northeast Thailand,* Royal Forest Department (RFD) and Forest Genetic Resources Conservation and Management Project (FORGENMAP), Bangkok, Thailand

Hattemer, H.H. (1997) 'Concepts and requirements in the conservation of forest genetic resources', in B. Valdés, V.H. Heywood, F.M. Raimondo and D. Zohary (eds) *Conservation of the Wild Relatives of European Cultivated Plants, Bocconea,* vol 7, pp329–343

Heywood, V.H. (2005) 'Master lesson: conserving species *in situ* – a review of the issues', *Planta Europa IV Proceedings,* http://www.nerium.net/plantaeuropa/proceedings.htm, accessed 10 May 2010

Heywood, V.H. (2008) 'Challenges of *in situ* conservation of crop wild relatives', *Turkish Journal of Botany,* vol 32, pp421–432

Heywood, V.H. and Dulloo, M.E. (2005) In Situ *Conservation of Wild Plant Species – A Critical Global Review of Good Practices,* IPGRI Technical Bulletin, no 11, FAO and IPGRI. IPGRI, Rome, Italy

Hutton, J.M. and Leader-Williams, N. (2003) 'Sustainable use and incentive-driven conservation: realigning human and conservation interests', *Oryx,* vol 37, pp215–226

Iriondo, J.M. and De Hond, L. (2008) 'Crop wild relative *in situ* management and monitoring: the time has come', in N. Maxted, B.V. Ford-Lloyd, S.P. Kell, J.M. Iriondo, M.E. Dulloo and J. Turok (eds) *Crop Wild Relative Conservation and Use,* pp319–330, CAB International, Wallingford, UK

Jarvis, D., Myer, L., Klemick, H., Guarino, L., Smale, M., Brown, A.H.D., Sadiki, M., Sthapit. B. and Hodgkin, T. (2000) *A Training Guide for* In Situ *Conservation On-Farm: Version 1,* International Plant Genetic Resources Institute (IPGRI), Rome, Italy

Jarvis, D. I. and Hodgkin, T. (eds) (1998) 'Strengthening the scientific basis of *in situ* conservation of agricultural biodiversity on-farm: options for data collecting and analysis', Proceedings of a workshop to develop tools and procedures for *in situ* conservation on-farm, 25–29 August 1997, IPGRI, Rome.

Kanowski, P. (2001) '*In situ* forest conservation: a broader vision for the 21st century', in B.A. Thielges, S.D. Sastrapradja and A. Rimbawanto (eds) In Situ *and* Ex Situ *Conservation of Commercial Tropical Trees,* pp11–36, Faculty of Forestry, Gadjah Mada University and International Tropical Timber Organization, Yogyakarta, Indonesia

Lefèvre, F., Barsoum, N., Heinze, B., Kajba, D., Rotach, P., de Vries, S.M.G. and Turok, J. (2001) *In Situ* Conservation of *Populus nigra,* EUFORGEN Technical Bulletin, IPGRI, Rome, Italy

Lindenmayer, D.B. and Franklin, J.F. (2002) *Conserving Forest Biodiversity: A Comprehensive Multiscaled Approach,* Island Press, Washington, DC

Masundire, H. (2004) 'Preface' in Shepherd, G., *The Ecosystem Approach: Five Steps to Implementation,* IUCN – The World Conservation Union, Gland, Switzerland and Cambridge, UK

Mathews, S. (2005) 'Imperial imperatives: ecodevelopment and the resistance of adivasis of Nagarhole National Park, India', *Law, Social Justice and Global Development (LGD),* http://www.go.warwick.ac.uk/elj/lgd/2005_1/mathews, accessed 14 May 2010

Maxted, N., Ford-Lloyd, B.V. and Hawkes, J.G. (1997) 'Complementary conservation strategies', in N. Maxted, B.V. Ford-Lloyd and J.G. Hawkes (eds) *Plant Genetic Conservation, The* In Situ *Approach*, Chapman and Hall, London

Moore, S.A. and Wooller, S. (2004) 'Review of landscape, multi- and single species recovery planning for threatened species', World Wide Fund for Nature (WWF) – Australia

Müller-Starck, G. (1995) 'Protection of genetic variability in forest trees', *Forest Genetics,* vol 2, pp121–124

Müller-Starck, G. (1997) 'Protection of variability in forest tree populations: an overview', in B. Valdés, V.H. Heywood, F.M. Raimondo and D. Zohary (eds) *Conservation of the Wild Relatives of European Cultivated Plants, Bocconea,* vol 7, pp323–327

Namkoong, G. (1986) 'Genetics and the forests of the future', *Unasylva,* vol 38, no 152, pp2–18

Palmberg-Lerche, C. (1993) 'International programmes for the conservation of forest genetic resources', in *Proceedings of the International Symposium on Genetic Conservation and Production of Tropical Forest Seed,* ASEAN/CANADA Forest Tree Seed Centre, Muak Lek, Thailand

Palmberg-Lerche, C. (2002) 'Thoughts on genetic conservation in forestry', *Unasylva,* vol 209, no 53, pp57–61

Poulsen, J. (ed) (2001) *Genetic Resources Management in Ecosystems,* Report of a workshop organized by CIFOR for the SGRP CIFOR, Bogor, Indonesia, 27–29 June 2000. Centre for International Forestry Research (CIFOR), Bogor, Indonesia for CGIAR SGRP, Rome, Italy, http://www.cifor.cgiar.org/publications/pdf_files/grme.pdf

Reid, W. (1990) 'Eastern black walnut: potential for commercial nut-producing cultivars', in J. Janick and J.E. Simon (eds) *Advances in New Crops,* pp327–331, Timber Press, Portland, OR, USA

Rogers, D.L. (2002) '*In situ* genetic conservation of Monterey pine (*Pinus radiata* D. Don): information and recommendations', Report No. 26, University of California Division of Agriculture and Natural Resources, Genetic Resources Conservation Programme, Davis, CA

Rubenstein, K.D., Heisey, J.P., Shoemaker, R., Sullivan, J. and Frisvold, G. (2005) *Crop Genetic Resources: An Economic Appraisal,* Economic Information Bulletin Number 2, United States Department of Agriculture (USDA), Washington, DC

Ruiz Pérez, M. and Arnold, J.E.M. (eds) (1996) *Current Issues in Non-Timber Forest Products Research,* Centre for International Forestry Research (CIFOR), Bogor, Indonesia

Sheppard, V., Rangeley, R. and Laughren, J. (2005) *Multi-Species Recovery Strategies and Ecosystem-Based Approaches,* World Wide Fund for Nature (WWF) – Canada, http://assets.wwf.ca/downloads/wwf_northwestatlantic_assessmentrecoverystrategies.pdf, accessed 14 May 2010

Simberloff, D. (1998) 'Flagships, umbrellas, and keystones: is single-species management passé in the landscape era?', *Biological Conservation,* vol 83, no 3, pp247–257

Smith, R.D. and Maltby, E. (2003), *Using the Ecosystem Approach to Implement the Convention on Biological Diversity: Key Issues and Case Studies,* Ecosystems Management Series No. 2, IUCN – The World Conservation Union, Gland, Switzerland and Cambridge, UK

Tan, A. and Tan, A.S. (2002) '*In situ* conservation of wild species related to crop plants: the case of Turkey', in J.M.M. Engels, V. Ramantha Rao, A.H.D. Brown and M.T. Jackson (eds) *Managing Plant Genetic Diversity,* pp195–204, CAB International, Wallingford, UK

Thomson, L., Graudal, L. and Kjaer, E.I. (2001) 'Selection and management of *in situ* gene conservation areas for target species', in DFSC and IPGRI (eds) *Forest Genetic Resources Conservation and Management, Vol 2*, pp5–12, IPGRI, Rome, Italy

Western, D. and Pearl, M. (eds) (1989) *Conservation for the Twenty-First Century,* Oxford University Press, New York, NY, USA

Woodruff, D.S. (1989) 'The problems of conserving genes and species', in D. Western and M. Pearl (eds) *Conservation for the Twenty-First Century*, pp76–88, Oxford University Press, New York, NY, USA

WWF (2004) *How Effective are Protected Areas?* Preliminary analysis of forest protected areas by WWF – the largest ever global assessment of protected area management effectiveness. Report prepared for the Seventh Conference of the Parties of the Convention on Biological Diversity, February 2004. World Wide Fund for Nature (WWF) – International, Gland, Switzerland

Part II

Conservation Planning

This part describes the planning processes and preparatory actions that are needed before practical conservation action begins. It also covers the tools that may be used to assist in establishing priorities and decision-making.

Chapter 4

Planning for CWR Conservation and Partnership Building

Biodiversity managers often underestimate the commitment, human resources and time necessary to develop trusting relationships that lead to collaboration between communities, other government agencies, businesses and conservationists (Hesselink et al, 2007).

Aims and purpose

The field of CWR *in situ* conservation, like other areas in biodiversity conservation, is susceptible to limited collaboration and subsequent, ineffective planning and implementation. Some reasons for this have already been highlighted and discussed in Chapters 1 and 3, and are also mentioned elsewhere in this manual. Having to operate in a time-bound project context constitutes one major challenge, while the lack of a traditional culture of collaboration between the agriculture, forestry and conservation sectors represents another. Addressing this existing disconnect and bridging such gaps is surely one of the foremost challenges limiting the success of CWR *in situ* conservation. This part aims to provide the reader with information and guidance for consideration when planning partnerships or collaborations to ensure the effective coordination and implementation of the CWR *in situ* conservation planning process and to highlight why such collaborations are important.

Introduction

Conservation does not just happen; it is the result of a planning process that includes a series of initiatives and policy decisions operating within a particular context – a strategic process of setting priorities and goals. The process may be organized at a national, regional or local level and financed in a wide variety of

Most CWR projects will contain a variety of activities and components, not all of them directly relevant to actual physical, on-the-ground conservation activities. Given project staff and partner tendencies to focus on the often easier tasks of data collection, documentation and public awareness, it is critical that the overall aims and goals of *in situ* conservation are clearly articulated. The most serious shortfall of a project can be the failure to appreciate, until late in the planning process, the importance of the conservation components or the sequence in which they need to be carried out and what *in situ* conservation of target species (as opposed to area conservation) entails. To prevent this, it may be advisable to establish a dedicated conservation committee at an early stage. Certainly, a global or regional project should have a technical advisory committee established early in the process to clarify these issues and a conservation inception workshop held to determine a common understanding of the technical steps involved in the *in situ* process.

Adapted from the UNEP/GEF CWR Project Technical Advisory Committee.

ways. It will involve a number of different agencies and will affect many stakeholders who may or may not be directly involved or consulted. The process must address an equally diverse range of activities – developing national action plans, conservation prioritization, data collection, adapting and developing management plans, community participation, education and public awareness – spanning the natural and social sciences skills spectrum (see Chapter 15). It operates within a timescale and requires considerable financial and human resources. Poor planning and consideration of the *in situ* conservation process and context can lead to a waste of valuable resources, a haphazard approach to the activities involved and a failure to achieve the expected conservation goals.

Planning will require partnership among a diverse range of actors, which may include local and national government agencies, national and international non-governmental organizations, academia, donor organizations, the private sector and local and indigenous communities (the topic of local and indigenous communities is dealt with in detail in Chapter 5 and highlighted in Figure 4.1). It is advisable that due attention is paid to the task of partnership building between these stakeholders at the outset (see the section below, 'planning for partnership'). Each discrete group will bring potential benefits to the partnership, but they will also come with their own interests, perspectives and expectations. It is the task of the conservation manager and planner to harness these divergent views for the greater good of the partnership and CWR conservation. This is a skill that few are adequately prepared for, and although often discussed in biodiversity and development circles, scant attention is given to mobilizing effective partnerships or providing the capacity building necessary to achieve this. Despite the complexity and challenges, effective planning and partnerships can lay the foundation for successful CWR *in situ* conservation by harnessing the enthusiasm, skills and resources of those working in this area, building on their strong interest in protecting this valuable global resource.

Figure 4.1 *When establishing partnerships, it is important to enter into consultation and dialogue with indigenous and local communities at an early stage. This is the main topic of Chapter 5*

Source: Danny Hunter

Context for planning – requirements of sponsoring agency (national or international) and timescales

> There is a strong case to be made for improved evaluation and learning by donor agencies involved in CWR *in situ* conservation. While most initiatives have been project-driven, there has been little attempt at any real organizational learning arising from this, other than the usual project evaluation exercises. To date, no effort has been made to undertake a strategic meta-type evaluation of multiple, related projects (even in cases where there is a common donor). We now have examples of national, regional and international projects, and there is much to be shared within and between agencies. Such analysis would generate key lessons learned and improved good practices that would influence future project interventions that are more tailored towards the long-term nature of conservation, such as more use of south-to-south capacity development.

Currently, most CWR *in situ* conservation projects have been sponsored by grants from agencies such as the Global Environment Facility (GEF) and with governmental approval and some degree of financial or in-kind support. They fall within traditional project implementation and funding cycles which introduce challenges for long-term conservation planning. As well as having a limited timescale, these projects usually have a specific geographic focus and often involve working at particular locations. In addition, countries may participate as part of a regional or global initiative, which adds another level of complexity to the process. By their nature, these grants must follow the detailed goals, requirements and restrictions of the sponsoring agency(ies) and are strictly monitored with onerous reporting requirements which can compromise conservation actions. In addition to GEF, the European Union and FAO have also provided support for CWR conservation projects in the past, and international NGOs such as the World Wide Fund for Nature also support CWR-related activities, although to a lesser extent. Unfortunately, there are few other agencies that actively support CWR conservation. In the case of GEF-supported projects, other institutions, in addition to the GEF, include the implementing agency (which may be UNEP, UNDP, FAO or the World Bank), the national government(s) and their relevant ministries and agencies, the executing agencies (in the case of the UNEP/GEF CWR Project, Bioversity International). Global and regional GEF projects also offer opportunities to collaborate with international partners, which, in the case of the UNEP/GEF CWR Project, include FAO, IUCN, BGCI and UNEP-WCMC. Some form of an international steering committee involving these various actors is required to provide guidance and oversight to activities that should be described in detail in the project's terms of reference.

Collaborating with international partners provides a much needed opportunity to attract technical expertise to a project as well as the possibility of co-financing, a compulsory requirement for GEF projects. When looking for possible international partners, it is critical to clearly define where and how their involvement will be required and to determine the most appropriate agency for the task. If co-financing is a requirement, you need to ensure the agency is committed to meeting its contribution.

A major constraint for any CWR *in situ* conservation project is the timescale. By its nature, CWR *in situ* conservation is a long-term approach: not only does it require considerable time for project preparation, but its success (or failure) may not be evident for 5–10 years, or even longer, after the initiation of activities. Indeed, as noted in Chapter 10, a conservation management or recovery plan may take many years to achieve and have short-term, medium-term and long-term goals of 30 years or more. Likewise, monitoring the success of CWR *in situ* interventions may be open-ended. On the other hand, funding for such activities, if obtained through grants, tends to be time-bound, limited to 3 to 5 years and usually without the possibility of renewal. This is why it is important to clearly convey the long-term nature of the project when preparing a CWR *in situ* project proposal for sponsors. It is also the reason why, ultimately, the responsibility for *in situ* conservation of CWR should be assumed by the state or by an international agreement. Further, it is critical that some form of overarching CWR national action plan or strategy is put in place, if this is not already the case (see Chapter 6).

Obviously, donor agencies such as the GEF are not in a position to provide long-term funding to specific conservation projects. However, until new sustainable funding mechanisms, as proposed in Chapter 3, are identified, this is the reality and the situation needs to be dealt with as effectively as possible. Typically, donor agencies look upon projects as short-term interventions, which will demonstrate localized impacts and in which partners, such as national governments and NGOs, will find value and seek to scale up and sustain. While challenging and rather idealistic, partnerships established for conservation planning can play a unique role in this situation. A partnership can help accurately identify financial needs and explore avenues that may sustain long-term conservation activities after the donor funding ends. This would include identifying funding gaps, sources and opportunities, as well as developing strategies to address these. Clearly, having an effective partnership in place, even in a project context, can assist with long-term planning and issues of sustainability surrounding CWR *in situ* conservation actions. This is more likely if the aforementioned national CWR action plan is in place.

Despite their different contexts, biodiversity frameworks and government structures, the CWR Project countries have formed an effective working partnership and have acquired unique experience in one of the most difficult areas of agrobiodiversity conservation. The attention of the GEF should be drawn to this and sympathetic consideration given to any proposals made for continuing this work into the future so that the effectiveness of the approaches developed in the project can be fully tested and applied by other countries.

Source: UNEP/GEF CWR Project Technical Advisory Committee.

Figure 4.2 *Beatriz Zapata Ferrufino (Bolivia) explaining CWR plans. National project coordinators and focal points have a large responsibility for consulting widely and explaining the project or programme to stakeholders*

Source: Bioversity International

Implications for national planning

A key challenge working in a project- or donor-driven context, with a focus on disbursement of funds and achievement of milestones and outputs, is the difficulty this presents in terms of the long-term nature of CWR *in situ* conservation, the need for organizational capacity development and for mainstreaming CWR conservation into relevant national programmes and strategies.

The constraints and challenges described for *in situ* conservation require a strategic, comprehensive and inclusive planning process. Adequate planning brings many benefits to enhancing CWR conservation (see Box 4.1).

A lead agency or organization with a mandate and capacity to plan and coordinate CWR *in situ* conservation activities will need to be identified. It is also likely that a national focal point for CWR *in situ* conservation will need to be identified within this agency. It will be the task of the mandate agency and the national focal point to facilitate bringing together relevant stakeholders and putting in place an appropriate process to undertake the planning and implementation of the range of activities necessary for successful CWR *in situ* conservation (see Figure 4.2). The national focal point will be responsible for articulating the objectives, goals and resources of the project and for ensuring that relevant stakeholders have a clear understanding of this information.

National focal points will be required to spend significant amounts of time consulting widely and publicizing a project or programme. This will include private and public meetings to describe the project and its goals and objectives, the type of partners sought, how to get involved, roles, responsibilities and obligations, and contacts for further information. This is more difficult than it may seem and it is important to not raise expectations unrealistically.

Given the level of complexity and multi-stakeholder nature of the task, it may be necessary to establish a national steering committee (see Box 4.2), which will have the overall responsibility for national planning and decision-making. The committee should have membership from as many relevant stakeholder groups as possible and detailed terms of reference should be provided. This may require formal or informal agreements, as outlined below, depending on the national context. It must be stressed that there will be pressure placed on national focal

Box 4.1 The benefits of planning

- decision-making is based on a clear understanding among all relevant stakeholders of the project, its goals and objectives and the resources available;
- roles and responsibilities are assigned and agreed;
- improved use of financial, staff and organizational resources;
- increased transparency and accountability;
- improved communication;
- being better placed to take advantage of opportunities;
- enhanced commitment and ownership.

Box 4.2 Steering the process

Given the complexity of CWR *in situ* conservation and the wide range of relevant institutional interests, it will be important to have a national coordinating mechanism or a national steering committee to oversee the planning and implementation process. In Bolivia, prior to the implementation of the UNEP-GEF CWR Project, a national steering committee with the role of guiding and monitoring project progress was formed. Representatives included senior decision-makers from the following institutions: the General Directorate of Biodiversity from the Vice-Ministry of Environment and Natural Resources; Unit of Production and Technology of the Vice-Ministry of Agriculture; the National Protected Areas Service; the Confederation of Indigenous Peoples of Bolivia; the Instituto de Ecología UMSA; and the seven national executing partner institutions of the CWR Project itself.

Source: Beatriz Zapata Ferrufino, National Project Coordinator, Bolivia

points and mandate agencies to meet the demands of all stakeholders. This must be carefully managed in an open and transparent manner and a national steering committee is well placed to balance priorities (this is a topic dealt with in detail in relation to species and locations in Chapter 7, see also Box 4.3).

More importantly, the committee should have linkages to, and be in communication with, other national biodiversity planning and reporting committees and processes so that CWR conservation receives national attention and recognition, which hopefully will translate into greater mainstreaming, political support and resources.

It is unlikely that such a national committee would be capable of undertaking the planning and coordination of all national activities. This will depend on the geographical size of the country, the political and institutional culture, diversity of agencies and stakeholders, and overall national capacity and resources. In many

Box 4.3 Whose priority counts?

As Chapter 7 illustrates, the task of prioritization of target CWR species for conservation action is an important, yet challenging, one. It is a task that will require consultation and negotiation with a wide range of stakeholders and institutions to reach consensus on a methodology, to ensure that relevant data is made available and to secure stakeholder and institutional commitment to follow-up actions. Each agency will have its preferred species and corresponding expertise, but this must be balanced against other criteria. In Armenia, Bolivia, Sri Lanka and Uzbekistan the process of prioritization sometimes took up to two months and involved a total of 97 experts from 27 different national organizations, including government departments, research institutes, universities, genebanks, herbaria, botanic gardens, indigenous peoples' organizations and non-governmental organizations.

instances, it will be necessary to initiate sub-committees that plan and coordinate activities addressing a particular geographical location or a thematic technical area, such as a sub-committee set up to develop a national CWR action plan, adapt a protected area management plan for CWR conservation or to prioritize target species for conservation actions (see Box 4.3).

Another key role for a national steering committee is to oversee the development of a national communication plan (see Chapter 16) and a national capacity development plan (see Chapter 15), both of which should be linked to any national CWR action plan or strategy (see Chapter 6).

Planning for partnership

This chapter has already referred to the difficulties involved in facilitating effective partnerships for conservation. Such impediments, and proposals for bridging them, have been described for disciplines strictly within the natural sciences (Golding and Timberlake, 2003; Lowry and Smith, 2003), as well as for disciplines from across the natural and social sciences spectrum (Mascia et al, 2003; Campbell, 2005). Despite the historical and complex reasons for these disconnects, which are beyond the scope of this manual, it is important to know that, with attention to planning and detail, progress can and must be made.

What is partnership?

Building partnerships for CWR *in situ* conservation is about working with others to achieve what cannot be achieved by individuals or individual institutions alone. The diversity of partner organizations involved in the UNEP/GEF CWR Project is already highlighted in the acknowledgements section and demonstrates clearly the scope for involvement. A partnership is a special kind of relationship, in which people or organizations combine their resources to carry out a specific set of activities. Partners work together for a common purpose and for mutual benefit. Different people, organizations and sectors have a wide range of resources and skills to offer each other in this regard. A good partnership should offer effective coordination, minimize duplication and make the best use of available resources; but, most importantly, it should ensure that everyone benefits from their involvement. It should also identify opportunities for collaboration with other initiatives relevant to CWR conservation. Building partnerships differs from 'networking' or 'public relations' in that partnerships are about in-depth relationships, involving a

Before embarking on partnership consider:

- What level of participation is required?
- What dangers/risks are involved?
- What are the potential benefits?

few carefully selected targets and having specific, practical goals as opposed to simply communicating a message or information. They also tend to be based upon informal, collaborative agreements or formal contracts such as memoranda of understanding, but such agreements will largely depend on the context.

The task of planning and implementing a partnership should involve wide consultation and effective communication between potential partners, strong commitment from all involved and, ideally, control of local decision-making on activities and resources. The potential benefits and pitfalls that may arise in partnership are many, but if planned and managed properly, the advantages greatly outweigh these difficulties.

Who can partnerships be built with?

The range of stakeholders involved in CWR *in situ* conservation is extensive, especially as activities will be area-based and those with an interest in the area concerned will normally need to be included (see Box 4.4).

Importantly, the conservation of CWR involves two major sectors that traditionally do not work together – agriculture and biodiversity conservation. This, in itself, presents an added challenge to the task of forming effective partnerships for CWR *in situ* conservation. There are already many techniques and methodologies for identifying and engaging potential stakeholders; these will not be elaborated here. Instead, the reader is directed to such tools and resources highlighted at the end of this chapter and also described in Chapter 5.

Box 4.4 Guidelines on identifying your key stakeholders

Key individuals who will play a role in a national strategy for CWR *in situ* conservation might include:

- political leaders and senior policy-makers;
- senior biodiversity, environment and agriculture decision-makers;
- heads of relevant organizations and institutes;
- national and local policy planners;
- scientists and researchers;
- protected area managers;
- project management staff;
- field technicians;
- university lecturers and postgraduate students;
- communications and public awareness specialists;
- extension and outreach specialists;
- information analysts and managers;
- training specialists; and
- community and indigenous leaders and groups.

When facilitating partnerships consider:

- common interests;
- common goals;
- reputation, both nationally and internationally;
- level of expertise;
- past track record, including past achievements/problems;
- proposed partner already working in similar area;
- clear objectives of what to achieve;
- what is in it for the partners;
- power relations with other sectors and actors;
- experience and attitudes towards other partners;
- receptivity to public opinion;
- what drives partners/limits them/enables them; and
- their interests/revenues/rewards.[1]

To assess if the context is conducive to partnership the following checklist questions can be posed:

- Where is the drive or motivation for this partnership coming from?
- How do you expect the partnership to address the problem?
- Will the partners be able to achieve more together than they would working on their own or individually?
- Is the partnership based on partners' differences rather than their similarities?
- What are the main strengths that each partner brings to the partnership?
- Are there gaps in strengths or skills that might be filled by another partner not yet identified?
- What do partners expect from the partnership?
- What do partners fear from the partnership?
- What can the partnership do to avoid, reduce or deal with these fears?
- Are there any problems or conflicts between partners before the partnership commences?
- Do the partners gain access to additional funds and resources that neither could access on their own?
- Will this access be on an equal basis?
- Will the partnership build a sense of local ownership?
- Will the partnership help sustain CWR *in situ* conservation actions?[2]

Planning the partnership

The development of a partnership should not be rushed and must be carefully nurtured if many of the above pitfalls are to be avoided. Roles and responsibilities will have to be clearly articulated and understood and these may need to be formalized in the appropriate manner. Three basic ingredients of a partnership need to be considered and negotiated (see Box 4.5).

Box 4.5 Partnership planning checklist

- Focus of the partnership
  - Define the objective (project, activity, product) of the partnership.
  - Define the time (period) and place involved.
  - If necessary, make sure it is clear what is *not* the objective of the partnership.
  - Define the limits of the partnership (a partnership does not mean complete involvement in each other's activities).
- Organization of the partnership
  Many of the challenges involved in partnerships can be managed through planning, but to complement this, you may want to establish formal or informal collaborative agreements to avoid misunderstandings and conflicts. These may include:
  - informal agreements, verbal agreement, guiding principles;
  - formal agreements (e.g. memorandum of understanding);
  - contracts (formal and legal).
- Rights and obligations of each partner
  Administrative, financial and legal issues involved will have to be openly discussed and agreement reached on such issues as:
  - financial inputs, material inputs;
  - access to resources;
  - sharing of information and benefits;
  - sharing of unexpected costs;
  - publicity and communication strategy;
  - financial accounting and liability aspects;
  - work plans, milestones, roles and responsibilities;
  - monitoring and reporting requirements.

Source: adapted from 'The Partnership Toolbox', WWF and other WWF partnership tools (see Further sources of information)

The importance of communication in the partnership

A partnership will bring together a variety of stakeholder interests, motives and objectives. Balancing these in a fair and open manner is one of the important challenges in managing a partnership. It is best to promote clear and open communication, right from the beginning, about partners' motives and desired benefits in order to provide a firm basis for a good partnership. The majority of problems that arise in partnerships can be traced to poor communication or lack thereof. At the planning stage, it is useful for the partnership to consider developing a communications strategy which should also incorporate aspects of external communication and advocacy for the partnership in general, not just internal communication between partnership members. Developing and maintaining clear communication channels between the partners will help build trust, maintain focus and momentum, and ensure that everyone shares in the partnership's successes. The subject of communication is dealt with in detail in Chapter 16.

The experiences of the UNEP/GEF CWR Project partnership

This global partnership was established to improve the *in situ* conservation of CWR and to use the experience of doing so as a platform to create and test tools to enable others to use similar methods. Throughout the project, all partners not only sought to improve matters within the target countries, but also to contribute to global knowledge about CWR and their conservation and use. Chapter 1 has already highlighted the considerable complexities of *in situ* conservation and the acute dilemma posed by climate change. This was the challenging context in which the project and partnership was implemented.

Bioversity International, in collaboration with Armenia, Bolivia, Madagascar, Sri Lanka and Uzbekistan, and the international organizations BGCI, FAO, IUCN and UNEP-WCMC, set out to establish a broad-based partnership to improve the conservation and sustainable utilization of these important resources, maximizing the use of existing information and conservation resources to protect CWR species occurring within these specific countries, through establishing further effective partnerships among relevant national agencies and individuals, and adding to the information base by carrying out original research on the distribution and uses of and threats to those populations.

The partnership was essential to overcome many of the national political, administrative and infrastructural obstacles limiting conservation efforts, and it provided a collaborative framework to target the effective *in situ* conservation of CWR. Most importantly, the partnership provided an interdisciplinary and apolitical platform for information gathering and sharing and for the development of national and international data resources, which are now available for other countries to use and employ.

The partnership included almost 60 national and international agencies essential to the complex and multidisciplinary nature of CWR *in situ* conservation planning and action (see acknowledgements). Planning, implementation and monitoring was carried out through a series of local and national committees, coordinated and guided by Bioversity International through an international steering committee made up of representatives from all participant countries and international organizations. A three-person technical advisory committee provided overall technical direction. At the national level, the partnership brought together individuals from universities, herbaria, government departments of agriculture, environment and biodiversity, protected areas administrations, local and indigenous community groups, NGOs, extension and outreach agencies, botanic gardens, natural history museums and research agencies.

The main advantage of the partnership was that it assembled and integrated the multidisciplinary expertise necessary to meet the complex challenge of *in situ* conservation of CWR (see Box 4.6). The agencies and organizations essential to this process traditionally had little history of working together; the partnership enabled them to do so, with great effectiveness. This, in itself, is a significant achievement. Despite their different cultures and contexts, and biodiversity and

Box 4.6 What did the UNEP/GEF CWR Project partnership achieve?

Because there are thousands of known CWR in the five countries and resources are limited, prioritization is vital. The partnership encouraged each country to consult widely and to negotiate with a diverse range of stakeholders and institutions to reach consensus on priority taxa and to agree on methodologies, to ensure that relevant data were made available and to secure stakeholder and institutional commitment to follow-up conservation actions. As a result, CWR species from 36 different genera were prioritized for action, including ecogeographic assessments. More than 310 CWR species were Red List assessed according to IUCN guidelines, and Bolivia published the first IUCN Red List specifically dedicated to CWR. This is probably the largest set of such assessments undertaken for CWR and represents a major contribution. Furthermore, the partnership worked closely with protected area authorities to develop species management plans for CWR in selected protected areas and put in place a series of important national action plans and strategies. This partnership has substantially expanded the previously limited body of knowledge on *in situ* CWR conservation in developing countries and used a series of innovative communication and outreach products to enhance awareness and understanding of CWR. Further, information and knowledge generated within the partnership has been consolidated in a series of national information systems which are, in turn, linked to a global CWR portal.

In addition, the partnership:

- created important synergies and facilitated sharing and learning through south–south and north–south exchanges;
- enhanced the capacity of individuals, organizations and communities to support CWR *in situ* conservation;
- linked national partners to the best and most up-to-date science by including relevant international partners in the fields of information management, conservation actions and legal and policy review and analysis; and
- strengthened linkages to utilization by undertaking evaluation of selected CWR species with potential for crop improvement.

government structures, the countries have formed an effective working partnership and have acquired unique experience in one of the most difficult areas of agrobiodiversity conservation, which must be undertaken with a long-term view. For this reason, it was crucial to incorporate responsibility for conservation of CWR into national biodiversity and plant genetic resources strategies. Having done so, the countries and the partnership are now well placed to act as hubs for CWR conservation in their regions.

Sources of further information

This chapter benefited immensely from the excellent partnering tools developed by the World Wide Fund for Nature (WWF). *The WWF Partnership Toolbox*, and other partnering tools available from WWF, are useful starting points for resources and guides for establishing, nurturing and monitoring partnerships. Website: http://assets.wwf.org.uk/downloads/wwf_parthershiptoolboxartweb.pdf.

The *Conservation Action Planning (CAP) Handbook*, developed by The Nature Conservancy, is a simple, straightforward and proven approach for planning and implementing conservation projects. The CAP Handbook is available to download from the internet and contains a variety of chapters including Step 1: Identify People Involved. Website: http://conserveonline.org/workspaces/cbdgateway/cap/practices/index_html

Tuxhill, J. and Nabhan, G. P. (2001) *People, Plants and Protected Areas: A Guide to In Situ Management*, Earthscan, London, UK.
This book has a useful chapter on 'who is involved?' when it comes to routine, on-the-ground conservation activities and who you need to be working with to ensure conservation of useful plants in their native habitat. The chapter discusses the various reasons why you cannot expect to achieve successful *in situ* conservation without fully engaging the relevant stakeholders. Some of the discussion is related to Chapter 5 on engaging with local and indigenous communities.

Biodiversity Conservation: A Guide for USAID Staff and Partners provides basic information about designing, managing and implementing biodiversity conservation programmes or activities. This publication includes a chapter on: Involving Stakeholders. Website: http://pdf.usaid.gov/pdf_docs/PNADE258.pdf

The *Effective Engagement* web pages of the Department of Sustainability and Environment, Australia have three useful downloadable documents: 'An Introduction to Engagement'; 'The Engagement Planning Workbook' and 'The Engagement Toolkit'. Website: http://www.dse.vic.gov.au/dse/wcmn203.nsf/Home+Page/8A461F99E54B17EBCA2570340016F3A9?open

Partnerships Online Guide which includes step-by-step guides for creating effective partnerships. Website: www.partnerships.org.uk/

The Partnering Toolbook, written by Ros Tennyson and produced in cooperation with the Global Alliance for Improved Nutrition (GAIN), the United Nations Development Programme (UNDP) and the International Atomic Energy Agency (IAEA), provides a concise overview of the essential elements that make for effective partnering. English and Spanish versions are available for download. Website: http://www.undp.org/partners/business/partneringtoolbook%5B1%5D.pdf

The Partnering Initiative works with individuals, organizations and systems to promote and develop partnerships for sustainable development – between business, government and civil society – and has a number of publications and resources available on their website: www.thepartneringinitiative.org/

(Links last checked 29 May 2010)

Notes

1. Adapted from 'The Partnership Toolbox', WWF and other WWF partnership tools (see Further sources of information).
2. Adapted from 'The Partnership Toolbox', WWF and other WWF partnership tools (see Further sources of information).

References

Campbell, L.M. (2005) 'Overcoming obstacles to interdisciplinary research', *Conservation Biology*, vol 19, pp574–577

Golding, J.S. and Timberlake, J. (2003) 'How taxonomists can bridge the gap between taxonomy and conservation science', *Conservation Biology*, vol 17, pp1177–1178

Hesselink, F. Goldstein, W., van Kempen, P.P., Garnett, T. and Dela, J. (2007) *Communication, Education and Public Awareness: A Toolkit for National Focal Points and NBSAP Coordinators,* Convention on Biological Diversity (CBD) and International Union for Conservation of Nature (IUCN)

Lowry, P.P. and Smith, P.P. (2003) 'Closing the gulf between botanists and conservationists', *Conservation Biology*, vol 17, pp1175–1176

Mascia, M.B., Brosius, J.P., Dobson, T.A., Forbes, B.C., Horowitz, L., McKean, M.A. and Turner, N.J. (2003) 'Conservation and the social sciences', *Conservation Biology*, vol 17, pp649–650

Chapter 5

Participatory Approaches for CWR *in situ* Conservation

Although the role of local people has not figured highly in most examples of in situ *conservation of rare and endangered species ... when we deal with species which have an economic or social value or otherwise impinge on the interests of local communities, such an approach is no longer tenable* (Heywood and Dulloo, 2005).

Aims and purpose

Participatory approaches present many opportunities for CWR *in situ* conservation, in addition to the positive contribution they can make to the social and economic empowerment of often marginalized groups. These approaches, though, also present immense challenges to scientists and their organizations, which often have limited understanding and capacity to support participatory methods effectively. Certainly, in the CWR conservation community, practitioners have had limited exposure to such approaches and techniques, compared to their counterparts in the on-farm community, and there is an almost complete lack of published information on approaches that might be replicated elsewhere.

This chapter focuses on these challenges and opportunities by introducing the concept of community participation and participatory approaches applicable to CWR *in situ* conservation planning and action. It is *not* meant as an exhaustive account of the many methods and tools of participation. The literature and internet abound with information on participatory approaches and tools (and how to use them), which have been successfully applied in other contexts and which can be readily applied to CWR conservation. The chapter guides the reader through some general information on participatory approaches, provides relevant examples and refers to the available resources as described at the end of this chapter. More importantly, the chapter aims to encourage an understanding of the development of participatory approaches, what participation involves and its role in various conservation settings. By doing this, it is hoped that it will make CWR practitioners more aware of the opportunities for such community-based approaches. In the context of this manual, the term 'community' refers to local and indigenous communities. While there are similarities and parallels between participation and partnerships (see Chapter 4), for the purpose of this manual, 'participation' refers to working with communities to achieve conservation and socio-economic goals and involves an element of community empowerment, whereas 'partnership' refers to agreements and working arrangements entered into with other key stakeholder groups, largely for the purpose of CWR *in situ* conservation planning. The chapter concludes by highlighting the importance of biocultural diversity conservation and the potential for collaboration with recent initiatives such as community conserved areas (CCA) and indigenous bio-cultural heritage areas (IBCHA).

Participatory processes are demanding. Those involved must be aware of this reality. There will be many different perspectives and interpretations of purposes and goals which must be discussed and debated. Role reversals and attitudinal change will be required as will new ways of learning. There will also be important resource issues to consider with regards to the significant capacity development required, as well as the funds necessary to support community consultation and engagement. Participation should not be seen as an expedient for convenient implementation of activities. Empowerment, as well as conservation action, should be one of the goals. Common understanding and commitment to this must be established early on in the process.

Introduction

> *Multi-stakeholder processes and terms such as adaptive management, collaborative management, participation, citizen involvement, community based natural resource management, communities of practice, dialogue, interactive decision making and societal learning have proliferated in the natural resources management literature* (Hesselink et al, 2007).

Local and indigenous communities in biodiversity-rich countries have been closely linked to their natural environments for millennia. Often, they have intimate knowledge about habitats and their wild plant species, including wild relatives. This may include knowledge of their sustainable management. In many instances, this intimacy has been disrupted by conventional conservation approaches (United Nations, 2009). The latter part of the 20th century has witnessed a re-examination of some of these approaches to biodiversity conservation, with a growing recognition of the need to enhance the role of local and indigenous communities in the management of their environments and resources. While this may present win–win situations for those involved, it is a process that generates many challenges and potential pitfalls and requires long-term commitment.

Community-based natural resource management (CBNRM)

CBNRM models represent a shift from centralized to more devolved approaches to management which work to strengthen locally accountable institutions, enabling local communities to make better decisions about the use of land and resources. A recent review by the International Institute for Environment and Development (IIED) of the impact of CBNRM approaches has highlighted some notable ecological, economic and institutional achievements. While CBNRM is identified as an important strategy in meeting the goals of various international targets such as the CBD, some important challenges remain.

Source: Roe et al, 2009; http://www.iied.org/pubs/display.php?o=17503IIED

Challenges aside, community participation presents key opportunities for those involved in CWR conservation. Working closely with local communities can facilitate data gathering (see Chapter 8) and provide insights into CWR and indigenous knowledge such as ethnobotanical knowledge on uses, understanding of the distribution of CWR, patterns of the use of CWR and potential threats (see Box 5.1 and Figure 5.1).

Participatory approaches allow opportunities for local and indigenous communities to be involved in planning and partnerships (see Chapter 4). Scientists and organizations can work with communities to strengthen the

Figure 5.1 *Collecting information on wild yams during a consultation with a community bordering Ankarafantsika National Park, Madagascar*

Source: Danny Hunter

Box 5.1 Participatory assessment of utilization of wild plants by local communities in Armenia

To preserve its wealth of globally important agrobiodiversity, in 1981, the Minister Councils of the Armenian Soviet Socialist Republic designated the south-eastern side of Yerevan city a protected area. Occupying an area of approximately 89ha, Erebuni State Reserve is situated in close proximity to a highly urbanized area, bordering the villages of Hatsavan and Voghchaberd, and the Erebuni district of Yerevan city. The reserve is rich in biodiversity and is home to 292 species of vascular plants, representing 196 genera from 46 families. Among these are over 40 species of wild relatives of wheat (*Triticum*), rye (*Secale*) and barley (*Hordeum*).

Despite sustained conservation efforts, the proximity of Yerevan city to the protected area is putting severe pressure on the distribution of wild plants, which are being collected for food and medicinal purposes to be sold in the city markets. Traditionally, wild plants contribute between 10 and 15 per cent to the average Armenian diet, yet due to overharvesting, they are becoming increasingly scarce. Plant collectors frequently trespass within the protected area to harvest and meet the increasing demand for wild crops. The phenomenon is becoming so widespread that many species of plants existing in the area have been included in the Red Data Book of Threatened Plants of Armenia.

In community consultations, lack of awareness of the importance of CWR as repositories of genetic diversity was identified as the major factor influencing overharvesting. For this reason, the UNEP/GEF CWR Project implemented a series of workshops and working groups in 2007, meeting with local community representatives, followed by surveys of residents of the communities, to gather information about the collection, use and conservation status of a range of wild plants. Meetings also provided local communities with the opportunity to learn more about the benefits and importance of conserving these valuable species. Discussions highlighted that rural communities, and mostly women from these communities, continue collecting a variety of wild plants for use in local dishes and for medicinal purposes.

The participatory process, carried out over a one-year period, has revealed the need to train local communities on the correct utilization of particular plant species. This holds particularly true for women, who continue to be the main source of knowledge about wild plants in Armenia, knowledge that has been passed down from generation to generation and continues to the present day. Furthermore, if the conservation efforts being made by the Erebuni State Reserve are to continue in the long term, it is essential that the surrounding local communities are engaged and aware of the benefits of conserving CWR in their natural environments and the threats posed to their well-being by overharvesting. To this end, participatory approaches must be sought, whenever possible, to improve cooperation with local communities to enhance CWR conservation.

Source: Naire Yeritsyan, UNEP/GEF CWR Project, Armenia

Figure 5.2 *Presenting research findings and other information back to communities is an important part of the participatory process*

Source: Danny Hunter

management of habitats and CWR species both inside (see Chapter 9) and outside (see Chapter 11) protected areas. Capacity can be developed so that communities and grassroots organizations are involved in the implementation of national action plans (see Chapter 6) and management plans (see Chapter 10), including species and habitat monitoring (see Chapter 13). Danielsen et al (2009) describe varying levels of local participation in natural resource monitoring which can be applied to CWR conservation. At the same time, working closely with communities presents opportunities to communicate knowledge on the importance of CWR and to raise awareness and build support for CWR conservation (see Chapter 16 and Box 5.1 and Figure 5.2). This can be matched with the appropriate community-based capacity development necessary to undertake related tasks (see Chapter 15).

A remarkable example of a participatory approach is that of farmers in Nepal who were able to improve rice crops by crossbreeding wild and local varieties through a participatory plant breeding programme facilitated by local organizations (Sthapit, 2008), thus demonstrating ways of strengthening the link between CWR conservation and utilization. A detailed account of participatory approaches and tools, such as the development of community biodiversity registers for on-farm conservation (many of which are applicable to the *in situ* context), is given by Friis-Hansen and Sthapit (2000). There has been further progress in participatory approaches to on-farm conservation compared with *in situ* conservation of agrobiodiversity in natural landscapes.

Community participation can help countries implement the CWR conservation actions necessary to meet their obligations and targets, as set out in international agreements and conventions such as the Convention on Biological Diversity (CBD) and the International Treaty on Plant Genetic Resources for Food and Agriculture (ITPGRFA). Also important are the opportunities that community participation in CWR *in situ* conservation offers in contributing to the poverty reduction and social and economic empowerment targets proposed by the Millennium Development Goals (MDG), especially MDG 1 and 7.

Nonetheless, it needs to be stressed that participatory approaches will generate many challenges for scientists who may be used to working with conventional, quantitative research approaches. Most natural scientists are usually not experienced with the attitudes, skills and behaviours considered necessary for participatory approaches. *To ensure an effective participatory process, it is good practice to seek out those social (and natural) scientists in your organization or others with extensive skills and experience in using participatory methods and tools, and facilitating participatory approaches with local and indigenous communities.* It is also good practice to review what other national conservation programmes and projects are doing in order to build on the lessons learned to ensure that the research team is adequately sensitized to the objectives, needs and demands of using a participatory approach. This will also help to identify who to contact for advice and guidance on engaging community groups and organizations (see Boxes 5.2 and 5.3).

What is participation?

Participation is a somewhat ambiguous term that enjoys a high level of popularity in strategy, policy and project documents but is not always accompanied by a similar level of understanding in terms of what it actually means or an appreciation of what is involved. Because of this ambiguity, participation is open to interpretation and variability in practice. Many typologies of participation have been described, such as that of Jules Pretty and colleagues (described in Bass et al, 1995). These are useful as a way of categorizing levels of, or commitment to, participation with the recognition that participation, by default, is not necessarily a good thing. Most typologies of participation describe a continuum of participation from passive to active such as that illustrated in Table 5.1. At its most effective, participation can lead to situations where communities gain control over decision-making and actions, as well as resources through a process of empowerment and self-initiated mobilization.

The goals and objectives of the planned CWR conservation intervention will determine the level and extent of participation required. *It is not always necessary to strive for a level of participation which equates with community autonomy or mobilization, but the work should in some way enhance community empowerment.*

Box 5.2 Involve local and indigenous communities early

Although the Confederation of Indigenous Peoples of Bolivia (CIDOB) were included at the outset (in 2004) as a member of the National Steering Committee of the UNEP/GEF CWR Project in Bolivia, it was more in an advisory role rather than as a partner for executing or undertaking specific project activities. However, CIDOB, as a member institution, was able to play an active role and lobby for the involvement of indigenous peoples. It also played an important role in advising the Vice-Ministry of Environment, Biodiversity and Climate Change (VMABCC), the Project Coordination Unit and the other national partner institutions, to respect and recognize the rights of indigenous peoples over their traditional knowledge associated with CWR, which included securing prior informed consent of indigenous peoples when considering the inclusion of information on traditional knowledge associated with CWR and ethno-botanical studies into the databases of the project's National Information System.

Eventually, with the support of the General Director for Biodiversity and Protected Areas (Indigenous Guaraní), CIDOB was given an executing role in the project in 2007. In December 2006, representatives of CIDOB and the Directorate-General for Biodiversity and Protected Areas – Vice-Ministry of Environment, Biodiversity and Climate Change (DGBAP-VMABCC) held a series of meetings to inform CIDOB on the project scope and the topics on which CIDOB could work. This eventually led to CIDOB carrying out the following activities on CWR species for three genera (*Arachis, Theobroma, Annona*), together with four other national partner institutions of the CWR Project. The activities included:

- systematization of information for inclusion into the institutional database on CWR of CIDOB as a part of NIS;
- creation of distribution maps in community lands;
- ecogeographic surveys and specimen collection in the field, in areas of species distribution and community lands;
- development of public awareness materials; and
- organization of outreach activities in the sub-central lowlands.

The clear lesson learned from CIDOB's involvement in executing CWR activities was the need to identify such activities at the early project design stage. Funds were already committed to other partners and the small amount that could be allocated to CIDOB was too limited to achieve major impact in community lands.

Despite this, the collaboration did achieve important outcomes. Prior to the project, little was known about the issue of *in situ* conservation of CWR by CIDOB as the national indigenous peoples' organization and also by the indigenous peoples on whose community lands many CWR species are found. The project was able to address this issue by strengthening capacity within CIDOB through support of an indigenous technician, the building of linkages to scientific organizations and the considerable sharing of information and knowledge related to CWR that took place. There was also a strong commitment by scientific researchers in the project to explain complex issues in non-technical language, which helped facilitate networking and an increased awareness among scientists of the rights of indigenous peoples over their traditional knowledge and natural resources. The issue of CWR and their conservation are now on the natural resource management agenda of CIDOB.

Such was the interest of CIDOB in these activities that they developed a stand-alone project proposal for the *in situ* conservation of CWR in community lands in an effort to continue the work of the project.

Source: Beatriz Zapata Ferrufino, National Coordinator, UNEP/GEF CWR Project, Bolivia

Box 5.3 Checklist for developing an effective consultation process

Begin consultation at the earliest stage possible of project design

Before commencing project design, consider how communities will be involved in this process and determine the best avenues to secure their engagement.

Prior to visiting local communities and villages, seek permission from community members. Share with them the motivation and purpose of the proposed research and explain the benefits of providing local knowledge and resources. Visit different community groups (e.g. women's groups, farmer associations) and hold meetings to share information about the project. Ensure that information is accessible by the community and is presented in a transparent manner. During community visits, identify local representatives who can serve as future contacts for the development of an agreement outlining project objectives and activities.

After obtaining local permission to undertake research, engage communities in the entire research process. Collect information about the location, population size and community members' interests, concerns and perceptions. You need to fully understand the local context and ensure the project addresses local needs.

Explain to communities their roles and responsibilities, including the activities to be conducted and the impacts these may have on community practices (limits to areas of use or specific species, presence of outsiders, etc.). Respect local traditions, culture and traditional knowledge, working to include community members as much as possible. Explore avenues to overcome language barriers and cultural differences which may hinder the success of the project. A relationship of trust must be built with local communities.

Build the confidence of communities

Community involvement should be at the centre of the project. Make sure to engage communities in the earliest stages of project design; be certain that no one is excluded. It is important to identify and involve traditional decision-making authorities within communities, as well as to encourage the participation of marginalized groups such as women and children. Offer support to these groups and others to ensure their voices are heard. Respect local customs and traditions and provide adequate information for communities to make informed decisions.

Identify stakeholders and their rights over land, natural resources and associated knowledge

To meet the needs of project stakeholders, it is important to identify:

- indigenous groups and local communities directly or indirectly affected;
- landowners and holders of resource rights where research will be conducted;
- authorities with jurisdiction over locations and activities, including local, state and national agencies;

- key persons with knowledge of the cultural, social and economic context of the communities where research will be conducted;
- individuals and authorities with the power to influence the project in a positive or negative manner; and
- community groups to be involved, including women, elders and youth – pay close attention to ensuring the participation of women as they may not hold formal positions in the community, but do bring a unique and important perspective to the table. Separate consultations with women may be required.

Agree on acceptable logistic and administrative frameworks for consultation

Formulate a plan outlining measures for communication and information exchange and access and identify capacity building needs of the communities. Raise awareness among communities to be sure they know their legal rights and their authority to influence the research process. Determine if interpreters are required and identify such support, as needed.

Develop and finalize the project work plan and timeframe for implementation in line with community suggestions and preferences. Seek advice from communities in terms of the most appropriate forums for consultations (e.g. workshops, informal discussions, video presentations). Informal discussions are often useful to identify different needs for consideration, which may not be raised by community members in formal settings or in front of a public audience. Jointly identify themes for, and agree on, the frequency of meetings throughout the life of the project. Ensure that joint decisions are clarified, being careful to consider various views and opinions. Finally, establish a mechanism to review the effectiveness of community consultations and identify accessible means to resolve conflicts which may arise during the project.

Source: adapted from Laird and Noejovich (2002) *Biodiversity and Traditional Knowledge*, Earthscan

Although some writers make it sound as though there is a separate 'participatory' research method, this is misleading. The idea of participation is more an overall guiding philosophy of how to proceed, than a selection of specific methods. So, when people talk about participatory research, participatory monitoring and participatory evaluation, on the whole they are not discussing a self contained set of methodologies, but a situation whereby the methods being used have included an element of strong involvement and consultation on the part of the subjects of the research. Not all methods are equally amenable to participation.

Source: Pratt and Loizos, 1992

Table 5.1 *A typology of participation*

Passive participation	People participate by being told what is going to happen or has already happened. It is a unilateral announcement by an administration or project management without listening to people's responses.
Participation in information giving	The information being shared belongs only to external professionals. People participate by answering questions posed by extractive researchers using questionnaire surveys or such similar approaches. People do not have the opportunity to influence proceedings, as the findings of the research are neither shared nor checked for accuracy.
Participation by consultation	People participate by being consulted, and external agents listen to views. These external agents define both problems and solutions and may modify these in the light of people's responses. Such a consultative process does not concede any share in decision-making, and professionals are under no obligation to take on board people's views.
Participation for material benefits	People participate by providing resources such as labour, in return for food, cash or other material incentives. Much on-farm research falls into this category, as farmers provide the fields but are not involved in experimentation or the process of learning. It is very common to see this called participation, yet people have no stake in prolonging activities when incentives end.
Functional participation	People participate by forming groups to meet predetermined objectives related to the project, which can involve the development or promotion of externally initiated social organization. Such involvement tends not to be at early stages of project cycles or planning, but rather after major decisions have already been made. These institutions tend to be dependent on external initiators and facilitators, but may become self-dependent.
Interactive participation	People participate in joint analysis, which leads to action plans and the formation of new local institutions or the strengthening of existing ones. It tends to involve interdisciplinary methodologies that seek multiple objectives and make use of systematic and structured learning processes. These groups take control/ownership over local decisions, so people have a stake in maintaining structures or practices.
Self-mobilization	People participate by taking initiatives independent of external institutions to change systems. Such self-initiated mobilization and collective action may or may not challenge existing inequitable distributions of wealth and power.

Source: Bass et al, 1995

Participatory approaches and methods – a brief history

The history of the systematic use of participatory methods can be traced back to the late 1970s with the introduction of a new research approach called *rapid rural appraisal* (RRA) which quickly became popular with decision-makers in development agencies, including NGOs. A criticism of the RRA approach was that it was 'extractive' and the role of local communities was limited to providing information, while the power of decision-making about the use of this information remained in the hands of outsiders. During the 1980s, NGOs working closely with communities further refined RRA approaches and developed what is known as *participatory rural appraisal* (PRA). While using similar methods and tools, the underlying philosophy and purpose changed: while RRAs led to situations of extracting information, often in a single event, PRAs were designed to follow the peoples' own concerns and interests and to build a process of involvement that would lead to actions and capacities to intervene and address such concerns. Thus, it enhanced a community's own capacities for analysing their circumstances of living, their potentials and their problems in order to actively decide on changes and action. These shifts towards interactive, mutual learning are now reflected in *participatory learning and action* (PLA), an approach and terminology commonly used by teams working in development and conservation, involving many of the elements and tools of RRA and PRA.[1] Some of the participatory tools and methods that can be used constructively for CWR *in situ* conservation planning and action are listed in Box 5.4.

The list included in Box 5.4 is by no means exhaustive and the reader is referred to the information at the end of this chapter for more detailed descriptions, many of which include advantages and disadvantages, on how to use these tools and others that may be relevant.

Before getting started, however, *it will be useful to ask the following questions* to stimulate thinking and guide decision-making during the formulation of the participatory intervention:

- Why is a participatory approach necessary?
- What experience and skills in participatory approaches exist in my organization?
- What experience and skills exist in other partner organizations?
- Who might make up the team for a participatory approach?
- Is there a need for additional training in participatory approaches for team members?
- Are the communities that need to be involved, well defined?
- Does my organization already have existing relations with the proposed community?
- Do other collaborating national organizations have existing relations with the proposed community?
- Has the participatory process and planning involved the community from an early stage?

Box 5.4 Participatory tools and methods to consider

Brainstorming – quick, easy way of generating ideas and information with groups of people.

Review of secondary data – often performed, although the emphasis on previous data can lead to erroneous interpretations.

Direct observation – observation related to What? When? Where? Who? Why? How?

Do it yourself – role reversal used to gain an insider's perspective. Community members are encouraged to become the 'experts' and teach the researcher how to perform daily tasks and activities.

Participatory mapping and modelling – community members draw or model past or current situation using local materials. Researchers gain an understanding of land-use patterns and changes, agricultural practices and resource distribution by asking questions on the picture/model. This approach has recently been developed further to include participatory GIS and 3-D modelling.

Transects, group treks and guided field walks – a walking tour is carried out through an area of interest with a local guide to learn about the area's geography and identify problems and solutions.

Seasonal calendars – set up with local materials showing monthly variations and seasonal constraints in rainfall, labour, income, expenditures, debt, harvesting periods, etc. This can help identify opportunities for action.

Daily activity profiles – the daily activities (tasks and time taken to complete them) of community members can be explored based on age and gender.

Semi-structured interviewing – this technique involves informal interviews that follow set questions, but which allow new topics to be explored as the interview develops.

Permanent-group interviews – groups exploiting the same resource are interviewed together to identify collective problems and solutions (e.g. people using a same forest source).

Timelines – major community events are dated and listed to help communities and outsiders understand cycles and reasons for change and take measures for future action.

Local histories – a similar exercise as timelines, but provides more detailed account of changes. This can be used for crops, wild resource changes, population changes, health trends, etc.

Local researchers and village analysts – training local people to collect, analyse, use and present data.

Venn diagrams – overlapping circles that help visualize the relationship between people, communities or institutions.

Participatory diagrams – people are encouraged to display their knowledge on pie and bar charts and flow diagrams.

Wealth and well-being rankings – this technique involves asking people to rank cards representing individuals or households from rich to poor or from sick to healthy. It can be used to cross-check information and produce a benchmark against which future development interventions can be measured or evaluated.

Direct-matrix pair-wise ranking and scoring – a tool used to assess local perceptions on different topics, ranging from value of resources to wealth. People are asked to rank and compare individual items, using their own categories and criteria, by raising hands or placing representative objects on a board. For example, trees can be ranked from best to worst for their properties as a source of fuel and fodder.

Matrices – tools for gathering information and facilitating discussions. For example, a problem–opportunity matrix could have columns with the following labels: soil type, land use, cropping patterns and available resources; and rows with the following labels: problems, constraints, local solutions and initiatives already tried.

Traditional management systems and local-resource collections – this tool can be used to learn about local biodiversity, management systems and taxonomies.

Portraits, profiles, case studies and stories – insightful descriptions of problems and how they are dealt with can be obtained by recording case studies and how household conflicts were resolved.

Key probes – questions addressing a key issue are posed to different interviewees and the answers compared. The question might be something like 'If my goat enters your field and eats your crops, what do you and I do?'

Folklore, songs, poetry and dance – local folklore, songs, dance and poetry are analysed to provide insight into values, history, practices and beliefs.

Futures possible – people's expectations are sounded as they are asked how they foresee the future and to predict the different scenarios if action for a specific problem is or is not taken.

Diagrams exhibition – diagrams, maps, charts and photos of the research activity are displayed in a common area to share information and promote discussion. The tool can provide a further means of cross-checking information and may inspire other community members to take part in research activities.

Shared presentations and analysis – participants are encouraged to share their findings with other community members and outsiders, providing a further opportunity for cross-checking information and obtaining feedback.

Night halts – interactions with community members are greatly facilitated when the researcher lives in the village during the study, as it allows for early morning and evening discussions, when community members tend to have more leisure time.

Short questionnaires – useful if conducted late in the research process and are topic-specific.

Field report writing – key findings are recorded and summaries made of diagrams, models and maps produced during the study, as well as the process involved in creating them. (Check that community has consented to data leaving the village.)

Self-correcting field notes – field notes help the investigator focus on achievements, lessons learned and outstanding activities. Regular revisits to the field notes help the researcher correct any mistakes and identify problems and solutions.

Survey of community members' attitudes toward participatory process – community members are asked to voice their expectations regarding the participatory activities. Their feedback helps improve the process and techniques, and maintain realistic expectations.

Source: Grenier, 1998

It is now widely accepted that local people need to share in the benefits derived from protected areas, and this is best achieved through their playing a role in the management and protection of such areas. This is now reflected in the protected areas work of WWF and UNESCO's Man and the Biosphere Programme (MAB) and other agencies.

Source: adapted from Heywood and Dulloo, 2005

Background to participation in conservation planning

A recent global survey and comparative case study analysis highlights that conservation professionals and managers of biosphere reserves now regard participation as one of the most important success factors for management. However, a separate study, using case studies from selected protected areas using participatory approaches in their formal structure, points out that it does not necessarily always translate into economic benefits for local people.

Source: adapted from Stoll-Kleemann and Welp (2008) and Galvin and Haller (2008)

Conservation planning, like its counterpart in agriculture and rural development, has often employed 'top-down' and centrally planned approaches that have a primary objective of biodiversity conservation and pay little attention to the needs or aspirations of local communities. Often, it was felt that any form of community involvement actually compromised this objective (Pimbert and Pretty, 1995). Conservation planning has not been served well by these 'command-and-control' strategies of the past, often perpetuating the poverty, inequality and power structures that hinder the realization of biodiversity conservation and sustainable development in the first place. As a result of the many lessons learned from this history, community participation is now regarded as fundamental to the attainment of the economic, political, social and environmental objectives that underpin conservation, while 'exclusionary conservation' is not considered sustainable (Kothari, 2006a). This has led to a paradigm shift, from 'ecology first' to 'people first' perspectives (O'Riordan and Stoll-Kleeman, 2002) in conservation planning and management. Such shifts in practice offer considerable opportunities for innovative approaches to CWR *in situ* conservation both inside (see Box 5.5) and outside (see Box 5.6) protected areas. Others refer to this shift as the move away from the 'preservation approach' – trying to isolate and maintain biodiversity in natural parks by excluding indigenous and local communities – towards a more 'biocultural systems approach' – allowing human activity as part of the process and thereby rendering a much more successful conservation strategy.

Box 5.5 Community participation in developing a management plan for wild yams in the National Park of Ankarafantsika, Madagascar

The UNEP/GEF CWR Project's work on wild yams in Madagascar is both exciting and innovative and highlights the challenges and conflicts faced in trying to promote *in situ* conservation in protected areas of a resource of considerable value and use by local communities living inside or bordering the park (there are around 58 small administrative units inside or bordering the national park). Overharvesting of wild yams, erosion and poverty in these communities are inter-related. The project in Madagascar has successfully facilitated a participatory process in developing a management plan that will allow local communities to sustainably harvest and manage these wild relatives. The management plan seeks to reduce highlighted threats and issues that negatively impact on biodiversity conservation in the park. Prior to the project, the national park authority's (National Association for the Management of Protected Areas in Madagascar – ANGAP) policies and regulations were not seen as favourable to local communities who have been harvesting wild yams inside the park for generations. They are an important source of food in times of scarcity (rice) and also sold to generate income. Wild yams are very much seen as an important component of villager's identities. Their ancestors have always harvested, eaten and sold wild yams. ANGAP now has plans to scale up this process to other national parks in the country. The effort and dedication involved in the mainstreaming of CWR conservation into management plans is all too often underestimated by the CWR community. Working directly with local communities through this process requires an even greater intensity of commitment (Figure 5.3).

Source: Jeannot Ramelison, UNEP/GEF CWR Project National Project Coordinator, Madagascar

Figure 5.3 *Working closely with local communities who depend on wild relatives for food or other needs is vital for the development of successful management interventions*

Photo: Danny Hunter

Box 5.6 The Potato Park, Peru

Six Quechua communities in Peru worked closely together, with the Asociación ANDES and other organizations, for several years to establish a 'Parque de la Papa' (the Potato Park). The Potato Park is a centre of diversity for a range of important Andean crops in addition to the potato, including quinoa and oca. The park represents a community-based agrobiodiversity-focused conservation area – also described as community conserved areas (CCAs) and indigenous biocultural heritage areas (IBCHAs) – and is home to a diversity of Andean crop landraces as well as CWR, along with many other species regularly harvested from the wild for food, medicine, and cultural and spiritual purposes. The park is also home to a host of endemic plant species. The park aims to ensure sustainable livelihoods of indigenous communities by relying on local resources to create alternative livelihoods while using customary laws and institutions to facilitate the effective management, conservation and sustainable utilization of biodiversity and ecosystems. The Potato Park, which is not an official protected area, and IBCH areas, in general, represent unique opportunities for CWR conservation practitioners to engage with communities and grassroots organizations to ensure that CWR issues and concerns are integrated into plans. Such work also presents opportunities for linkages between protected and agricultural landscapes to facilitate expected CWR species migration under climate change.

Recently, the UNEP/GEF CWR Project collaborated with Asociación ANDES on capacity building and CWR conservation. In 2009, the Asociación ANDES hosted an international training workshop on 'Design and Planning of Agrobiodiversity Conservation Areas' in Cuzco, Peru, at the request of a delegation of farmers and researchers from Ethiopia who were considering a similar concept for an Ensete park. The UNEP/GEF CWR Project was able to work with colleagues at the Asociación ANDES to ensure that resources and materials on CWR *in situ* conservation were included in the training. The workshop resulted in the Joint Declaration on Agrobiodiversity Conservation and Food Sovereignty, which draws attention to the importance of CWR in community conserved areas.

Source: adapted from Argumedo (2008) and Argumedo and Stenner (2008)

Let the locals lead

To save biodiversity, on-the-ground agencies need to set the conservation research agenda, not distant academics and non-governmental organizations.

Source: Smith et al, 2009

Biocultural diversity conservation: An opportunity for CWR *in situ* conservation

As already alluded to, there is a growing body of knowledge on indigenous management systems of *in situ* plants and crops which shows that local people possess a great diversity of sustainable and localized conservation-oriented knowledge and practices. This field of biocultural diversity conservation (Leakey and Slikkerveer, 1991; Adams and Slikkerveer, 1996) is rapidly emerging as a highly dynamic and integrative approach to understanding the links between nature and culture, and the interrelationships between humans and the environment from the local to global scale (Maffi and Woodley, 2010), providing opportunities that are certainly worth exploring in regard to enhancing the *in situ* conservation of CWR. Such approaches rightly point to the need for integrating human values and needs in conservation strategies (Maffi and Oviedo, 2000; Maffi, 2004). Several other authors have highlighted models of low-intensity mosaic usage of the environment and its resources by local communities for positive and equitable biodiversity conservation outcomes, including Altieri and Merrick (1987); Alcorn (1991, 1994, 1995); Toledo (2001); Carlson and Maffi (2004).

Global commitment and support in recent years for enhanced community participation in biodiversity conservation has led to the emergence of community conserved areas (CCAs, see Box 5.6), recently described as the most exciting conservation development of the 21st century (Kothari, 2006b). Although most CCAs, which are dealt with in more detail in Chapter 11, fall within the definition of Category V protected areas, they do not necessarily have this designation in practice and may also not be identified as part of the national protected area network. CCAs, most of which address agrobiodiversity conservation, both wild and domesticated, have been defined (see Kothari, 2006b) as:

> *Natural and modified ecosystems with significant biodiversity, ecological and related cultural values, voluntarily conserved by indigenous and local communities through customary laws or other effective means.*

CCAs contain three essential elements:

- communities closely tied to ecosystems and/or species through cultural, livelihood, economic or other important links;
- community-based management decisions leading to the conservation of habitats, species and ecosystem services; and
- communities as the prime actors in decision-making and implementation of actions.

Kothari (2006b) identifies two general types of CCA with implications for sustainability:

- **Strong types** are usually internally originated and driven, fully backed by local practice and culture, strongly supported by other stakeholders (e.g.

NGOs) and with the community entitled to some form of ownership rights recognized by the national policy framework.
- **Weak types** are usually externally originated and driven, poorly supported by NGOs and do not secure long-term ownership rights.

Importantly, CCAs include mosaics of natural and agricultural ecosystems containing significant biodiversity value and are managed by farming and rural communities. This can help synergize links between agricultural biodiversity and wildlife and gene flow and migration, and represents an exciting prospect for future community-based work in CWR conservation.

Sources of further information

The **Center for People and Forests** (RECOFTC) has one of the most useful CBNRM online sources of information available. There is a wealth of downloadable manuals and publications available. Of particular interest is their manual for 'Participatory Management of Protected Areas'. The manual on 'Facilitation Skills' will also be useful for anyone working with participatory approaches. Website: http://www.recoftc.org/site/index.php?id=392

Chambers, R. (2002) *Participatory Workshops: A Sourcebook of 21 Sets of Ideas and Activities*, Earthscan, London, UK.

Community Empowerment is a website dedicated to strengthening communities through participation, which includes a useful set of downloadable modules. Website: http://www.scn.org/cmp/

The **Community Planning website** has clear advice on a whole range of ways and tools to get people involved. Website: http://www.communityplanning.net/index.php

The **FAO Participation Website** brings together a broad cross-section of stakeholders interested in participatory approaches and methods in support of sustainable rural livelihoods and food security. It also provides a wealth of resources and field tools for successful participation. Website: http://www.fao.org/participation/default.htm (last accessed 7 October 2010).

FAO (1990) *The Community Toolbox: the Ideas, Methods and Tools for Participatory Assessment, Monitoring and Evaluation in Community Forestry*. Website: http://www.fao.org/docrep/x5307E/x5307e00.HTM.

Friis-Hansen, E. and Sthapit, B. (2000) *Participatory Approaches to the Conservation and Use of Plant Genetic Resources*, IPGRI, Rome, Italy. Although dealing with the on-farm context, there is a lot of useful information in this book on participatory approaches, many of which are applicable to CWR conservation. There is a chapter that provides a brief review of participatory tools and techniques.

Louise Grenier (1998) *Working with Indigenous Knowledge: A Guide for Researchers*, International Development Research Centre (IDRC). Website: http://www.idrc.ca/en/ev-9310-201-1-DO_TOPIC.html

The online **Guide to Effective Participation** offers information on partnerships and participation, theory to practice including toolkits, ideas and other downloadable resources. Website: http://www.partnerships.org.uk/guide/index.htm

IIED Participatory Learning and Action is the world's leading series on participatory learning and action approaches and methods. Website: www.planotes.org/

IGNARM (Network on Indigenous Peoples, Gender and Natural Resource Management) shares experiences and knowledge within the field emerging at the intersection between indigenous peoples, gender and natural resource management. Website: http://www.ignarm.dk/

The **IUCN Indigenous and Community Conserved Areas** website contains many resources including a worldwide database and publications. Website: http://www.iucn.org/about/union/commissions/ceesp/topics/governance/icca/index.cfm

Lockwood, M., Worboys, G.K. and Kothari, A. (2006) *Managing Protected Areas: A Global Guide*, Earthscan, London, UK. This includes detailed chapters dealing with community conserved areas and collaboratively managed protected areas.

Martin, G. (2004) *Ethnobotany; A Methods Manual,* Earthscan, London, UK. Chapters 1, 4 and 8 contain useful information on participatory approaches.

Parque de la Papa (The Potato Park). Website: http://www.parquedelapapa.org/eng/03parke_01.html

Participatory Approaches: A Facilitator's Guide. Website: http://community.eldis.org/.59c6ec19/

Pretty, J., Guijt, I., Thompson, J. and Scoones, I. (2003) *Participatory Learning and Action: A Trainers Guide*, IIED. It is the standard reference on participatory PLA training and tools and is designed for both experienced and new trainers with an interest in training others in the use of participatory methods, whether they are researchers, practitioners, policy-makers, villagers or trainers.

Terralingua, an international non-profit organization, maintains a useful community of practice portal for exchange and sharing of information on biocultural diversity. The portal is an online companion to the book *Biocultural Diversity Conservation: A Global Sourcebook* (Earthscan, 2010). Website: http://www.terralingua.org/bcdconservation/

Tuxhill, J. and Nabhan, G.P. (2001) *People, Plants and Protected Areas: A Guide to In Situ Management,* Earthscan, UK. This book has a useful chapter 'Working with local communities', which provides detailed background information on the rationale for involving local communities in conservation. The chapter also has much information on participatory information gathering tools, the materials required, the advantages and disadvantages, and protocols for implementing. It includes information on preparing for community meetings.

[Links last checked on 28 May 2010]

Note

1. FAO Participation website: www.fao.org/particpation/default.htm (last accessed 7 October 2010).

References

Adams, W.M. and Slikkerveer, L.J. (eds) (1996) *Indigenous Knowledge and Change in African Agriculture*, Studies in Technology and Social Change No. 26, TSC Programme, Iowa State University, Ames, Iowa, USA

Alcorn, J.B. (1991) 'Ethics, economies and conservation', in M. Oldfield and J.B. Alcorn (eds) *Biodiversity: Culture, Conservation and Ecodevelopment*, Westview Press, Boulder, CO, USA

Alcorn, J.B. (1994) 'Noble savages or noble state? Northern myths and Southern realities in biodiversity conservation', in V.M.E. Toledo, *Ethnoecologica*, vol 1, no 3

Alcorn, J.B. (1995) 'Ethnobotanical knowledge systems: A resource for meeting rural development goals', in D.M. Warren, L.J. Slikkerveer and D. Brokensha (eds) *The Cultural Dimension of Development: Indigenous Knowledge Systems*, IT Studies in Indigenous Knowledge and Development, Intermediate Technology Publications, London, UK

Altieri, M.A and Merrick, L.C. (1987) '*In Situ* conservation of crop genetic resources through maintenance of traditional farming systems', *Economic Botany*, vol 41, no 1

Argumedo, A. (2008) 'The Potato Park, Peru: Conserving agrobiodiversity in an Andean Indigenous Biocultural Heritage Area', in T. Amend, J. Brown, A. Kothari, A. Phillips and S. Stolton (eds) *Protected Landscapes and Agrobiodiversity Values*, vol 1 of *Protected Landscapes and Seascapes*, pp45–48, International Union for Conservation of Nature (IUCN) and Deutsche Gesellschaft für Technische Zusammenarbeit (GTZ), Kasparek Verlag, Heidelberg, Germany

Argumedo, A. and Stenner, T. (2008) *Association ANDES: Conserving Indigenous Biocultural Heritage in Peru*, Gatekeeper Series 137a, International Institute for Environment and Development (IIED)

Bass, S. Dalal-Clayton, B. and Pretty, J. (1995) *Participation in Strategies for Sustainable Development*, Environmental Planning Issues No 7, International Institute for Environment and Development (IIED)

Carlson. T.J.S. and Maffi, L.(2004) *Ethnobotany and Conservation of Biocultural Diversity*, Advances in Economic Botany Series, vol 15, New York Botanical Garden Press, Bronx, NY, USA

Danielsen, F., Burgess, N.D., Balmford, A., Donald, P.F., Funder, M., Jones, J.P., Alviola, P., Balete, D.S., Blomley, T., Brashares, J., Child, B., Enghoff, M., Fjeldså, J., Holt, S., Hübertz, H., Jensen, A.E, Jensen, P.M., Massao, J., Mendoza, M.M., Ngaga, Y., Poulsen, M.K., Rueda, R., Sam, M., Skielboe, T., Stuart-Hill, G., Topp-Jørgensen, E. and Yonten, D. (2009) 'Local participation in natural resource monitoring: A characterization of approaches', *Conservation Biology*, vol 23, pp31–42

Friis-Hansen, E. and Sthapit, B. (2000) *Participatory Approaches to the Conservation and Use of Plant Genetic Resources*, International Plant Genetic Resources Institute (IPGRI), Rome, Italy

Galvin, M. and Haller, T. (eds) (2008) *People, Protected Areas and Global Change: Participatory Conservation in Latin America, Africa, Asia and Europe*, Perspectives of the

Swiss National Centre of Competence in Research (NCCR) North-South, vol 3, University of Bern, Geographica Bernensia, Bern

Grenier, L. (1998) *Working with Indigenous Knowledge: A Guide for Researchers*, International Development Research Centre (IDRC), Ottawa, Canada

Hesselink, F., Goldstein, W., van Kempen, P.P., Garnett, T. and Dela, J. (2007) *Communication, Education and Public Awareness: A Toolkit for National Focal Points and NBSAP Coordinators*, Convention on Biological Diversity (CBD) and International Union for Conservation of Nature (IUCN)

Heywood, V.H. and Dulloo, M.E. (2005) In Situ *Conservation of Wild Plant Species – A Critical Global Review of Good Practices*, IPGRI Technical Bulletin, no 11, FAO and IPGRI, International Plant Genetic Resources Institute (IPGRI), Rome, Italy

Kothari, A. (2006a) 'Community conserved areas', in M. Lockwood, G. Worboys and A. Kothari (eds) *Managing Protected Areas: A Global Guide*, Earthscan, London, UK

Kothari, A. (2006b) 'Community conserved areas: Towards ecological and livelihood security', *Parks*, vol 16, no 1, pp3–13

Leakey, R.E. and Slikkerveer, L.J. (1991) 'Origins and development of indigenous agricultural knowledge systems in Kenya, East Africa', *Studies in Technology and Social Changes*, no 19, Iowa State University, Ames, Iowa, USA

Maffi, L. (2004) 'Maintaining and restoring bio-cultural diversity: The evolution of a role for ethnobotany', in T.J.S. Carlson and L. Maffi (eds) *Ethnobotany and Conservation of Biocultural Diversity*, Advances in Economic Botany Series, vol 15, New York Botanical Garden Press, Bronx, NY, US

Maffi, L. and Oviedo, G. (2000) *Indigenous and Traditional Peoples of the World and Ecoregion Conservation*, WWF/Terralingua, Gland, Switzerland

Maffi, L. and Woodley, E. (2010) *Biocultural Diversity Conservation: A Global Sourcebook*, p304, Earthscan, London, UK

O'Riordan, T. and Stoll-Kleemann, S. (2002) *Biodiversity, Sustainability and Human Communities: Protecting Beyond the Protected*, Cambridge University Press, UK

Pimbert, M. and Pretty, J. (1995) *Parks, People and Professionals: Putting Participation into Protected Area Management*, United Nations Research Institute for Social Development (UNRISD), International Institute for Environment and Development (IIED) and World Wide Fund for Nature (WWF)

Pratt, B. and Loizos, P. (1992) *Choosing Research Methods: Data Collection for Development Workers*, Development Guidelines No 7, Oxfam, Oxford

Smith, R.J., Verissimo, D., Leader-Williams, N., Cowling, R.M. and Knight, A.T. (2009) 'Let the locals lead', *Nature*, vol 462, pp280–281

Sthapit, B. (2008) 'Blurring the line between farmer and breeder', *Geneflow*, p32, Bioversity International, Rome, Italy

Stoll-Kleemann, S. and Welp, M. (2008) 'Participatory and integrated management of biosphere reserves', *Gaia*, vol 17/S1, pp161–168

Toledo, V.M. (2001) 'Biocultural diversity and local power in Mexico: Challenging globalisation', in L. Maffi (ed) *On Biocultural Diversity*, Smithonian Institution, Washington, DC

United Nations (2009) *State of the World's Indigenous Peoples*, Department of Economic and Social Affairs of the United Nations Secretariat, United Nations, New York, NY, US

Chapter 6

Developing National CWR Strategies and Action Plans

The Convention on Biological Diversity calls for each Party to develop a National Biodiversity Strategy and Action Plan (NBSAP) to guarantee that the objectives of the Convention are undertaken at all levels and in all sectors in each country (CBD, 2010).

Importance and purpose

Under Article 6 of the Convention on Biological Diversity (CBD), the parties are required to develop national strategies, plans or programmes for the conservation and sustainable use of biodiversity. Guidance on their preparation was given in the national biodiversity guidelines published by the United Nations Environmental Programme (UNEP), the World Resources Institute (WRI) and the International Union for Conservation of Nature (IUCN) (Miller and Lanou, 1995). Such strategies can be considered a call to action and set a national direction for biodiversity conservation. A sample survey showed that most countries' biodiversity strategies and action plans do not specifically refer to CWR or even to the *in situ* conservation of targeted species but, such is the importance of CWR, that it is clearly desirable for countries to develop a separate national strategy and action plan for their conservation and sustainable use. On the other hand, some countries have developed national plant conservation strategies in response to the Global Strategy for Plant Conservation (GSPC). In such strategies, CWR are included in several targets; CWR are specifically covered by target 9 of the European Strategy for Plant Conservation (ESPC).

Prior to the UNEP/GEF CWR Project, very few countries had developed a CWR strategy or included one on CWR in their national biodiversity strategy and action plans, so there are few country examples that offer guidance. The one exception is Turkey, which produced the National Plan for *In Situ* Conservation

Box 6.1 Main objectives and expectations of the National Action Plan for *In Situ* Conservation of Plant Genetic Diversity in Turkey

- The Turkish National Plan for *in situ* conservation of plant genetic diversity is the first example of its kind in the world. It could serve as an example for other countries.
- The implementation of the National Plan for *in situ* conservation of selected (target) species of the wild relatives of herbaceous and woody plants and important forest trees will provide efficiency and continuity in conservation programmes in Turkey by establishing gene management zones (GMZs) for target species throughout the country.
- Since the GMZs are accepted as one of the most effective ways of *in situ* conservation, allowing the evolutionary changes and continuity of genetic diversity in target species in the National Plan, the alternatives for the selection criteria, management responsibility and policy for GMZs, as well as the methods for utilization of genetic material from GMZs will be also developed for target species with special requirements.
- The basic purposes of all environmental actions are to prevent environmental problems before they occur, and to sustain the quality and quantity of the biotic and abiotic components in ecosystems. With the implementation of the National Plan, the plant genetic resources which are seriously threatened by various environmental problems will be efficiently conserved and managed *in situ*.

Source: Albayrak (2004)

of Plant Genetic Diversity in Turkey (Kaya et al, 1997) as an output of the World Bank/GEF-sponsored project '*In Situ* Conservation of Genetic Biodiversity' (Tan and Tan, 2002) (see Box 6.1).

Why develop a strategy?

Given the importance of CWR, a national strategy is needed to provide a coherent and coordinated approach to their conservation and utilization. Further, the many challenges highlighted and addressed elsewhere in this manual, such as lack of collaboration across sectors, absence of policy and legislative reforms, lack of technical expertise and limited finances demand a strategic approach. To implement the strategy, a plan of action is needed to implement future coordinated actions to achieve its goals. It may also be used by countries to meet the targets they have committed to under international agreements such as the CBD and its GSPC and other global strategies such as the Global Strategy for Conservation and Use of CWR (see below).

A national CWR strategy or action plan should seek to:

- ensure coordination of planning and implementation of CWR conservation so that collaboration occurs and activities are harmonized between the relevant stakeholders and actors involved;
- institutionalize the practice of CWR conservation by embedding it in national planning mechanisms supported by relevant policy, legislative and financial measures;
- promote the public awareness and understanding of the importance and value of CWR and their conservation; and
- provide a mechanism for reporting on progress towards targets and plans agreed under other agreements – e.g. the CBD.

The experience gained during the UNEP/GEF CWR Project has shown very clearly the value of a CWR national strategy. The preparation by the countries of their strategies has been a valuable exercise, highlighting the need for greater coordination and collaboration between ministries, agencies and institutions, and for improved partnerships and more effective planning across sectors and thematic areas. The Project has helped draw attention to the importance of CWR, both nationally and globally, and the growing threats that they now face. Furthermore, a national strategy can be a useful instrument to help secure funding for CWR in a climate of financial difficulty and competition from other demands. It might also assist a country to better align its CWR activities with other relevant international initiatives such as the GSPC and the International Treaty on Plant Genetic Resources for Food and Agriculture (ITPGRFA).

Once a strategy has been prepared and approved, an action plan for its implementation will need to be developed. The action plan is likely to be phased over a period of years, according to the availability of resources and finances. Full implementation is likely to take many years in countries with numerous CWR.

As already highlighted, most countries have well-established arrangements for the preparation and implementation of National Biodiversity Strategies and Action Plans (NBASPs), as required by the CBD. Some government departments also have considerable experience in developing strategies and action plans for thematic topics, as do other national organizations. It should be noted that NBSAPs and related reporting to the CBD is generally carried out by a country's ministry of environment, whereas CWR often fall under the responsibility of the ministry of agriculture. A CWR strategy can therefore help to bring these different sectors together.

Considerable expertise already exists in most countries; it is good practice to seek out this expertise, as well as other resources and tools which may be relevant to the CWR strategy and action planning process.

Who should be involved?

The relevant government department or agency(ies) with a mandate or responsibility for CWR conservation may wish to establish a working group or task force to oversee the drafting of the national CWR action plan. Countries should also consider designating a national focal point for CWR; this person would be responsible for coordinating CWR-related activities, including the design and implementation of the national action plan for CWR. It is important that the working group includes individuals from other relevant agencies and sectors if there is to be ownership of the final action plan. Cross-sector support and buy-in will also be a key to the success of the action plan; such support is necessary to ensure that actions are integrated into relevant agency work plans and budgets.

The agency may wish to employ a consultant to prepare an initial draft of the national CWR action plan if no in-house expertise exists.

Guidelines for preparation

In the absence of previous examples of national strategies for the conservation of CWR, little published guidance on how they may be produced is available. However, as the preparation of a national action plan for CWR conservation and use was one of the outputs of the UNEP/GEF CWR Project, the subsequent action plans/strategies for Armenia, Bolivia, Madagascar, Sri Lanka and Uzbekistan represent a unique resource. These are discussed below. In addition, a major component of the CWR Project was the development of a national information system on CWR (as well as an international information system); this represents a major source of information for use in preparing a national strategy/action plan. Similarly, if a national CWR database has already been constructed for a country, it will contain much of the information needed for inclusion in the strategy.

The preparation of national CWR strategic action plans was one of the main objectives of the draft Global Strategy for the Conservation and Use of CWR proposed by the PGR Forum Project and First International Conference on Crop Wild Relatives Conservation and Use in 2005 (Heywood et al, 2008).

A national strategy for the conservation and sustainable use of CWR may be prepared as:

- a free-standing document, as in the case of Armenia, Bolivia and Uzbekistan;
- incorporated into the country's National Biodiversity Strategy and Action Plan, as in the case of Sri Lanka; or
- included in a country's plant genetic resources strategy, as in the case of Madagascar, where the process of outlining a national strategy for CWR is ongoing. It has been agreed that CWR will be integrated into Madagascar's National Management Strategic Plan for Forest Phytogenetic Resources, which is under revision.

Box 6.2 Components and actions for preparing a CWR national strategy/action plan

- Provide the background context for CWR:
  - state of biodiversity conservation in the country;
  - international agreements entered into, relevant to CWR, e.g. CBD, ITPGRFA, Global Plan of Action for Conservation and Sustainable Utilization of PGRFA, GSPC;
  - national legal framework relevant to CWR;
  - national biodiversity strategy and action plan;
  - national and international information system on CWR;
  - lientification of stakeholders.
- Compile a national inventory of CWR and lists of other potential economically important target species – forestry species, medicinal/aromatic plants, indicating their conservation status (where known).
- Review existing national data sources on CWR, with regard to their current state of conservation:
  - their occurrence in protected areas;
  - any *in situ* actions affecting them (including recovery plans);
  - their representation in genebanks.
- From the national inventory, select a list of priority species of CWR for which conservation action is proposed, either *in situ* or *ex situ* or both.
- For the priority species, make a baseline assessment of their ecogeographic status and undertake a threat assessment.
- Undertake a gap analysis to establish where gaps exist in conservation measures.
- For priority species, outline proposals for *in situ* conservation action (including threat management), both within protected areas, preferably as a network of genetic reserves, and outside currently protected areas.
- For priority species for which *ex situ* conservation is required, make proposals for their sampling and storage in national or international genebanks, botanic gardens or other long-term facilities.
- Make proposals for other actions to protect CWR outside protected areas, such as easements, incentive-based schemes or micro-reserves.
- Make proposals for complementary conservation.
- Determine the policy framework changes needed.
- Review adequacy of existing legislative and determine what further action, if any, is required.
- Assess budget and funding issues and develop a financing plan.
- Make proposals for ensuring national awareness of the importance of conserving and using CWR sustainably, preferably within the framework of a communications strategy.
- Devise a capacity development plan.
- Arrangements for implementation of the strategy and allocation of management responsibilities.

There is no single, right way to prepare a national CWR strategy, but the key elements are provided in Box 6.2. An outline scheme and further information for the development of a national CWR strategy are presented by Stolton et al (2006). Given that, in most countries, an array of different strategies and action plans, national reports and assessments on various aspects of biodiversity and conservation have been produced, every effort should be made to draw on these and avoid duplication of effort.

Provide the background context for CWR

As noted above, most countries will have already prepared a number of strategies, action plans or other instruments that document the state of biodiversity. Some of these will correspond to reporting requirements, such as national reports under international treaties or other agreements that have been entered into, including the CBD, ITPGRFA, Global Plan of Action for Conservation and Sustainable Utilization of PGRFA and GSPC. There may also be useful background information in past and current country reports submitted for the *State of the World Report on Plant Genetic Resources* and in the consolidated report itself. In addition, there may be regional agreements with similar reporting requirements – for example, in Europe, the European Union Habitats Directive and the Council of Europe Bern Convention. National biodiversity or agrobiodiversity institutes will also hold relevant information. While these documents may not specifically mention CWR, they will provide much background on the species that might be identified as CWR and the areas in which they occur.

Data sources for the national inventory

The backbone of a national CWR strategy is the inventory or listing of CWR. In very few cases, such a list will already exist as in the case of Armenia, where a catalogue of the wild relatives of food crops was prepared by Gabrielian and Zohary (2004). The main source of data for the inventory will normally be the national Flora(s). For most countries, one or more standard Flora(s) exist: these are the Floras generally acknowledged by botanists in the country or region as the most reliable sources of information on plants occurring there and, consequently, those that are the most widely used. Lists of standard Floras for Europe are given by Tutin et al (1964–1980; 1993) and for the Mediterranean Region by Heywood (2003); a guide to the standard Floras of the world has been compiled by Frodin (2001). In addition, many countries have a published or online checklist of existing Flora.

Unfortunately, a number of countries do not have a comprehensive Flora or even a catalogue. In such cases, the cooperation of local taxonomists should be sought. For example, in the case of the five UNEP/GEF CWR Project countries, Bolivia, which has an estimated 20,000 species (Ibisch and Beck, 2003), does not have a complete Flora nor a recent checklist; the last listing of the ferns and flowering plants of Bolivia was that of Foster's *Catalogue* (Foster, 1958). On the other hand, a handbook of the economic plants of Bolivia was published by the

Box 6.3 A Catalogue of the Vascular Plants of Madagascar

The Vahinala project aims to bring together information on all native and naturalized vascular plant species in Madagascar, evaluating the available taxonomic literature and specimen base for each taxon. The project will result in the 'Catalogue of Vascular Plants of Madagascar', comprising an online database and, eventually, a printed version. The project is led by the Missouri Botanical Garden in collaboration with numerous institutional and individual partners. The harmonized list of accepted species is nearing completion; it aims to have evaluated all genera and compiled distributional, ecological, and conservation status information for all accepted species by the end of 2010.

Source: Missouri Botanical Garden, St Louis, USA

Bolivian botanist, Cárdenas, in 1969 (Cárdenas, 1969), which has proved to be a useful source of information on CWR. A checklist of the Bolivian Flora is under preparation in association with the Missouri Botanical Garden and the New York Botanical Garden. Other partial sources of information include the 'Checklist of New World Grasses'[1] and the 'Preliminary Checklist of the Compositae of Bolivia,' published in 2009.[2]

Likewise, there is no comprehensive Flora of Madagascar, estimated to have at least 9500 species, although the 99 sections of the *Flore de Madagascar et des Comores*, which commenced in 1936 have been published. The Vahinala project, based at the Missouri Botanical Garden, USA, plans to produce the 'Catalogue of the Vascular Plants of Madagascar' (see Box 6.3). The aim is to create a practical, up-to-date, online synthesis of the flora of Madagascar for a diverse group of users, including systematists working on Malagasy plants, ethnobotanists and natural products chemists, natural resource and protected areas managers, conservation scientists and government agencies. It is now possible to prepare this catalogue only because baseline taxonomic data on all names applied to Malagasy plants have already been compiled over the past 25 years into the TROPICOS database (see below).

In the case of Armenia, extensive studies of higher vascular plants have been carried out since the 1950s and have culminated in the production of nine volumes of the 'Flora of Armenia' (Takhtajan, 1954–2001) documenting dicotyledonous vascular plants. A further two volumes on monocotyledons are expected to be published. However, in comparison to these two groups, others are not well studied. At present, not all groups of Armenian flora (lower and higher plants) are equally well known – those most studied are fungi and flowering plants (Plant Genetic Resources in Central Asia and Caucasus: http://www.cac-biodiversity.org/arm/arm_biodiversity.htm).

Another invaluable source of information is herbarium material. Most countries have a national herbarium or one or more major herbaria, as well as university and local herbaria. These herbaria vary enormously in the scope and number of collections they hold. The two herbaria in Madagascar (Parc de

Botanique et Zoologique de Tsimbazaza and Centre National de la Recherche Appliquée au Développement Rural), both in Antanarivo, each hold about 40,000 specimens, the National Herbarium of Bolivia in La Paz houses 100,000 specimens, while another 150,000 specimens are housed in other Bolivian herbaria. The main herbarium in Sri Lanka, at the Royal Botanic Gardens, Peradeniya, contains 130,000 specimens, while the main herbarium in Armenia, at the Institute of Botany of the National Academy of Sciences, Yerevan, has 500,000. Finally, the herbarium of the Scientific Production Centre (SPC) Botanika of the National Academy of Science, Uzbekistan has over 1 million specimens.

Some of the major world herbaria have massive collections and may contain material that is highly relevant for the study of other countries' CWR. For historic reasons, there may be more of a particular country's material in foreign herbarium collections than in those of the country itself, given that much of the plant exploration and collection of herbarium specimens and other material was undertaken by botanists from other countries before appropriate national institutions were established. For example, the herbarium of Phanérogamie at the Muséum National d'Histoire Naturelle, Paris, with 8 million specimens, is of major importance for the study of the Madagascar flora. The major herbaria such as those of the Royal Botanic Gardens, Kew; Natural History Museum, London; Botanischer Garten und Botanisches Museum Berlin-Dahle; New York Botanical Garden; Missouri Botanical Garden, St Louis; and the Central National Herbarium of the Botanical Survey of India, National Botanic Garden, Howrah, all of which have 1 to several million specimens with special emphasis on particular geographic regions or individual countries other than the host country, may be consulted. However, access to such collections to obtain data on CWR may be difficult due to the costs involved, though some information may be accessible electronically.

Herbarium specimens, through their label data, can provide valuable information on the distribution, abundance and, to some extent, the ecology and conservation status of CWR. Obtaining data from herbarium specimens can, however, be time-consuming and laborious and there are many pitfalls. Two of the most serious downfalls include the incorrect naming of material and the use of names which differ from those employed by the standard Floras or checklists, leading to confusion and misunderstanding. Incorrect determination is usually difficult to detect without professional assistance; therefore, the help of taxonomists should always be enlisted. The problem of synonymy – the use of different names for the same plant – is a fact of life and, again, may require the services of a professional taxonomist to resolve. It is beyond the scope of this manual to go into further details in this regard.

In recent years, considerable progress has been made by herbaria across the world in digitizing herbarium material. Digitization involves the process of capturing data such as a plant species' name, the names of collectors and the date of collection, as well as other descriptive and ecological data obtained mainly from the specimen's label. The image of the specimen, itself, is then scanned and stored in digital form along with the aforementioned data. The process is not without its problems: in a project to digitize the herbarium specimen label data from the

Box 6.4 Major initiatives contributing to the digitization of herbaria collections

The Mellon Foundation is supporting work to digitize all the plant type specimens anywhere in the world, coupled with institutional initiatives: http://www.mellon.org/internet/grant_programs/programs/conservation#current.

These include:

The African Plants Initiative (API), an international partnership collaborating to produce an online database of scholarly information about African plants. The partnership currently (December 2009) includes 44 botanical institutions representing 20 countries in Africa, Europe and the US. http://www.aluka.org/action/doBrowse?sa=1&sa_set=1.

The Latin America Plant Initiative (LAPI) and the Global Plant Initiative (GPI) covers Mexico, Central America and the Caribbean, and all of South America. http://www.rbge.org.uk/science/herbarium/digitisation-of-collections/the-latin-american-plants-initiative-and-global-types-initiative.

For a case study of the East African Herbarium (EA) digitization process see: http://www.e-biosphere09.org/posters/H21.pdf.

The Royal Botanic Gardens, Kew, has made substantial efforts in this area in the last five years: http://apps.kew.org/herbcat/gotoProjects.do. This site contains links to many other initiatives that enhance digitization of records and which may be important sources of information, some of which is relevant to Madagascar. Kew has implemented an electronic catalogue for its herbarium specimen collections, known as HerbCat – a relational database that stores information about specimens including collection details (where, when and by whom) and naming history (what taxon has this specimen been assigned to now and previously, when and by whom). Other information such as the part of the plant collected, related material in Kew's collections, and any restrictions on the use of the specimen, are also recorded where appropriate. Each specimen is given a unique barcode and represented as a separate record in HerbCat. http://apps.kew.org/herbcat/navigator.do.

Botanical Research Institute of Texas, Fort Worth, USA, a preliminary survey showed that only 41 per cent of the specimens' labels could be translated into error-free, computer-readable text with off-the-shelf OCR (optical character recognition) software. The remaining 59 per cent of the labels were older, poorly hand-typed or handwritten, and could not be digitized by machines alone, and a system whereby humans could work with computers to transform label data had to be devised. Once digitized, the information can then be readily disseminated and made available to those who do not have direct access to the collections. Some of the major digitizing initiatives are given in Box 6.4.

No comprehensive global checklist or database of plant species exists although target 1 of the CBD Global Strategy for Plant Conservation aims to

Box 6.5 International Plant Names Index (IPNI)

IPNI is a list of plant names giving place of publication, storing around 1.5 million scientific plant names. Comprising data from three hitherto separate indexes (Index Kewensis, Gray Card Index and the Australian Plant Name Index), IPNI is the result of collaboration between the Royal Botanic Gardens, Kew, the Harvard Herbaria and the Australian National Herbarium, Canberra. IPNI data is copyright protected under the Plant Names Project. Website: www.ipni.org

Box 6.6 The Catalogue of Life

The aim of Species 2000 and Integrated Taxonomic Information System (ITIS) Catalogue of Life is to become a comprehensive catalogue of all known species of organisms on earth. The 2010 edition comprises some 1,257,735 species from 77 databases, representing approximately two-thirds of the world's known species. Species 2000 and ITIS teams peer-review databases, select appropriate sectors and integrate the sectors into a single coherent catalogue with a single hierarchical classification. Two products have thus far been published by the Catalogue:

- Species 2000 and ITIS Catalogue of Life: 2010 Annual Checklist
  The Annual Checklist is published each year as a fixed edition that can be cited and used as a common catalogue for comparative purposes by many organizations; http://www.catalogueoflife.org/annual-checklist/2010.
- Species 2000 and ITIS Catalogue of Life: Dynamic Checklist
  The Dynamic Checklist is a virtual catalogue operated on the internet and available both for users and as an electronic web-service at http://www.catalogueoflife.org/dynamic-checklist. The Dynamic Checklist harvests taxonomic sectors and associated strands of hierarchical classification dynamically from the source databases across the internet. The Dynamic Checklist is presently less extensive than the Annual Checklist because fewer taxonomic sectors have been connected so far. It differs in concept from the Annual Checklist in that (i) the taxonomic records may be updated and the catalogue changed more frequently than in the Annual Checklist, and (ii) the Dynamic Checklist contains additional regional species checklists (such as the Regional Checklist – Europe, effectively a Pan-European Species Checklist) not included in the Annual Checklist.

Source: http://www.catalogueoflife.org/

produce a working list by 2010.[3] Major databases and information systems such as GBIF, TROPICOS, IPNI (see Box 6.5), the Catalogue of Life (see Box 6.6) and the electronic Plant Information Centre (see Box 6.7) are also important resources. In addition, there are countless regional, national or local databases and information systems relating to particular areas. For an increasing number of

Box 6.7 The electronic Plant Information Centre (ePIC)

The ePIC is a major resource discovery project to provide a single point of search across all Kew's major specimen, bibliographic and taxonomic databases on the internet. In addition, Kew plans to add digital images and electronic documents into the available resources, and to develop links to external sites with complementary information. The website will be developed through successive releases, with additional data and features being made available at each one. The main components of ePIC are the website; software to enable the cross-database searching and provide ancillary services; hardware to store the data and support the website; and the data itself.

Source: Royal Botanic Gardens Kew, http://epic.kew.org/index.htm, accessed 21 August 2009

Box 6.8 World Checklist of Monocotyledons

A database of accepted names, synonyms, geographical distribution and life forms for monocot plants. Currently, the checklist includes roughly 65,000 accepted taxa in 78 families. When complete, it will include approximately 80,000 accepted taxa for all monocot families. Generic concepts follow *Vascular Plant Families and Genera*. Citation of authors follows *Authors of Plant Names* and terminology for life forms is based on the Raunkier system (1934). Geographical distribution is comprised of a generalized statement in narrative form, and TDWG Level 3 codes. Website: www.kew.org/wcsp/monocots

Source: The Royal Botanic Gardens, Kew

families, global taxonomic databases exist and can be found through normal search engines. Examples are the International Legume Database and Information Service (ILDIS) and the World Checklist of Monocotyledons (see Box 6.8).

Data standards

A major difficulty in working with taxonomic, ecological and geographical information is the lack of consistency, not just in terminology – something that has been addressed in developing the CWR Global Portal – but in the ways names of plants and the literature about them (books and journals) are cited, the application of geographical terms and so on. These issues have been addressed by the Biodiversity Information Standards (TDWG, formerly known as Taxonomic Database Working Group), an international not-for-profit group that develops standards and protocols for sharing biodiversity data. Standards are available from the TDWG website and some of them, especially the earlier ones, are relevant for developing a national CWR catalogue (see Box 6.9). In particular, the so-called Darwin Core (often abbreviated as DwC) is now increasingly adopted

Box 6.9 TDWG standards

The following earlier TDWG standards[4] may be relevant to the preparation of a national CWR strategy:

- Economic Botany Data Collection Standard;
- Plant Occurrence and Status Scheme: Status and Categories;
- Plant Names in Botanical Databases Best Current Practice;
- Authors of Plant Names;
- World Geographical Scheme for Recording Plant Distributions;
- XDF – A Language for the Definition and Exchange of Biological Data Sets;
- Botanico-periodicum-huntianum/supplementum;
- Index Herbariorum. Part I: The Herbaria of the World: Status and Categories;
- International Transfer Format for Botanic Garden Plant Records;
- Floristic Regions of the World: Status and Categories; and
- Taxonomic Literature, ed. 2 and its Supplements.

Source http://www.tdwg.org/standards/

by bioinformatics projects. Darwin Core is a body of data standards that consists of a glossary of terms aimed at facilitating the discovery, retrieval, and integration of information about organisms, their occurrence in nature in space and time, as documented by observations, specimens and samples, and related information housed in biological collections (http://rs.tdwg.org/dwc/). The Simple Darwin Core [SIMPLEDWC] is 'a specification for one particular way to use the terms – to share data about taxa and their occurrences in a simply structured way and is probably what is meant if someone suggests to "format your data according to the Darwin Core"' (http://rs.tdwg.org/dwc/terms/simple/index.htm).

Data sources on CWR conservation

It is important to obtain data on which CWR occur in a country's protected areas, if at all possible. Inventories of the plants occurring in protected areas are sometimes published in protected area management plans or in scientific literature and may be available from the managers of the protected areas. Unfortunately, inventories are lacking or incomplete for the majority of protected areas. With regards to the UNEP/GEF CWR Project areas, the Erebuni Reserve in Armenia has a vascular flora of some 1800 CWR, according to unpublished data from M. Grigoryan cited in Khanjyan (2004), who also quotes approximate figures for other protected areas in the country. A list of species growing in the reserve is also given as an Annex to the Erebuni State Reserve Management Plan.

The project 'Plant and Vertebrate Animal Species Reported from the World's Protected Areas',[5] which aimed to provide databases containing documented, taxonomically standardized species inventories of plants and animals reported from the world's protected areas, was initiated by the Information Centre for the Environment (ICE), in cooperation with the United States Man and the

Biosphere program (US MAB), the Man and the Biosphere (MAB) programme of UNESCO, the National Biological Information Infrastructure, the US National Park Service, and the Biological Resources Discipline of the United States Geological Survey (USGS). The project, however, is still a work in progress.

Information on any actions being taken to manage or conserve CWR species' populations occurring in protected areas should be recorded when available. Again, such information may be available from protected area management plans (often officially published by the state), from the scientific literature or from conservation agencies or non-governmental organizations (NGOs).

Likewise, information on the existence of accessions of CWR in national and local genebanks, botanic gardens and arboreta should be recorded. Accessions may be held in genebanks or collections in other countries and in international genebanks such as those of the Consultative Group on International Agricultural Research (CGIAR) centres; United States Department of Agriculture (USDA), Fort Collins, USA; Leibniz Institute of Plant Genetics and Crop Plant Research (IPK), Gatersleben, Germany; Commonwealth Scientific and Industrial Research Organisation (CSIRO), Australia; Vavilov Institute, St Petersburg, Russia; and the Brazilian Agricultural Research Corporation (EMBRAPA), Brazil. Information on *ex situ* collections may be obtained from the *ex situ* Collection Database, which is a component of the Food and Agriculture Organization of the United Nations' (FAO) World Information and Early Warning System of Plant Genetic Resources for Food and Agriculture (WIEWS).[6] It contains summary records of plant genetic resource holdings (more than 5 million accessions belonging to more than 18,000 species) reported by more than 1500 national, regional or international genebanks. Passport and phenotypic information for many *ex situ* collection holdings (including those from the CGIAR international collections, the European catalogue of genebank holdings and the USDA-ARS GRIN collections) will become available through a single portal (called the *Genesys* portal) to be launched in early 2011 as a result of a collaborative project between Bioversity International, the Global Crop Diversity Trust and the Secretariat of the International Treaty on Plant Genetic Resources for Food and Agriculture. The PlantSearch database, maintained by Botanic Gardens Conservation International (BGCI), can be used to identify plants in *ex situ* collections of botanic gardens. Currently (May 2010), it contains over 575,000 records.

Published information on the conservation of CWR is scarce and the review of information sources by Thormann et al (1999), although somewhat outdated, is a useful resource.

Selecting a list of priority species of CWR

Many countries will have extensive lists of CWR, but resources will be limited and it will not be cost-effective to undertake conservation actions for all CWR, or even a great many. Therefore, it is necessary to undertake a process whereby CWR species can be prioritized. This topic is dealt with in detail in Chapter 7. A national action plan and strategy for CWR should elaborate a list of CWR (the long list) and then prioritize these into those that will be the subject of conservation action in the short, medium and long term. This should be supported by a detailed plan as to what kinds of conservation activities will be applied to the species on these lists. A national strategy is not only about the few selected or prioritized species that any single project can hope to deal with. Instead, the strategy needs to indicate which of the listed species will be targeted, what will be the timeframe, how many species can be afforded protection over this period, what kinds of actions can be undertaken inside and outside protected areas, and so forth.

Baseline assessments of ecogeographic status and threats

Before conservation actions on a priority or target species can be undertaken, as much information as possible about it needs to be gathered in order to make informed decisions and set effective goals for conservation, a topic which is dealt with in detail in Chapter 8.

Conservation gap analysis

Gap analysis was initially put forward as a technique for conservation evaluation, aimed at identifying areas where selected elements of biodiversity are under-represented. Conservation planners regularly use the gap analysis technique to identify biodiversity that is not adequately conserved in protected areas or by other conservation approaches (Stolton et al, 2006). It is a technique that can be used to evaluate current gaps in CWR *ex situ* and *in situ* conservation; this topic is dealt with in Chapter 8.

Proposals for *in situ* conservation action inside and outside protected areas

This topic is dealt with in Chapters 7, 9, 10, 11 and 13.

Proposals for complementary conservation actions, including ex *situ* conservation

This topic is dealt with in Chapter 12.

Review of policy framework for CWR conservation

At the international level, the conservation and sustainable use of CWR are addressed in both the agriculture and environment sectors through the ITPGRFA and CBD. At the national level, it is important to undertake an analysis of relevant national policy documents, such as the national biodiversity and conservation strategies and national biodiversity action plans, to review their relevance to CWR conservation. Where it is weak it will be necessary to draft and promote necessary revisions to national policy. For a summary of the steps that may be involved in developing such a policy framework see Laird and Wynberg (2002).

Review of legal framework for CWR conservation

Most countries have a legislative basis for biodiversity conservation, which includes laws that are related to CWR conservation and use. It is important to review the national legal framework and assess whether it is suitable for PGR conservation, including CWR, and whether it is consistent with international agreements such as the ITPGRFA and CBD. The steps involved in drafting and implementing institutional policy are outlined by Laird and Wynberg (2002).

Box 6.10 Reviewing national legislation on plant genetic resources in Bolivia

Within the framework of the UNEP/GEF CWR Project, and with legal support from FAO, the government of Bolivia reviewed the adequacy of its legislation targeting the protection of plant genetic resources for food and agriculture and CWR. Results from the legal framework review showed that although the sustainable use of natural resources and the conservation of biodiversity were regulated to some extent by Bolivian legislation, particularly Decision 391 that regulates access to genetic resources in Andean countries, no specific legislation was in place for plant genetic resources or for the *in situ* conservation of CWR. Recommendations ensuing from the report suggested that new international priorities set by the ITPGRFA and the CBD be streamlined into Decision 391. The report further highlighted the need to improve national legislation to facilitate access to PGRFA and proposed that a new law be drafted, regulating the conservation, study, evaluation and use of CWR, taking into account traditional knowledge associated with CWR and the safeguarding of farmers' rights in indigenous communities. It also recommended Bolivia's ratification of the ITPGRFA.

Following workshop recommendations, the Bolivian government agreed to consider the issue of streamlining international priorities stemming from the CBD and the International Union for the Protection of New Varieties of Plants (UPOV) Convention into national legislation, and committed to developing a study on Bolivia's food security dependence on PGRFA species included in Annex I of the ITPGRFA. To build consensus at the institutional level for the ratification of the ITPGRFA, a workshop was organized bringing together Bolivian stakeholders involved in plant genetic resources' management. The aim of the workshop was to inform stakeholders of the benefits and obligations linked to the signing of the ITPGRFA and draft a set of recommendations to present to relevant government authorities as a basis for decision-making on the document's ratification. Although aware of the benefits that could stem from signing the ITPGRFA and gaining access to foreign PGRFA material, Bolivia has not yet ratified the agreement. The endorsement of the ITPGRFA remains a politically sensitive issue, particularly regarding plant genetic resources ownership, benefit-sharing mechanisms and farmers' rights, which are not, according to relevant stakeholders, clearly defined in the ITPGRFA.

Source: Beatriz Zapata Ferrufino, National Project Coordinator of Bolivia for the UNEP/GEF CWR Project

Assessment of budget and funding issues

Issues of finance and budgets are covered briefly in Chapter 4. Close attention must be given to budgets and financial support. Most countries will not allocate specific budgets for CWR conservation actions and sources of potential support are limited. This is why it is important that the national action plan has political support and agency buy-in. This might ensure that the action plan is integrated into the relevant agency annual work plan and budgetary mechanisms.

Arrangements for implementation of the national action plan

There are a number of important and challenging issues, many of them cross-cutting, which need serious consideration for the successful implementation of the action plan. Many of these are dealt with in detail elsewhere in this manual. The planning process and the importance of effective partnerships and participation for successful conservation actions are dealt with in Chapters 4 and 5. These chapters also provide information on collaborative agreements, identification of stakeholders and the allocation of management roles and responsibilities. The successful implementation of any action plan will depend on identifying what national capacity already exists and the current gaps that need to be addressed; this topic is dealt with in detail in Chapter 15 and should be addressed through the development and implementation of a capacity development plan. Likewise, communication, public awareness and education are all critical, yet complex and challenging, issues. These topics are covered in Chapter 16 and should be considered in the context of a well-developed communication strategy.

Summary of CWR national strategies/action plans of the UNEP/GEF CWR Project countries

The five UNEP/GEF CWR Project countries have approached the preparation of a national CWR strategy or action plan in different ways and without the benefit of prior agreed guidelines.

The outline of the CWR conservation national action plan for Armenia is given in Box 6.11.

In the case of Uzbekistan, work on the national strategy and national action plan went through various stages. At a national meeting of project partners, it was decided that the strategy would consist of nine chapters and annexes; responsibility for preparing these was shared among partner institutions, according to their interests. The chapters were put together and delivered to the experts of the technical advisory group and a draft version of the strategy was circulated among directors of the following government organizations for comment: Uzbek Research Institute of Plant Industry; Republican Scientific Production Centre on Ornamental Gardening and Forestry; Scientific Plant Production Centre 'Botanica' of the Academy of Sciences of Republic of Uzbekistan; Institute of Market Reform; Main Department of Forestry, Research Institute of

Box 6.11 Outline of the Crop Wild Relatives Conservation National Action Plan for the Republic of Armenia

Executive Summary

1. Conservation of CWR varieties in Armenia
1.1 *In situ* conservation
1.1.1 *In situ* conservation of CWRs in SPAs
1.1.2 *In situ* conservation of CWR outside of SPAs
1.2 *Ex situ* conservation
1.3 International agreements and cooperation national legal frameworks
1.4 National Legal Framework
1.5 Biodiversity National Strategy and Action Plan
1.6. Stakeholders related to CWR
1.6.1 Ministry of Nature Protection of RA (MoNP)
1.6.2 Ministry of Agriculture (MoA)
1.6.3. Ministry of Economy (MoE)
1.6.4 State Regional Administrative Bodies (Marz Administrations)
1.6.5 Local Self Administrative Bodies (LSAB)
1.6.6 Scientific Educational Institutions
1.7 Current status of CWR conservation
1.8 Use of CWR
1.9 Threats
1.10 CWR Information System
1.10.1 CWR International Information System
1.10.2 CWR National Information System

2. National Goals and Objectives

Literature
Annexes
N1 Schedule for the Implementation of the National Action Plan of the Republic of Armenia for the Conservation of Crop Wild Relatives (2007–2011)
N2 Crop Wild Relatives Species and Family Quantitative Distribution Growing in the Territory Of Armenia
N3 List of Endemic Crop Wild Relatives of the Republic of Armenia

Horticulture, Viticulture and Wine Production; and Ministry of Agriculture and Water Resources. After being further reviewed by a meeting at the Institute of Genetics and Experimental Plant Biology, publication was recommended. An implementation schedule has been agreed.

Sri Lanka appointed a team of stakeholders to discuss the best way to develop a national action plan for conservation of CWR. During a workshop held for this purpose, stakeholders unanimously agreed that the development of a separate national action plan for CWR conservation would be futile as the country is already burdened with numerous conservation action plans and has limited adequate capacity to implement these. Therefore, the stakeholders suggested that CWR conservation should be included in other selected conservation action plans already being developed by the authorities. Accordingly, *in situ* conservation was included as a priority area in Sri Lanka's national action plan for biodiversity conservation (2007 addendum) and in provincial biodiversity conservation action plans (Southern, North-Western, Central). In all these action plans, CWR are recognized as an important component of biodiversity that should be given priority in conservation.

In the case of Bolivia, it was decided to employ a consultant to prepare a national strategy for the conservation and use of the CWR and a corresponding plan of action (Elaboración de una Estrategia Nacional para la conservación, uso y aprovechamiento de los parientes silvestres de cultivos de Bolivia y su respectivo Plan Nacional de Acción).

Issues and problems (legal, scientific, technical and logistical) encountered by Armenia, Bolivia, Madagascar, Sri Lanka and Uzbekistan in preparing national CWR strategies

The main problems that arose during the preparation of the national strategies were mainly to do with (1) the fact that no prior models or experience could be called upon; (2) the need to involve or consult with many different national institutions that normally do not work together; (3) the lack of institutions specialized in conservation and monitoring, especially at the species level; (4) the generally low level of appreciation of the importance and issues involved in CWR conservation.

Further sources of information

Brehm, J.M., Maxted, N., Ford-Lloyd, B.V. and Martins-Loução, M.A. (2007) 'National inventories of crop wild relatives and wild harvested plants: case-study for Portugal', *Genetic Resources and Crop Evolution*, vol 55, pp779–796.

GEF, UNEP and CBD (2007) *The Biodiversity Planning Process: How to Prepare and Update a National Biodiversity Strategy and Action Plan*, Module B-2, Version 1, July 2007.

Hagen, R.T (1999) *A Guide for Countries Preparing National Biodiversity Strategies and Action Plans*, UNDP-BPSP, United Nations Development Programme-Global Environment Facility, New York, NY, USA.

Kaya, Z., Kun, E. and Güner, A. (1997) *National Plan for* In Situ *Conservation of Plant Genetic Diversity in Turkey*, Milli Egitim Basimevi, Istanbul.

Miller K.R. and Lanou, S.L. (1995) *National Biodiversity Planning: Guidelines Based on Early Experiences Around the World*, WRI, UNEP and IUCN, http://archive.wri.org/publication.cfm?id=2667&z=?; http://pdf.wri.org/nationalbiodiversityplanning_bw.pdf

Thormann, I., Jarvis, D., Dearing, J. and Hodgkin, T. (1999) 'International available information sources for the development of *in situ* conservation strategies for wild species useful for food and agriculture', *Plant Genetic Resources Newsletter*, no 118, pp38–50.

Notes

1. *Catalogue of New World Grasses* (CNWG) is an ongoing project led by agrostologists from five US and South American institutions to create a database, using TROPICOS, and link all nomenclature, types, synonymy, current taxonomy and distribution for grasses occurring from Alaska and Greenland to Tierra del Fuego (http://mobot.mobot.org/W3T/Search/nwgc.html).
2. http://www.kew.org/science/tropamerica/boliviacompositae/index.html
3. Latest estimates (May 2010) suggest that the list will be 85 per cent complete by 2010, with some progress made on the remaining 15 per cent.
4. These are technically called 'prior standards' and while they are not currently being promoted by TDWG, are widely used.
5. http://www.ice.ucdavis.edu/bioinventory/bioinventory.html (accessed 21 August 2009)
6. http://apps3.fao.org/wiews/wiews.jsp

References

Albayrak, A. (2004) Biodiversity of Turkey. http://people.exeter.ac.uk/rwfm201/cbbia/downloads/Van04/AlbayrakVancouver.2.pdf

Cárdenas, M. (1969) *Manual de Plantas Económicas de Bolivia*, 2nd edition, Los Amigos del Libro, Cochabamba, Bolivia

CBD (2010) *National Biodiversity Strategies and Action Plans (NBSAPs)*, Secretariat of the CBD, Montreal

Foster, R.C. (1958) 'A catalogue of the ferns and flowering plants of Bolivia', *Contr. Gray Herb*, vol 184, pp1–223

Frodin, D.G (2001) *Guide to Standard Floras of the World*, 2nd edition, Cambridge University Press, Cambridge

Gabrielian, E. and Zohary, D. (2004) 'Wild relatives of food crops native to Armenia and Nakhichevan', *Flora Mediterranea*, vol 14, pp5–80

Heywood, V. (2003) 'The future of floristics in the Mediterranean region', *Israel Journal of Plant Sciences*, vol 50, ppS.5–S.13

Heywood, V.H., Kell, S.P. and Maxted, N. (2008) 'Towards a global strategy for the conservation and use of crop wild relatives', Chapter 49, in N. Maxted, B.V. Ford-Lloyd, S.P. Kell, J.M. Iriondo, M.E. Dulloo and J. Turok (eds) *Crop Wild Relative Conservation and Use*, pp657–666, CAB International, Wallingford, UK

Ibisch, P.L. and Beck, S. (2003) La diversidad biológica: espermatófitas, in P.L. Ibisch and G. Mérida (eds) *Biodiversidad: La Riqueza de Bolivia*, pp103–112, Editorial Fundación Amigos de la Naturaleza (FAN), Santa Cruz

Kaya, Z., Kun, E. and Güner, A. (1997) *National Plan for* In Situ *Conservation of Plant Genetic Diversity in Turkey*, Milli Egitim Basimevi, Istanbul

Khanjyan, N. (2004) *Specially Protected Nature Areas of Armenia*, Tigran Mets, Ministry of Nature Protection of the Republic of Armenia, Yerevan

Laird, S.A. and Wynberg, R. (2002) 'Institutional policies for biodiversity research', in S.A. Laird (ed), *Biodiversity and Traditional Knowledge: Equitable Partnerships in Practice,* Earthscan, London, UK

Miller, K.R. and Lanou, S.M. (1995) *National Biodiversity Planning: Guidelines Based on Early Experiences around the World,* World Resources Institute, Washington, DC, United Nations Environment Programme, Nairobi, Kenya and The World Conservation Union, Gland, Switzerland

Stolton, S., Maxted, N., Ford-Lloyd, B., Kell, S.P. and Dudley, N. (2006) *Food Stores: Using Protected Areas to Secure Crop Genetic Diversity,* World Wide Fund for Nature (WWF) Arguments for Protection Series, WWF International, Gland, Switzerland

Tan, A. and Tan, A.S. (2002) '*In situ* conservation of wild species related to crop plants: the case of Turkey', in J.M.M. Engels, V. Ramantha Rao, A.H.D. Brown and M.T. Jackson (eds) *Managing Plant Genetic Diversity,* pp195–204, CAB International, Wallingford, UK

Takhtajan, A. (ed) (1954–2001) *Flora Armenii* (*Flora of Armenia*), vols 1–10, National Academy of Sciences of Armenian SSR, Nauka, Moskva-Leningrad

Thormann, I., Jarvis, D., Dearing, J. and Hodgkin, T. (1999) 'Internationally available information sources for the development of *in situ* conservation strategies for wild species useful for food and agriculture', *Plant Genetic Resources Newsletter,* no 118, pp38–50

Tutin, T.G., Heywood, V.H., Burges, N.A., Moore, D.M., Valentine, D.H., Walters, S.M. and Webb, D.A. (eds) (1964–1980) *Flora Europaea,* vols 1–5, Cambridge University Press, Cambridge

Tutin, T.G., Burges, N.A., Chater, A.O., Edmondson, J.R., Heywood V.H., Valentine, D.H., Moore, D., Walters, S.M. and Webb, D.A. (eds) (1993) *Flora Europaea, Vol 1: Psilotaceae to Platanaceae,* 2nd edition, Cambridge University Press, Cambridge

Chapter 7

Selection and Prioritization of Species/Populations and Areas

Setting conservation priorities is not an easy or comfortable undertaking (K.A. Saterson, 1995).

The nature of the problem

The amount of resources, both human and financial, available for conservation is insufficient to satisfy all the demands being made. CWR are no exception and actions to conserve them have to compete with other biodiversity conservation activities. As a consequence, some form of triage or priority setting has to be applied. Furthermore, as already noted, in many countries, the number of CWR identified will be so large that it would not be feasible to prepare management plans and monitoring regimes for all of them, nor would it be cost-effective to do so, even assuming finance was made available. As indicated above (in Chapter 6), in preparing a national CWR conservation strategy and action plan, some form of selection should be used so that the candidate species can be placed in different priority categories and appropriate forms of genetic conservation applied to them. These may range from population and habitat recovery programmes, conservation plans with various levels of management intervention, through conservation statements to simply monitoring the status of the CWR populations. In some cases, no formal genetic conservation may be possible and alternative arrangements may be made that limit the threats to them or their habitats (see Chapter 11).

The selection of areas in which conservation of CWR is to be undertaken may be straightforward, for example when a CWR consists of a small population(s) geographically restricted to a small area(s), whether protected or not. Or, it may be complex, as in the case of variable species comprising many populations and with an extensive geographical distribution within the country (and sometimes also in adjacent countries). In recent years, various methods for reserve selection have been proposed, but these are primarily aimed at designing a protected area system that includes the maximum representation of biodiversity. Such considerations are well beyond the scope of this manual.

Selection of priority CWR species

Methodology and criteria

There is no precise or agreed methodology for selecting the species or populations that should be given priority as targets for *in situ* conservation and much will depend on local requirements and circumstances. In practice, the selection made will be influenced by the priorities and mandate of the institution or agency involved commissioning the conservation actions (Ford-Lloyd et al, 2008). Thus the species chosen and the actions proposed by agricultural or forestry staff in a country will most likely differ from those made by conservationists, conservation biologists, ecologists or taxonomists. For example, the CWR of economically important crops might well be given high priority, as was the case in the UNEP/GEF CWR Project where Sri Lanka based their selection of priority CWR for conservation primarily on the importance of the crop; or priority may be given to those CWR that are most threatened or endangered, but such an approach is to oversimplify a complex situation. In the absence of an agreed set of criteria, the UNEP/GEF CWR Project countries adopted different sets of criteria based on the knowledge, experience and interests of those involved in the exercise.

Commonly used criteria are listed in Box 7.1. Because there are so many possible factors that might be taken into account, a multilayer approach may be adopted and a scoring system may be used as in the case of Armenia (see below).

A proposal for applying scientific criteria to establish priorities using indicators was made by Flor et al (2006) (see Box 7.2) at the PGR Forum workshop on Genetic Erosion and Pollution Assessment Methodologies. For each criterion, indicators can be assigned and then values can be attributed to the indicators (see Box 7.3).

In addition, pragmatic considerations that may influence the choice of taxa include:

- the likelihood of conservation success and sustainability;
- the relative monetary costs of conservation actions;
- being taxonomically well known and unambiguously delimited;
- being readily available and easy to locate and sample;
- its biological characteristics (e.g. breeding system).

For further information on the various criteria mentioned above, see Maxted et al (1997) and Brehm et al (2010).

Box 7.1 General criteria for selecting target species

A scoring system could be applied to each of the questions below, with some having more weight than others depending on the objective of the strategy.

- What is the actual or potential use of the target species? Is it a CWR, medicinal plant, forest timber tree, fruit tree, ornamental, forage etc.? Can the species be used for habitat restoration or rehabilitation?
- What is the current conservation status of the target species?
- Is the species endemic, with a restricted range or is it widely distributed?
- Is the species experiencing a continuing decline in its occurrence?
- Is there evidence of genetic erosion?
- Does the species have some unique characteristics in terms of:
    a. ecogeographic distinctiveness;
    b. taxonomic or phyletic distinctiveness or uniqueness or isolated position;
    c. focal or keystone species;
    d. indicator species;
    e. umbrella species;
    f. flagship species?
- Does the species have cultural importance or is it in high social demand?
- Does the species occur in a protected area system or does it have some sort of legal or community protected status?

Source: adapted from Heywood and Dulloo, 2005

Box 7.2 Groups of criteria for priority setting

The criteria are grouped in five sets in order to reflect all of the variants that contribute to a taxon's status in terms of genetic importance in relation to its cultivated relatives.

Threat assesses the risk of extinction or any other threat to taxon viability while being an integral part of an ecosystem.

Conservation assesses the existence of programmes or conservation and management plans for the taxon.

Genetic assesses the genetic potential and the status in terms of taxon conservation when its importance as a plant genetic resource is attested.

Economic assesses the economic importance of the taxon.

Utilization assesses the social importance and the extent and frequency of traditional or other uses.

Source: Flor et al, 2006

Box 7.3 Examples of values applied to indicators

Group of criteria	*Criterion*	*Indicator*	*Valuation*
Threat	IUCN threat category	EW (Extinct in the wild)	13
		CR (Critically endangered)	11
		EN (Endangered)	9
		VU (Vulnerable)	7
		NT (Near threatened)	5
		LC (Least concern)	3
		DD (Data deficient)	1
Genetic	Gene pool	Primary gene pool	13
		Secondary gene pool	7
		Tertiary gene pool	3
		Unknown	0

Source: Flor et al, 2006

Conservation status and threat assessment

It is likely that in setting priorities, at some stage preference may be given to CWR that are threatened to some degree; this is usually expressed as their *conservation status or assessment*. What then is involved in evaluating the conservation status of a species? Essentially it is a process of assessing its current state in terms of its distribution and range, population size and numbers, genetic variation, the availability of habitat and the health of the ecosystem, the effects that any threats are having on its current maintenance and prospects for survival in the short, medium and long term.

It should be emphasized that we will rarely, if ever, know the exact population size or range of a species, because of measurement error and natural variation. Moreover, the information available for different species varies enormously and this has to be taken into account when using a set of rules or a framework for deciding on conservation status, which have to be applied irrespective of the amount and quality of the data. A simple approach to the interpretation of such rules, which treats the uncertainty associated with parameters in a precautionary manner, is provided by Burgman et al (1999).

The most commonly used system for assigning conservation status of species is that of the IUCN Red List programme. Red Books and Red Lists are intended both to raise awareness and to help to direct conservation actions. The goals of the IUCN Red List are summarized in Box 7.4 (see also IUCN, 2000). Attention is drawn to the comments on the role of global red lists at a local scale, a subject also addressed by Gardenfors et al (1999) who provide draft guidelines for the application of the IUCN Red List criteria at regional and national levels.

In 1994, a new set of rules was adopted by IUCN for assessing the conservation status of species in Red Lists and Red Data books (IUCN, 1994). Essentially,

Box 7.4 The Goals of the IUCN Red List

The formally stated goals of the Red List are: (1) to provide scientifically based information on the status of species and subspecies at a global level; (2) to draw attention to the magnitude and importance of threatened biodiversity; (3) to influence national and international policy and decision-making; and (4) to provide information to guide actions to conserve biological diversity.

To meet the first two of these goals, the classification system should be both objective and transparent; it therefore needs to be inclusive (i.e. equally applicable to a wide variety of species and habitats), standardized (to give consistent results independent of the assessor or the taxon being assessed), transparent, accessible (a wide variety of different people can apply the classification system), scientifically defensible and reasonably rigorous (it should be hard to classify species without good evidence that they really are or are not threatened). The application of a consistent system also has the benefit that changes in the list over time can be used as a general indicator of the changing status of biodiversity worldwide.

The third and fourth stated goals of the Red List mean that it needs to influence policy- and decision-makers: the challenge here is more complicated. Effective conservation actions generally take place nationally and locally and not at the global level. There are very few mechanisms to conserve species above the national level. Even the Convention on International Trade in Endangered Species of Wild Fauna and Flora (CITES) and the Convention on Biological Diversity (CBD), which are global agreements among countries, rely on implementation within countries for their effectiveness. The Red List is therefore intended to focus national and local conservation actions on the species that most need support. However, it is important to recognize that for various reasons the highest conservation priorities within countries or regions may not simply be the most threatened species found in that region. Certain species may be relatively secure within a politically defined area but nevertheless be at risk globally, whereas other species that are relatively secure globally may be at the edge of their geographic range and hence be highly threatened within a region. For this reason, the role of global red lists within countries must simply be to give shape and force to conservation planning and help set local actions in a global context. There are various ways in which countries might choose to use global information in their own assessments and so far IUCN has provided no more than general guidance.

Source: IUCN, 1996

a new, quantitative system replaced a set of qualitative definitions that had been in place since the early 1960s and were familiar and widely used in scientific, political and popular contexts as a means of highlighting the world's most threatened species. The development of the IUCN criteria took place over a period of five years and led to considerable debate and some controversy from the first proposals to formal adoption by IUCN. According to IUCN (2000), *the most fundamental feature of the new system is its intention to measure extinction risk, and not other factors, such as rarity, ecological role or economic importance that are commonly incorporated into conservation priority systems*. Attention is drawn to this as it is widely misunderstood.

Also, it needs to be stressed that global lists of threatened species do not provide a simple assessment of global conservation *priorities* among those species. As the IUCN clearly states (IUCN, 2000):

> *whilst a threat assessment is a necessary part of any conservation priority assessment, it is not on its own sufficient. Priority-setting should involve many other considerations. These might include assessments of the likelihood of successful remedial action for a species, of the wider benefits for biodiversity that will accrue from directed conservation actions (e.g. for other species within the region, the status of the habitat or ecosystem), and of political, economic and logistic realities. Under some circumstances additional factors are also incorporated in priority assessments, such as the evolutionary distinctiveness of the species ..., the status of existing protection measures, actual or potential economic value, ecological specialisations of particular note and the level of information on the species ...*

The current IUCN categories of threat[1] are given in Box 7.5 and Figure 7.1.

As noted above, the IUCN system of threats is primarily intended for making global assessments but has been widely adopted for national use by many

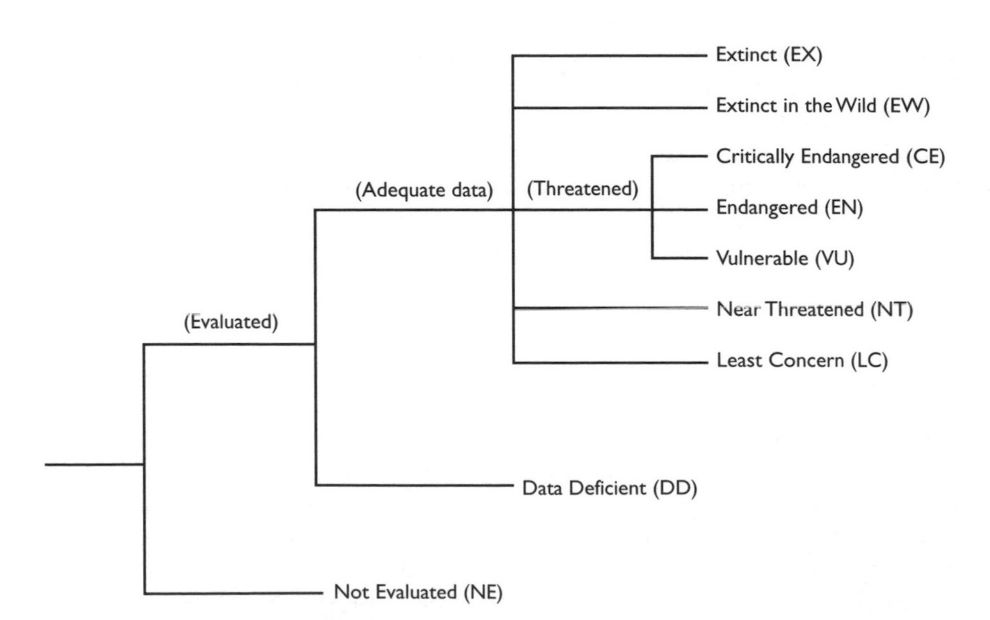

Figure 7.1 *Schema of the current IUCN categories of threat*

Box 7.5 IUCN categories of threat

Extinct (EX) – A taxon is Extinct when there is no reasonable doubt that the last individual has died. A taxon is presumed Extinct when exhaustive surveys in known and/or expected habitat, at appropriate times (diurnal, seasonal, annual) throughout its historic range, have failed to record an individual. Surveys should cover a timeframe appropriate to the taxon's life cycle and life form.

Extinct in the wild (EW) – A taxon is Extinct in the Wild when it is known only to survive in cultivation, in captivity or as a naturalized population (or populations) well outside the past range. A taxon is presumed Extinct in the Wild when exhaustive surveys in known and/or expected habitat, at appropriate times (diurnal, seasonal, annual) throughout its historic range have failed to record an individual. Surveys should be over a timeframe appropriate to the taxon's life cycle and life form.

Critically Endangered (CR) – A taxon is Critically Endangered when the best available evidence indicates that it meets any of the criteria A to E for Critically Endangered, and it is therefore considered to be facing an extremely high risk of extinction in the wild.

Endangered (EN) – A taxon is Endangered when the best available evidence indicates that it meets any of the criteria A to E for Endangered, and it is therefore considered to be facing a very high risk of extinction in the wild.

Vulnerable (VU) – A taxon is Vulnerable when the best available evidence indicates that it meets any of the criteria A to E for Vulnerable, and it is therefore considered to be facing a high risk of extinction in the wild.

Near Threatened (NT) – A taxon is Near Threatened when it has been evaluated against the criteria but does not qualify for Critically Endangered, Endangered or Vulnerable now, but is close to qualifying for or is likely to qualify for a threatened category in the near future.

Least Concern (LC) – A taxon is Least Concern when it has been evaluated against the criteria and does not qualify for Critically Endangered, Endangered, Vulnerable or Near Threatened. Widespread and abundant taxa are included in this category.

Data Deficient (DD) – A taxon is Data Deficient when there is inadequate information to make a direct, or indirect, assessment of its risk of extinction based on its distribution and/or population status. A taxon in this category may be well studied, and its biology well known, but appropriate data on abundance and/or distribution are lacking. Data Deficient is therefore not a category of threat. Listing of taxa in this category indicates that more information is required and acknowledges the possibility that future research will show that threatened classification is appropriate. It is important to make positive use of whatever data are available. In many cases great care should be exercised in choosing between DD and a threatened status. If the range of a taxon is suspected to be relatively circumscribed, and a considerable period of time has elapsed since the last record of the taxon, threatened status may well be justified.

Not Evaluated (NE) – A taxon is Not Evaluated when it has not yet been evaluated against the criteria.

Source: http://intranet.iucn.org/webfiles/doc/SSC/Redlist/RedListGuidelines.pdf (accessed 23 November 2010).

Table 7.1 *Assessing the conservation status of CWR*

	IUCN Red Listing	*Expert-based system evaluation*	*GIS-based assessment*
Advantages	Internationally recognized methodology Includes expert data	Based on field observations	Objective, standardized and repeatable
Disadvantages	Detailed information not always available Comparability (different levels of expertise)	Detailed information not always available Subjective Comparability (different levels of expertise)	Does not include species expertise (Reality check)

Source: Nelly de la Barra, presentation 'Assessing conservation status', delivered at the 5th ISC Meeting of the UNEP/GEF CWR Project, 1–6 December 2008, Cochabamba, Bolivia

countries. Other national or sub-national systems also exist, for example in Australia, the US and New Zealand. Many countries supplement the use of the IUCN system with additional criteria for particular requirements and circumstances.

The advantages and disadvantages of the IUCN system compared with other approaches when applied in Bolivia are summarized in Table 7.1.

Effective conservation of CWR involves the identification of the causes of threats to both the species and its habitat and the implementation of practices to manage them. *Threats or threatening processes* are those that may detrimentally affect the survival, abundance, distribution or potential for evolutionary development of a native species or ecological community.

The IUCN or other red listing systems, by definition, involve some degree of threat assessment, but in deciding on which CWR species should be selected for conservation action, a number of other factors may be taken into account. It should be noted, moreover, that threatened status is not so much a selection criterion as a filter that may be applied after other criteria have been employed. *Endangered status does not automatically qualify a CWR or any other species for selection for conservation action.* As IUCN points out:[2]

> *The category of threat is not necessarily sufficient to determine priorities for conservation action. The category of threat simply provides an assessment of the extinction risk under current circumstances, whereas a system for assessing priorities for action will include numerous other factors concerning conservation action such as costs, logistics, chances of success, and other biological characteristics of the subject.*

It should be noted that a taxon may require conservation action even if it is not listed as threatened. Indeed, a case can be made for conserving *in situ* samples of economically important CWR species that are widespread and not currently threatened. Examples are some major forest trees, many of which have extensive

Box 7.6 Genetic conservation of widespread species

... to effectively conserve the genetic resources of a widespread species several aspects of genetic variation need to be incorporated, i.e. identification of conservation genetic units through integration of patterns of quantitative and neutral genetic structure across multiple spatial scales. Once the organisation and dynamics of genetic diversity are described, an approach that assesses species case-by-case, taking into account unique factors such as recommended forestry practice and geopolitical distribution, should allow formulation of an effective strategy.

Source: Cavers et al, 2004

natural ranges and high levels of diversity within and between populations. An example is the widespread tropical tree *Cedrela odorata* L. for which Cavers et al (2004) bring together the results of previous studies on chloroplast, total genomic and quantitative variation and use the data to describe conservation units and assess their importance for resource management and policy recommendations (Box 7.6). Similar considerations may apply to some other widespread CWR such as *Brassica* crop relatives and leguminous fodder crop relatives.

Threat status and global change

The Intergovernmental Panel on Climate Change (IPCC) and many papers have drawn attention to the likely effects of global and, in particular, rapid climate change on species and their habitats (see Figure 7.2 and Box 14.1), a topic that is discussed in detail in Chapter 14. In the criteria used to assess the threat status of species, these effects have not so far been taken into account. For example, while the current IUCN Red List criteria are designed for classification of the widest set of species facing a diversity of threatening processes, they do not take accelerated climate change as such into consideration. IUCN (2008) does recognize the growing evidence that climate change will become one of the major drivers of species' extinctions in the 21st century and has listed five groups of traits that are believed to be linked to increased susceptibility to climate change:

- specialized habitat and/or microhabitat requirements;
- narrow environmental tolerances or thresholds that are likely to be exceeded due to climate change at any stage in the life cycle;
- dependence on specific environmental triggers or cues likely to be disrupted by climate change;
- dependence on inter-specific interactions likely to be disrupted by climate change;
- poor ability to disperse to or colonize a new or more suitable range.

So far, these have only been applied to a small number of taxa. It follows, therefore, that current Red List or other threat assessments of species can only be

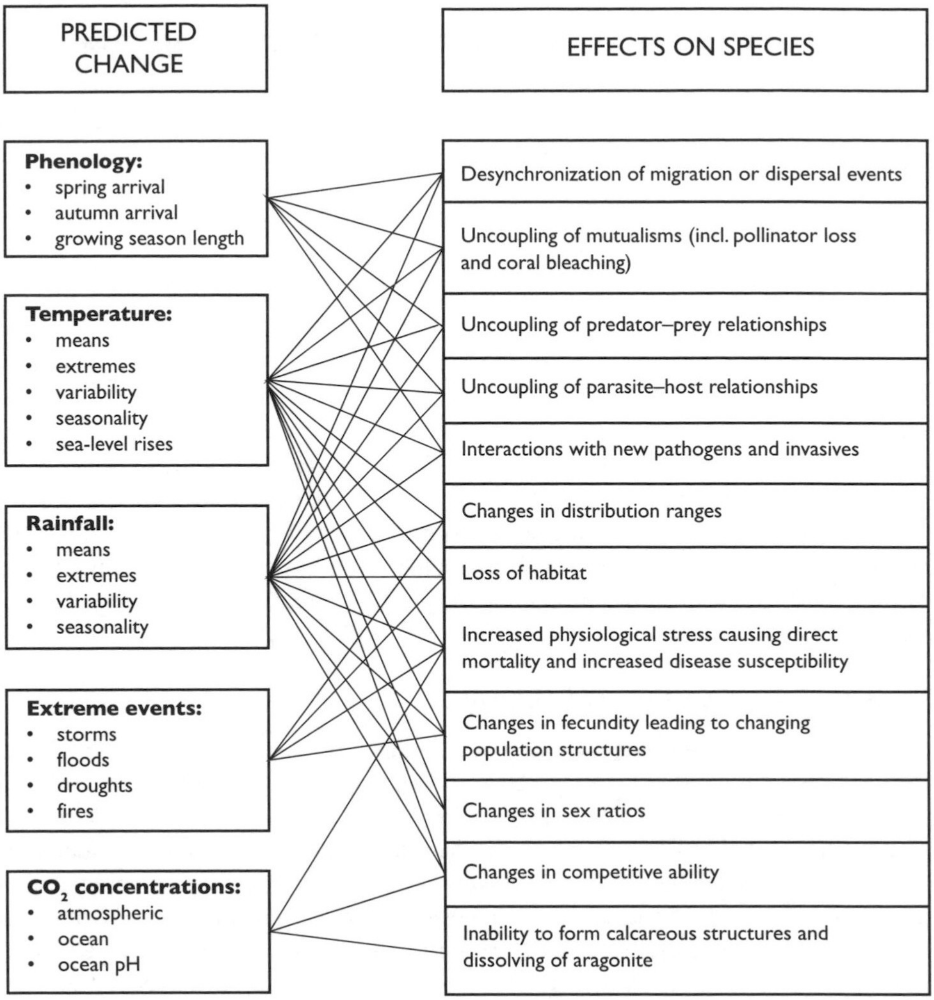

PREDICTED CHANGE

Phenology:
- spring arrival
- autumn arrival
- growing season length

Temperature:
- means
- extremes
- variability
- seasonality
- sea-level rises

Rainfall:
- means
- extremes
- variability
- seasonality

Extreme events:
- storms
- floods
- droughts
- fires

CO_2 concentrations:
- atmospheric
- ocean
- ocean pH

EFFECTS ON SPECIES

Desynchronization of migration or dispersal events

Uncoupling of mutualisms (incl. pollinator loss and coral bleaching)

Uncoupling of predator–prey relationships

Uncoupling of parasite–host relationships

Interactions with new pathogens and invasives

Changes in distribution ranges

Loss of habitat

Increased physiological stress causing direct mortality and increased disease susceptibility

Changes in fecundity leading to changing population structures

Changes in sex ratios

Changes in competitive ability

Inability to form calcareous structures and dissolving of aragonite

Figure 7.2 *Summary of some of the predicted aspects of climate change and examples of the effects that these are likely to have on species*

Source: Foden et al, 2008

regarded as valid in the short term and will all need to be reviewed and updated to take into account accelerated climate change and other aspects of global change if they are to continue to be used as an effective part of any triage system. There are, however, difficulties in incorporating climate change into the criteria and Akçakaya et al (2006) warn of the dangers of their misuse for this purpose. These issues are discussed by Foden et al (2008) who note that:

> *most assessments of species extinctions under climate change have been based on either isolated case studies or large-scale modelling of species'*

> *distributions. These methods depend on broad and possibly inaccurate assumptions, and generally do not take account of the biological differences between species. As a result, meaningful information that could contribute to conservation planning at both fine and broad spatial scales is limited.*

The possible impacts of climate change on CWR are discussed in Chapter 14.

The nature of threats

> *... any system that tries to summarize the complexity of threats to wild nature in a simple, categorical classification is bound to be imperfect* (Balmford et al, 2009).

Threats to CWR species and the communities in which they occur arise in various ways, many of them directly or indirectly as a result of human action. Various attempts have been made to develop classifications of direct threats to the various components of biodiversity, notably the schemes developed by the Conservation Measures Partnership (CMP, 2005) and the IUCN Species Survival Commission (IUCN 2005a, 2005b). In the belief that a single global comprehensive classification of threats and of the conservation actions needed to address them, Salafsky et al (2008) merged these two schemes into a unified classification of direct threats to biodiversity and a unified classification of conservation actions. The schemes are too complex to be reproduced here and the reader should refer to the original paper for details. They have been criticized as 'combining two key but sequential aspects of threat – the threat mechanism and its source – into a single and incomplete linear system' (Balmford et al, 2009), a criticism that has been countered by Salafsky et al (2009). These schemes should, in principle, be applicable to CWR, but so far have not been tested in such a context.

The main kinds of threats are:

- at population level: small subpopulations caused through fragmentation of habitat; low numbers in a population; narrow or small distributional range;
- changes in disturbance regime: for example, as a result of fragmentation and the consequent effects on dispersal and gene flow between isolated populations;
- fire: changes in components of fire regimes, including season, extent, intensity or frequency, inhibiting regeneration from seed or by vegetative reproduction; generally, inappropriate fire regimes lead to the competitive disadvantage of the threatened species against local and introduced species, or represent a future threat if fire recurs before plants are mature and seed is produced;
- threats of biotic origin: disease or predation, e.g. fungal disease; interactions with native species, e.g. allelopathy, competition, parasitism, feral grazing by

rabbits, goats, pigs, cattle, camels etc., including trampling by wild and feral animals and damage caused by rabbit warrens, pika tunnels;
- invasive alien species;
- threats due to development;
- threats due to contamination or pollution;
- indirect threats;
- potential accidents;
- global change (demographic, disturbance regimes, climatic).

Threats primarily due to human action include:

- habitat loss or destruction, degradation, modification or simplification as a result of land-use change such as clearing for agriculture (for crops and pastures, draining swamps and wetlands), forestry, plantations; housing and urban and coastal development; energy production and mining; agriculture edge effects (including herbicides, pesticides, drainage etc.);
- pollution;

Table 7.2 *Main threats to biodiversity in Armenia and their causes*

Threats	*Causes*
Loss of habitat	– agriculture – land appropriation – cattle breeding – drainage of marshes – forest logging – open mining – construction – recreation and tourism – hydroelectric engineering – decrease in level of lakes
Overexploitation of bio-resources (timber, medicines, fodder, fruits, nuts, fibres, oils)	– defective/incomplete legislation – incomplete control over use of resources – lack of inventory data and of bio-resources and quotas for their use – absence of a biodiversity monitoring system
Environmental pollution	– impact of industry – impact of agriculture – transport
Impact of alien invasive species	– deliberate introduction of species and natural introductions
Climate change	

Based on the Fourth Armenia National Report to the CBD in 2009

Table 7.3 *A summary of the major threats to biodiversity in Bolivia and their causes*

Threats	*Causes*
Loss of habitat	– Mainly caused by the expansion of agriculture (Baudoin and España, 1997). In 2008, the agricultural boundary was expanding at a rate of 300,000ha/year in Bolivia. – Opening of roads, establishment of pipelines and others related to the development process of urban expansion and centres of population (MDSP, 2001). – The replacement of forest by crops or livestock pasture and agricultural methods, such as the use of fire for regeneration of grasslands, are having major impacts on wildlife. The effects of these activities on the degradation of specific ecosystems such as savannas and cloud forests are evident (MDSP, 2001).
Degradation of habitat	– Fires and the expansion of other economic activities, such as forest overexploitation, mining and hydrocarbon exploitation (MDSP, 2001).
Impact of alien invasive species	– Competition for habitat, introduction of invasive alien species, introduction of new diseases, which affect both the flora and fauna, even in some cases to become pests for crops (Baudoin and España, 1997). – Introduction of goats in areas of the dry valleys of the Departments of La Paz, Cochabamba, Potosi, Chuquisaca, Santa Cruz and Tarija, which have generated an extensive loss of vegetation and the consequent destruction of habitat for wildlife (Baudoin and España, 1997).
Overexploitation of wildlife	– Overexploitation of species for consumption. – Overexploitation of species or products derived from them for trade, mainly for export.

Source: Wendy Tejeda Pérez, Technical Assistant and Beatriz Zapata Ferrufino, CWR Project Coordinator Proyecto UNEP/GEF 'Conservación *in situ* de parientes silvestres de cultivos a través del manejo de información y su aplicación en campo', 4 January 2009

- overexploitation for commercial, recreational, scientific or educational purposes;
- tourism and ecotourism;
- recreation (e.g. off-road vehicles).

A synopsis of the main threats to biodiversity in Armenia and their impacts is given in Armenia's Fourth National Report to the CBD (Table 7.2).

A synopsis of the main threats to biodiversity in Bolivia, most of which will affect CWR and their habitats, is given in Table 7.3; Table 7.4 for Madagascar.

Table 7.4 *The main threats to biodiversity in Madagascar*

Ecosystems	*Threats*	*Direct causes*	*Indirect causes*	*Consequences*
Agricultural ecosystems	Genetic erosion of agrobiodiversity	– Erosion and silting – Diseases – Lack of measures to conserve cultivars and seeds – Invasive species	– Poverty – Lack of scientific knowledge – Under-utilization of traditional and local knowledge – Unsustainable production methods – Lack of resources for management purposes	– Diminution of production rate – Food insecurity
Forest ecosystems	– Deforestation and forest degradation – Ecosystem fragmentation	– Agriculture expansion – Slash and burn and uncontrolled forest fires – Invasive species – Climatic change – Forest exploitation – Mining – Firewood collecting – Overexploitation of resources – Hunting, gathering and extraction	– Poverty – Usages and customs – Lack of good governance – Insufficient safeguard measures – Unsustainable means of consumption and production – Underestimation of value of biodiversity goods and services – Increase and density of population – Insufficiency of regulation mechanisms	– Impoverishment of ecosystems species richness – Disappearance of threatened species – Reduction of ecosystem services

Source: Fourth National Report to the CBD – Madagascar, 2009

Invasive alien species (IAS)

Globally, invasive alien species are acknowledged as one of the major threats to biodiversity, second only to habitat loss and degradation. In South Africa, for example, alien plant species are considered the single biggest threat to the country's biological biodiversity and now cover more than 10.1 million hectares, threatening indigenous plants.[3]

The term '*invasive*' is applied to alien plants that have become naturalized and are or have the potential to become a threat to biodiversity through their ability to reproduce successfully at a considerable distance from the parent plants and have an ability to spread over large areas and displace elements of the native biota. When they cause significant habitat transformation, leading to biodiversity loss and reduction in ecosystem services, they are often known as *transformers or transformer species*.

Information on invasive species may be obtained from:

- *Global Invasive Species Programme* (GISP)[4] which aims to facilitate and assist with the prevention, control and management of invasive species throughout the world.
- *GISP Global Strategy on Invasive Alien Species*[5] which highlights the dimensions of the problem and outlines a framework for mounting a global-scale response.
- *Global Invasive Species Information Network* (GISIN)[6] which was formed to provide a platform for sharing invasive species information at a global level, via the internet and other digital means.
- *Invasive Alien Species: A Toolkit of Best Prevention and Management Practices*[7] which provides advice, references and contacts to aid in preventing invasions by harmful species and eradicating or managing those invaders that establish populations.

Threats from IAS are likely to increase substantially in some regions as a consequence of global change (see Chapter 14). Examples of the effects of invasive species in the project countries are given in continuation. Although there are few examples so far of their effects on CWR and their habitats, it is highly likely that some of the areas in which CWR conservation will be proposed will be impacted.

Armenia

According to the Botanical Institute, there are over 100 invasive species that can cause damage to Armenia's natural ecosystems. A range of invasive species has been introduced to Armenia and some of them have expanded their ranges to the detriment of native species, and have resulted in population declines and disruptions of ecological relationships, affecting both biodiversity and agricultural systems. Among the most aggressive invasive plant species are *Xanthium*, *Cirsium*, and *Galinsoga parviflora*, while *Ambrosia artemisiifolia* has expanded its distribution by over 200km^2 within the last decade (ECODIT, 2009).

Box 7.7 Summary of the situation of invasive alien species (IAS) in Bolivia

Until 2007, the impact of invasive alien species on the biodiversity and the national economy had not been considered as a problem in Bolivia. The issue is not referred to in the National Strategy of Biodiversity Conservation of Bolivia, approved in 2001, which covers current national policy about the environment and agriculture.

A workshop on biological invasions, held in May 2007 in La Paz, highlighted the need to generate documented sources of information about the effects of invasive species on Bolivia's biodiversity. Subsequently, the Institute of Ecology, Universidad Mayor de San Andrés of La Paz, was given the responsibility of developing a system for collecting and organizing national information on invasive alien species, under the project 'Establishment in Bolivia of Data Bases on Invasive Alien Species, as part of the Inter-American Biodiversity Information Network – IABIN' (Rico, 2009).

According to Rico (2009), as of August 2009, the National Information System of Invasive Alien Species, contained information about invasive species of grass, acacia, pine, and eucalyptus. On the other hand, according to Fernández (2009), 17 species of alien invasive plants have been recorded and verified in three ecological zones of Bolivia: Altoandino: *Poa annua, Pennisetum clandestinum* and *Hordeum muticum*; Puna: *Pennisetum clandestinum, Taraxacum oficcinale, Medicago polymorpha, Trifolium pratense* and *Erodium circutarium*, and Dry Valley: *Pennisetum clandestinum, Rumex acetocella, Matricaria recutita, Taraxacum officinale, Atriplex suberecta, Medicago polymorpha, Spartium junceum, Dodonaea viscosa* and *Opuntia ficus-indica*.

Source: Wendy Tejeda Perez and Beatriz Zapata Ferrufino, December 2009

Bolivia

The issue of IAS in South America is enormous both in terms of the number and diverse range of species invading the continent, and of their impact on the health and livelihoods of all peoples of the region.[8] In Bolivia, however, little information is currently available but ten alien invasive species are reported by the Global Invasive Species Database: *Acacia melanoxylon, Ambrosia artemisiifolia, Leucaena leucocephala, Melia azedarach, Pittosporum undulatum, Rubus niveus, Cedrela odorata, Pisidium guajava, Arundo donax, Rottboellia cochinchinensis* (see Box 7.7).

Madagascar

About 49 invasive species have been recorded from Madagasacar: *Acacia dealbata, Acacia farnesiana, Acacia tortilis, Acanthospermum hispidum, Agave ixtli, Agave sisalana, Albizia lebbeck, Carica papaya, Cissus quadrangularis, Citrus aurantifolia, Citrus aurantium, Citrus medica, Clidemia hirta, Eichhornia crassipes, Erigeron albidus, Eucalyptus* spp., *Grevillea banksii, Lantana camara* var. *aculeate, Mimosa pigra, Mimosa pudica, Opuntia ficus-indica, Opuntia monacantha, Passiflora foetida, Passiflora incarnata, P. suberosa, Phoenix reclinata, Pinus patula, Pithecellobium dulce, Psidium guajava, Psidium cattleianum, Rubus moluccanus, Rubus rosifolius, Salvinia molesta*, *Solanum mauritianum*, *Syzygium jambos*, *Vangueria madagascariensis*, *Ziziphus jujube* and *Zizyphus spina-christi*.

The impact of invasive species on forest composition in Ranomafana National Park in south-eastern Madagascar, a global 'hotspot' of biodiversity, is serious. Common invasive trees and large shrubs established in south-eastern Madagascar include *Clidemia hirta* (Melastomacaceae), *Psidium cattleianum* Sabine (Myrtaceae), *Eucalyptus robusta* (Myrtaceae), *Lantana camara* (Verbenaceae) and *Syzygium jambos* (Myrtaceae,) and can dramatically alter the trajectory of forest succession. The impacts of the invasive species in the Park were compared inside and outside the Park. Studies based on paired transects inside and outside the boundaries of the Park and measuring and counting all the individuals over 1.5cm diameter showed that the percentage of non-native or invasive plants was significantly lower inside the Park as well as the diversity of utilitarian species. Therefore, it was assumed that protected areas play an important role in reducing the spread of invasive plants (Brown et al, 2009).

Detailed information on the extent of plant invasions in Madagascar and their effects are given by Bingelli (2003).

Sri Lanka

Twenty plant species (some of which are now domesticated) have already reached, or have high probability of reaching, invasive proportions in the country. In parts of the country, *Prosopis juliflora* is now a serious problem, where it has invaded agricultural and grazing land, protected areas and national parks. The national list of invasive species for Sri Lanka is presented in Table 7.5.

Uzbekistan

The decreased availability of downstream water and increased salinity levels have led to the shrinkage of wetlands and lakes by up to 85 per cent. Their loss is resulting in the widespread disappearance of native flora and fauna. As water availability declines, native plants are being replaced by invasive species more suited to the dry, saline environment. The following native species are reported as being invasive in Uzbekistan by the Global Invasive Species Database (http://www.gisp.org/): *Brassica elongata*, *B. tournefortii*, *Bromus rubens*, *Butomus umbellatus*, *Elaeagnus angustifolia*, *Erodium cicutarium*, *Hydrocharis morsus-ranae*, *Hypericum perforatum*, *Lepidium latifolium*, *Melilotus alba*, *Phalaris arundinacea*, *Populus alba*, *Tamarix ramosissima*, *Typha latifolia*.

Threat management

After the assessment of threat status, actions need to be taken for the control, mitigation or elimination of threats to target populations. A *threat management strategy* (sometimes known as a *threat abatement strategy*) needs to be developed as part of the conservation or recovery plan/actions (see Chapter 10). The strategy may contain protocols and guidelines directed at how best to abate, ameliorate or

Table 7.5 *National List of Alien Invasive Plants for Sri Lanka*

No.	Botanical Name	Status (Distribution)	
1	*Alstonia macrophylla* Wall. ex D.Don (Apocynaceae)	Degraded forests and forest edges in moist lowland	Provincial
2	*Annona glabra* (L.) (Annonaceae)	Coastal and inland	Provincial
3	*Clidemia hirta* (L.) D.Don (Melastomataceae)	Degraded forests in moist lowlands	Provincial
4	*Clusia rosea* Jacq.(Clusiaceae)	Mid-country moist, open and rocky areas, forest edges	Provincial
5	*Chromolaena odorata* (L.) King & Robinson (Asteraceae)	Road sides, waste ground in lowlands	National
6	*Dicranopteris linearis* (L.) (Gleicheniaceae)	Wastelands and fallow fields	Provincial
7	*Eichhornia crassipes* (Mart.) Solms. Laub (Pontederiaceae)	Inland stagnant water bodies	National
8	*Lantana camara* (L.) (Verbenaceae)	Open scrublands, waste ground	National
9	*Mikania cordata* (Burm.) Robinson (Asteraceae)	Secondary forests in moist regions up to 1000m	Provincial
10	*Miconia calvescens* DC. (Melastomataceae)	Degraded forests in sub-montane regions	Provincial
11	*Mimosa pigra* (L.) (Mimosaceae)	River banks and reservoir edges up to 1000m in moist regions	Provincial
12	*Panicum maximum* Jacq. (Poaceae)	Grasslands, open areas up to 1000m	Provincial
13	*Panicum repens* L.(Poaceae)	Grasslands, open areas up to 2000m	Provincial
14	*Pennisetum polystachyon* (L.) (Poaceae)	Grassland, fallow fields, roadsides up to 1100m	Provincial
15	*Pistia stratiotes* (L.) (Araceae)	Water bodies in wet zone and dry zone	National
16	*Pteridium aquilinum* (Dennstaedtiaceae)	Grassland and/or bare ground	National
17	*Salvinia molesta* D.Mitch. (Salviniaceae)	Inland stagnant water bodies	National
18	*Swietiena macrophylla* (Meliaceae)	In forests	
19	*Ulex europaeus* (Fabaceae)	Nuwara Eliya (Horton Plains)	Provincial
20	*Wormia suffruticosa* (Dilleniaceae)	Degraded forests and scrublands in wet lowlands	Provincial

Source: prepared by the First National Experts Committee on Biological Diversity of the Ministry of Environment, Sri Lanka, 1999*

* Since 1999, two other invasive plant species, *Alternanthera philoxeroides* (alligator weed) and *Parthenium hysterophorus* (congress weed), were recorded in Sri Lanka.

eliminate the impacts that threatening processes have on the target species or on the areas they occupy. Because threats may occur at any level from the landscape to individual populations, actions will need to be directed at the appropriate levels. The management of threats may involve a range of stakeholders and land managers (see Chapter 10). The agency or team responsible for designing and implementing the threat management strategy will need to coordinate actions and liaise with these stakeholders, such as protected area managers, other government agencies, local authorities, community members, conservation bodies and individuals.

Threat management has political, local and training dimensions and the success of threat management strategies may depend, to a large extent, on being able to establish effective community awareness and education programmes. The local community and landholders need to be made aware of the nature of the threats to the CWR and their habitats and how they might become involved in remedial measures.

Country experience and challenges

One of the problems found by some of the countries was disagreement between specialists from different fields as to which species should be given priority. As noted earlier, such differences of opinion between experts is to be expected, given their different interests and experience.

Armenia

In Armenia, during long discussions on choosing priority taxa, there were some disagreements between botanists from different fields. Wild relatives of cereals, pulses, vegetables and fruits were evaluated using unique criteria, specifically developed for each group. The main problem was the existence of biological and ecological differences among CWR families. It must be noted that each of these families' socio-economic characteristics were also evaluated and considered as very important for agronomy and economy.

Priority taxon selection

An evaluation method for three to five classes of crops to be selected for protection was devised as a result of meetings, debates and discussions held to consider particular crops and methods for their evaluation and selection. Botanists representing various fields were involved to ensure objectivity and transparency of the project and, on the basis of the chosen criteria, the crops were evaluated. As a result of the discussions, all CWR were divided into four key groups: cereals; pulses; vegetables; and fruits, berries and nuts. For each group a separate set of criteria was developed, paying special attention to the group's ecological, biological, economic and agricultural indicators/values. Despite the fact that this separation is a mechanical process, it allows one to bring together the groups that have similar qualities and at the same time allows a new strategy for priority taxa selection to be developed.

Box 7.8 List of priority CWR selected for Armenia

Cereals

Triticum boeoticum
Triticum araraticum
Triticum urartu
Aegilops tauschii
(Selection made by Estela Nazarova, Institute of Botany, National Academy of Sciences)

Pulses

Vavilovia formosa
Cicer anatolicum
Onobrychis transcaucasica
Trifolium pratense
(Selection made by Zirair Vardanyan, Institute of Botany, National Academy of Sciences)

Vegetables

Beta lomatogona
B. macrorrhiza
B. corolliflora
Asparagus officinalis
(Selection made by Andreas Melikyan, Armenian Agricultural Academy)

Fruits, berries and nuts

Pyrus caucasica (pear)
Armeniaca vulgaris
Amygdalus fenzliana
Malus orientalis
(Selection made by Eleanora Gabrielyan, Institute of Botany, National Academy of Sciences)

Leading professionals in various fields were involved in the selection process for each of the CWR. The main deciding factors were the same for all four groups – conservation status and gene sources – and were included in the list of criteria. A list of characteristics for each of the four type groups was developed by the editing and grouping method, using additional characteristics such as plant products, use as fodder, honey-yielding plants, environmental uses and food supplements.

In the list of the criteria for each of the species group, every indicator was evaluated on a 10-point system. Therefore, each list of evaluated criteria was applied to the corresponding group in order to select the priority species for protection (see Box 7.8). The cumulative number of points assigned to the particular species group is the summation of the points given to the indicators of the individual species. Subsequently, the species having the highest total points from

Table 7.6 *The conservation status and distribution of CWR selected as target species for Armenia*

Name	*Conservation status*	*In-country distribution*
Triticum araraticum	EN under B1ab (ii, iii, iv, v) +2ab (ii, iii, iv, v)	Yerevan and Darelegis floristic regions corresponding to Ararat, Kotayk, Vayots Dzor marz (administrative regions) and Yerevan city
Triticum boeoticum	EN under B1ab(ii, iii, iv, v) +2ab (ii, iii, iv, v)	Yerevan and Darelegis floristic regions corresponding to Ararat, Kotayk, Vayots Dzor marz (administrative regions) and Yerevan city
Triticum urartu	CR under B1ab(iii) +2ab(iii)	Yerevan floristic region, corresponding to administrative boundaries of Yerevan city
Aegilops tauschii	LC	Yerevan city, Tavush, Shirak, Lori, Kotayk, Ararat, Aragatsotn, Vayots Dzor, Armavir and Syunik marzes, corresponding to floristic regions of Shirak, Ijevan, Yerevan, Darelegis, Zangezur and Meghri
Beta lomatogona	EN under B1ab (i, ii, iii, iv) +2ab(i, ii, iii, iv)	Aragatsotn and Kotayk marzes (administrative regions)
Pyrus caucasica	LC	Lori, Tavush, Kotayk, Aragatsotn, Gegharkunik, Vayots Dzor, Syunik and Ararat marzes
Vavilovia formosa	EN under B1ab(iii) +2ab(iii)	Kotayk, Gegharkunik and Syunik marzes (administrative regions)

each group were selected as the priority species for conservation. The list includes 104 from approximately 250 CWR. Using the above method, seven species were selected for priority conservation (by the highest point score): *Triticum araraticum*, *Triticum boeoticum*, *Triticum urartu*, *Aegilops tauschii*, *Beta lomatogona*, *Vavilovia formosa* and *Pyrus caucasica*. The conservation status of the target CWR in Armenia is presented in Table 7.6.

It should be noted that none of the seven priority species listed above is endemic to Armenia.

Bolivia

During the period from 2000 to 2002, as part of the preparatory PDF-B phase of the UNEP-GEF CWR Project, Bolivia identified 53 genera of wild species important for food and agriculture, medicine and other uses as a part of their National Report (see Table 7.7). Twenty-two of the genera (in bold in Table 7.7) selected had already been the subject of a project to prepare an inventory of CWR in Bolivia, the outcome of which was an 'Atlas of Crop Wild Relatives'. The systemization of the information included in the atlas, was conducted with the support of

Table 7.7 *Genera of crop wild relatives in Bolivia*

Amaranthus	Cuphea	**Manihot**	Psidium
Anacardium	**Cyphomandra**	Nicotiana	Pseudoananas
Ananas	Dioscorea	**Oryza**	Rheedia
Annona	Euterpe	**Oxalis**	Rollinia
Arachis	**Gossypium**	**Pachyrhizus**	**Rubus**
Arracacia	Hevea	Passiflora	Saccharum
Bactris	**Hordeum**	Persea	**Solanum sect. Petota**
Bixa	**Ipomoea**	**Phaseolus**	Spondias
Canna	Ilex	Physalis	Swietenia
Capsicum	Inga	Piper	Theobroma
Carica	**Juglans**	Polymnia	Tripsacum
Chenopodium	**Lupinus**	Pouteria	Ullucus
Cinchona	**Lycopersicon**	Prunus	Vaccinium
Cucurbita			

the National Herbarium of Bolivia, the Museum of Natural History Noel Kempff Mercado, the Herbario Nacional Forestal Martín Cárdenas, Centro de Investigaciones de Pairumani Fitoecogéneticos, PROINPA and FAN. In addition, other national institutions from Argentina: Instituto Darwinion, Buenos Aires (SI); Universidad Nacional del Noreste, Corrientes (CTES); Herbarium of the Fundación Miguel Lillo, Tucuman (LIL); and Instituto Nacional de Tecnología Agropecuaria (INTA); from the United States: Missouri Botanical Garden, St Louis, Missouri (MO); New York Botanical Garden, New York (NY); National Herbarium, Washington, DC; Field Museum of Natural History, Chicago; and National Plant Germplasm System (NPGS); and from Brazil: National Center Genetic Resources (CENARGEN) were involved, as were three CGIAR centres, Centro Internacional de Agricultura Tropical (CIAT), Colombia; International Potato Centre (CIP), Peru; and the International Rice Research Institute (IRRI), the Philippines.

In June and August 2005, national workshops were held involving the eight national partner institutions of the CWR Project, DGBAP and Ecology Institute of UMSA, based in La Paz, Cochabamba and Santa Cruz to further prioritize this extensive list of 53 genera. As a result, the national research institutions of public universities from La Paz, Cochabamba and Santa Cruz, three genebanks, a national organization of indigenous peoples and a non-governmental organization dedicated to biodiversity conservation systemized information from different sources and identified 195 species of CWR from 17 genera (*Anacardium*, *Ananas*, *Annona*, *Arachis*, *Bactris*, *Capsicum*, *Chenopodium*, *Cyphomandra*, *Euterpe*, *Ipomoea*, *Manihot*, *Phaseolus*, *Pseudananas*, *Rubus*, *Solanum*, *Theobroma* and *Vasconcellea*) to be the primary focus for conservation activities during the full implementation of the project (see Table 7.8).

The taxon selection procedure used a number of sub-criteria under the following broad headings:

Table 7.8 *Priority CWR identified for Bolivia*

National Partner Institution	Genus	Spanish Common Name	English Common Name
Herbario Nacional de Bolivia (LPB)	*Euterpe*	Asaí	
	Bactris	Chima, palmito	Palm heart
	Theobroma	Cacao	Cocoa
	Anacardium	Cayú	Cayu
Centro de Biodiversidad y Genética (CBG-BOLV)	*Annona*	Chirimoya	Custard apple
	Rubus	Mora	Blackberry
	Cyphomandra	Tomate de árbol	Tree tomato
	Vasconcellea	Papaya	Papaya
Centro de Investigaciones Fitoecogenéticas de Pairumani (CIFP)	*Phaseolus**	Frijol	Beans
	Arachis	Maní	Peanut
	Capsicum	Ajíes	Chilli pepper
Museo de Historia Natural Noel Kempff Mercado (MHNNKM)	*Manihot**	Yuca	Cassava
	Ananas	Piña	Pineapple
	Pseudananas		
	*Ipomoea**	Camote	Sweet potato
Fundación para la Promoción e Investigación de Productos Andinos (PROINPA)	*Chenopodium*	Quinua, Cañahua	Quinoa
	*Solanum**	Papa	Potato

* crops listed in Annex I of the ITPGRFA

- potential use and economic, social and cultural importance;
- state of knowledge;
- inclusion in the International Treaty (ITGRFA).

Each sub-criterion was scored as either 1= low, 3= medium or 5= high. Each sub-criterion was also given a weighting (1 to 5) based on its overall importance as assessed by the national partners. The final tally for each sub-criterion was determined by multiplying the given score by the weighted value. Of the 53 genera, a final list of 17 were selected.

To further prioritize the many species that exist within the selected 17 genera, the national partner institutions selected the most threatened species for conservation. National partner institutions initially selected the species, from the 17 genera, which existed in protected areas before deciding on the target species.

The information generated by national research institutions on prioritized CWR species in Bolivia is available to the general public on the National Portal (www.cwrbolivia.gov.bo) and through the Global CWR Portal (www.cropwildrelatives.org).

In addition, during the period from 2006 to 2008, 195 CWR species were identified in Bolivia by the six national partner institutions (see Annex I).

Madagascar

The selection of the five priority taxa for conservation action was discussed with representatives of partner institutions involved in the implementation of the CWR Project and members of the Ministry in charge of the environment and forest resources as well as the Ministry of National Education and Research. They covered various fields of expertise in plant biology, such as taxonomy and systematics, botany and ecology, genetics and plant breeding, forestry and agronomy, and management of natural resources.

Based on the knowledge of the participants and development that had taken place in the CWR Project, a first list of eight CWR taxa were proposed as important: *Cinnamosma*, *Coffea*, *Dioscorea*, *Musa*/*Ensete*, *Oryza*, *Piper*, *Tacca* and *Vanilla*. *Musa* and *Ensete* were considered as congeneric. To reduce this list to five, the following selection criteria and value were used (also see Table 7.9):

- number of species occurring in Madagascar for each genus;
- the presence status of the species in each taxon (0 – introduced; 1 – naturalized; 3 – endemic);
- use of the taxon as food (0 – no; 3 – yes);
- contributions of species within the genus to food security (0 – no; 3 – yes);
- economic value of the crop relative (0 – low; 1 – mid; 3 – high);
- potential of the species as specific gene donor for crop improvement (0 – low; 1 – mid; 3 – high);
- level of threats to the taxon (unrated due to lack of data);
- availability of information (0 – high; 1 – mid; 3 – low), a lack of information is highly rated because the committee considered the CWR Project as an opportunity to gather information on the taxa.

Table 7.9 *Priority taxa selection in Madagascar*

TAXA	Number of species	Presence status (0-1-3)	Used as food (0-3)	Contribution to food security (0-3)	Economic value of the crop relative (0-1-3)	Gene donor potentiality (0-1-3)	Availability of information (0-1-3)	Total score
Cinnamosma	1	0	3	0	1	0	0	4
Coffea	60	3	0	0	3	3	1	10
Dioscorea	32	1	3	3	0	1	1	9
Musa and *Ensete*	3	3	0	0	3	3	1	10
Oryza	2	1	0	0	3	3	1	8
Piper	4	1	0	0	3	1	3	8
Tacca	11	3		3	0	0	1	7
Vanilla	6	3	0	0	3	3	1	10

In case of equal scoring, and so as to vary the plant types being represented, an additional criterion was applied – the category of use of the crop relative (aromatic, cereal, fruit, spice and tonic, tuber). Thus, the following taxa were selected: *Vanilla* as an aromatic plant; *Coffea* as a stimulant and tonic; *Dioscorea* as a tuber; *Musa/Ensete* as a fruit; and *Oryza* as a cereal.

The selection of the actual species on which conservation action would be carried out was only done after the ecogeographical surveys on the different species had been undertaken.

Sri Lanka

As noted above, in the absence of agreed criteria at the outset of the project, Sri Lanka based its selection of priority CWR for conservation primarily on the importance of the crop and, by default, gave priority to the wild relatives of the crops finally selected. This approach differed from that employed by the other UNEP-GEF CWR Project countries. As a result, five field crops were selected that represented a potential total of 33 CWR species.

Eighteen participants at a national workshop involving Agriculture, National Botanic Garden and Biodiversity Secretariat staff met to discuss and classify the important field crops in Sri Lanka. A list of 187 food crops was compiled of which only 103 were considered to be native to the South Asian region. As a next step, the most commonly grown crop species that were native to the region and which had known wild relatives in Sri Lanka, were selected from the list of 103 crops. This resulted in a core group of 31 crops with a corresponding total of 98 CWR. To further reduce the list of 31 crops the following list of criteria and values were used:

1. availability of wild relatives (1=many; 5=few);
2. degree of genetic erosion (1=high; 5=low);
3. potential crop improvement (1=high; 5=low);
4. presence status/endemism (1=high; 5=low);
5. geographical distribution (1=scanty; 5=well distributed);
6. current and potential economic value (1=high; 5=low);
7. multiple/combined value (1=high; 5=low);
8. traditional value (1=high; 5=low);
9. present state of conservation of wild relatives (1=neglected; 5=conserved);
10. availability of information (1=low; 5=high).

Other than criteria 1 and 4, the assessment for each crop was subjective and the final scores for each crop were decided through participant consensus. Following assessment using all criteria, a total of 14 crops with the lowest aggregate scores were selected, representing a potential total of 57 CWR (see Table 7.10).

Due to limited project resources, and the obvious fact that 57 CWR is too many to deal with in a five-year project, a decision was made to prioritize this list even further. An internal consultation within the project recommended that only

Table 7.10 *Priority CWR for conservation action – Sri Lanka*

Crop	Total score
Mangosteen (*Garcinia*)	12
Pepper	14
Cinnamon	16
Mango	16
Brinjal	18
Snake gourd	18
Rice	20
Banana	20
Okra	20
Green gram	22
Bitter gourd	24
Vanilla	24
Cardamom	32
Onion	34

five field crops be selected and that at least three of the five selected crops be from those included in Annex 1 of the ITPGRFA. The final decision to select the priority crops fell to the Director General of Agriculture, Director of the National Botanic Garden and the Director of the Biodiversity Secretariat. The final list included rice (*Ozyza*), banana (*Musa*) and cowpea (*Vigna)* as representatives from Annex 1 of the ITPGRFA, as well as pepper (*Piper*) and cinnamon (*Cinnamomum*), which were considered among the most economically important crops for the country. The importance of the final selected crops to the work of the different institutions in Sri Lanka was also a deciding factor. This final list represented a potential total of 33 wild relatives as priority CWR for Sri Lanka.

Uzbekistan

The approach for prioritizing CWR in Uzbekistan initially involved specialists from the Scientific Plant Production Centre 'Botanica' defining a list of genera including CWR that grow in Uzbekistan. They selected 48 genera and 70 species of CWR.

A further working group was organized with experts from five scientific research institutions (Institute of Genetics and Plant Experimental Biology; Scientific Research Institute of Fruit Growing, Viticulture and Winemaking; Scientific Plant Production Centre 'Botanica'; Scientific Research Centre for Ornamental Gardening and Forestry; and the Scientific Research Institute of Plant Industry), two universities (National University of Uzbekistan and Tashkent Agrarian University) and the Department of Forestry Management. The working group consisted of 30 specialists from the above-mentioned organizations; this group defined the criteria to further prioritize wild relative species for conservation action. The criteria were:

Table 7.11 *Selected genera and species for targeted CWR conservation – Uzbekistan*

Genus	Species	Genus	Species
1 *Aegilops*	*Aegilops crassa*	6. *Amygdalus*	*Amygdalus bucharica*
	Aegilops cylindrica		*Amygdalus communis*
	Aegilops juvenalis		*Amygdalus petunnikovii*
2 *Hordeum*	*Hordeùm bulbosum*		*Amygdalus spinosissima*
	Hordeum spontaneum		*Amygdalus vavilovii*
	Hordeum turkestanicum	7. *Pyrus L.*	*Pyrus korshinskyi*
	Hordeum leporinum		*Pyrus bucharica*
	Hordeum brevisubulatum		*Pyrus regelii*
3 *Allium*	*Allium pskemense*		*Pyrus vavilovii*
	Allium suvorovii	8. *Pistacia*	*Pistacia vera*
	Allium vavilovii	9. *Juglans*	*Juglans regia*
	Allium aflatunense	10. *Crataegus*	*Crataegus pontica*
	Allium oschaninii		*Crataegus turkestanica*
4 *Cucumis*	*Cucumis melo*	11. *Elaeagnus*	*Elaeagnus angustifolia*
5 *Malus*	*Malus sieversii*		*Elaeagnus orientalis*
	Malus niedzwetzkyana		

- cultural importance for mankind (socio-cultural importance of species in genera);
- use by the local people as a food source;
- local and national commercial importance;
- nearness to the centre of origin;
- diversity in habitat of the species;
- threat of species' extinction;
- importance for breeding;
- availability of information on the species.

Each genus on the list was scored by a '+' (if the criterion was important) or a '–' (if the criterion was not important). The maximum any genus could score was eight and the minimum was zero. From the initial list, 11 genera (representing 31 species of CWR) were selected (see Table 7.11).

At the final stage, the same scoring system was reapplied to the 31 remaining species, and the following CWR species were prioritized as a result: *Malus sieversii* (apple); *Allium pskemense* (onion); *Amygdalus bucharica* (almond); *Pistacia vera* (pistachio); *Juglans regia* (walnut); *Hordeum spontaneum, H. bulbosum* (barley – also listed in Annex 1 of ITPGRFA).

Malus sieversii, M. niedzweckiana, Allium pscemense, Amygdalus bucharica, A. petunnikova, A. spinosissima and *Pistacea vera* are endemic to Central Asia. *Hordeum spontaneum* and *H. bulbosum* are endemic to Uzbekistan.

Box 7.9 Conservation of walnut (*Juglans regia*) in Uzbekistan

Although the fruits of some other species of *Juglans* are edible, English or Persian walnut (*Juglans regia*) is the most horticulturally developed and widely cultivated species. Wild walnut populations in Uzbekistan grow in three isolated areas – in Western Tien Shan, Nurata and South Gissar – remote from each other by more than 200km. They occur in Ugam Chatkal State Natural National Park and in Nurata State Reserve, but are only partially protected. Uncontrolled cattle pasturing and harvesting is widespread in the reserves so that regeneration is not observed and the trees are of very old age. Ecosystems that include this species are partially or completely disturbed. The second tree layer is absent and the underneath layer is only partially conserved. The diversity of grass species is very poor because many have been eliminated, especially those that are grazed by cattle. Because of disturbances to the ecosystem, the walnut trees are almost completely affected by fungal diseases of the leaves and fruits. Recommendations for action to conserve the species in the wild include: strengthening the protection of areas containing walnut populations by restricting cattle grazing and fruit harvesting; improving and implementing existing legislation targeting the protection of CWR; creating walnut regeneration sites; involving local communities in conservation work; increasing public awareness on the importance of CWR conservation; and carrying out research to select genetic material for breeding purposes.

Box 7.10 *Malus sieversii* and the origin of the domestic apple

For many years, there has been a debate about whether *Malus domestica* evolved from chance hybridization among various wild species. Recent DNA analysis has indicated, however, that the hybridization theory is probably incorrect. Now, it appears that a single species, *Malus sieversii*, a wild apple native to the mountains of Central Asia in southern Kazakhstan, Kyrgyzstan, Tajikistan and Xinjiang, China, is the sole progenitor of most of today's domestic and commercial apples (Juniper and Mabberley, 2006). Leaves taken from trees in this area were analysed for DNA composition, which showed them all to belong to the species *M. sieversii*, with some genetic sequences common to *M. domestica*. Another recent DNA analysis (Coart et al, 2006), however, indicated that *Malus sylvestris* has also contributed to the genome of *M. domestica*. A third species that has been thought to have made contributions to the genome of the domestic apples is *Malus baccata*, but there is no hard evidence for this in older apple cultivars. The government of Kazakhstan and the United Nations Development Programme have established a conservation project and a protected reserve for *Malus sieversii* in the Zailijskei Alatau mountains. Fauna & Flora International (FFI) is working in Kyrgyzstan to save and restore one of the most highly threatened apple species, the Niedzwetzky apple (*Malus niedzwetzkyana*), as part of the Global Trees Campaign.

Box 7.11 Summary of CWR taxa selected by the project partners

Armenia – cereals: *Triticum boeoticum, Triticum araraticum, Triticum urartu, Aegilops tauschii*; pulse: *Vavilovia formosa*; vegetable: *Beta lomatogona*; fruits, berries and nuts: *Pyrus caucasica.*

Bolivia – *Annona, Rubus, Cyphomandra, Carica, Phaseolus, Arachis, Capsicum, Chenopodium, Solanum, Euterpe, Bactris, Theobroma, Anacardium, Manihot, Ananas, Ipomoea.*

Madagascar – rice (*Oryza*), *Ensete* (a wild relative of banana), vanilla (*Vanilla*), yam (*Dioscorea*), coffee (*Coffea*)

Sri Lanka – 5 wild species of rice (*Oryza*); 2 wild species of banana (*Musa*); 6 wild species of *Vigna*; 8 wild species of cinnamon (*Cinnamomum*); 8 wild species of pepper (*Piper*).

Uzbekistan – onion (*Allium*), apple (*Malus*), walnut (*Juglans*), pistachio (*Pistacia*), almond (*Amygdalus*), barley (*Hordeum* – 2 species).

It is estimated that around 90 per cent of the fruit and nut forests in Kyrgyzstan, Kazakhstan, Uzbekistan, Turkmenistan and Tajikistan have been destroyed over the past 50 years so that conservation of genetic resources of the species involved is a matter of high priority (see also Boxes 7.9 and 7.10).

A summary of the CWR selected by the UNEP/GEF CWR Project countries is given in Box 7.11.

Selection of priority areas

Protected areas can play a significant role in the conservation of agrobiodiversity, including CWR. The WWF report *Food Stores: Using Protected Areas to Secure Crop Genetic Diversity* (Stolton et al, 2006) (see Box 7.12) looks at how protected area managers can find which CWR species are present in the protected areas they manage and how they might adapt management practices to facilitate conservation of CWR and landraces.

The presence of populations of target species in an already existing protected area obviously confers an advantage in that, provided the conditions are suitable, the need for often lengthy and expensive negotiations in setting up a new protected area or reserve is obviated.

For further details on the selection of priority areas, the volume *Conserving Plant Diversity in Protected Areas* (Iriondo et al, 2008) is a useful resource, as is *Establishment of a Global Network for the In Situ Conservation of Crop Wild Relatives: Status and Needs* (Maxted and Kell, 2009).

Many CWR, probably the majority, occur outside protected areas and are found in a variety of natural and semi-natural habitats or even occur as weeds. The options for *in situ* conservation of CWR in such areas are reviewed in Chapter 11.

Box 7.12 Main conclusions of the *Food Stores* report

- Many of the centres of diversity of our principal cultivated plants are poorly protected.
- The role of protected areas in conserving crop genetic diversity could be greatly increased by better understanding of this issue within protected area organizations.
- The promotion of the conservation of crop genetic diversity within existing protected areas may further enhance the public perception of protected areas and help to ensure longer-term site security.
- There are already a few protected areas that are being managed specifically to retain landraces and CWR, and there are many more protected areas that are known to contain populations essential to the conservation of plant genetic resources.
- By conserving locally important landraces, protected areas can contribute to food security, especially for the poorest people.

Source: Stolton et al, 2006

Criteria for selection of areas

Selection of areas for *in situ* conservation of target species is quite different from designing a national system of protected areas that aim to include the maximum biodiversity possible or maintenance of ecosystem services. Extensive literature on reserve selection exists (e.g. Pressey et al, 1993, 1997; Balmford, 2002; Kjaer et al, 2004) and a review of genetic reserve location and design is given by Dulloo et al (2008). To a large extent, the areas for CWR conservation are self-defining by the presence in them of the target species as revealed by ecogeographical surveying (see Chapter 8). The issues here are more concerned with deciding how many populations and how much genetic variation is to be included and then whether the resultant area(s) required is ecologically viable and physically maintainable. The following criteria for locating genetic reserves have been suggested (cf. Dulloo et al, 2008):

- distribution pattern and abundance of the target species;
- level and pattern of genetic diversity of the target species' populations and presence of desirable alleles, if known;
- number of populations;
- number of individuals within the population;
- current conservation status;
- presence in protected areas or centres of plant diversity;
- accessibility;
- size of reserves;
- health and quality of the reserve;
- state of management of the reserve;
- political and socio-economic factors.

These and other factors that will influence the choice of reserve are discussed in continuation:

Size – Different species require reserves of different sizes. Generally, populations in larger areas are exposed to less risk of extinction: a larger population implies less vulnerability to inbreeding and stochastic factors and less negative influence of edge effects. On the other hand, the larger an area, the more likely it is to be at risk from invasive species and the larger an area and the lower its protection status (in terms of the IUCN classification of Protected Areas), the less likely the management of the area is to address the conservation needs of target species.

Boundaries, shape, integrity and context – The nature, location, state and effectiveness of the boundaries of a reserve are factors that need to be considered in choosing a protected area or reserve. If the range of biophysical conditions and habitats, and native organisms and ecosystems needed to maintain the ecological processes are not included within the boundaries set, then there is a risk of changes taking place in the disturbance regimes, ecological productivity and species dynamics, which could lead to a loss of species.[9] Natural boundaries are normally to be preferred to arbitrarily drawn ones.

Shape is a feature commonly associated with the selection of nature reserves: an irregular or elongated reserve has relatively more exposed areas so organisms may be more vulnerable to external threats, including invasion by alien species.

Integrity and context are two other issues of relevance. Internal roads, railways, power lines and fences are sources of fragmentation that create new borders with the undesired effects that these promote, including their role as pathways for invasive species. Biodiversity within the reserve is also influenced by the context of the countryside in which it occurs: it is not worth designing a reserve that is not incorporated into the surrounding environment or without considering land-use patterns at different scales.

Presence of invasive species – The presence of invasive species in the reserve can cause serious problems, especially when active measures (and a budget) are needed to control them. Their elimination or control may be an important component of management plans both for protected areas and for targeted species.

Sustainability – The sustainability of a protected area is a key concern and this will depend on a series of factors such as good governance, adequate finance and staffing. Many areas are what are termed 'paper parks', which have been designated but not properly implemented. Fewer than one-third of protected areas report having a full management plan (Ervin et al, 2008); in most cases, their biodiversity has not been adequately inventoried and many protected areas are inadequately protected, staffed or managed (WWF, 2004). While these are matters that are outside the responsibility of those undertaking *in situ* conservation of target species, they will clearly influence the choice of areas.

Box 7.13 Selection of genetic management zone sites for lychee (*Litchi chinensis*) in Vietnam

The selection of study sites proceeded in two steps. The first step was to identify genetically important areas (henceforth, referred to as 'genetic management zones' – GMZs) or 'hot spots' based on the following criteria:

- *presence and genetic diversity of target species;*
- *presence of endemic species;*
- *presence of high numbers of other economic species;*
- *overall floristic species richness;*
- *presence of high numbers of other economic species;*
- *containing natural and/or semi-natural ecosystems;*
- *presence of traditional agricultural systems; and*
- *protection status and/or existence of conservation-oriented farmers or communities that manage a number of species and cultivars ...*

The second step was to select specific sites and communities within the larger GMZs where socio-economic conditions indicate good feasibility for on-farm agrobiodiversity conservation activities. Several workshops, stakeholder consultations and numerous meetings between IAG, NGOs working in the GMZs, local institutes, and farmer groups aided this process. Visits were made to each site to assess community receptivity to sharing traditional knowledge and practices that promote in situ *conservation.*

Source: Thi Hoa et al, 2005

The criteria adopted for the selection of gene management zones or genetic reserves in Vietnam for lychee (*Litchi chinensis*) are described in Box 7.13.

It is likely that many protected areas will become vulnerable to the effects of global and, in particular, climate change and human population growth. This is discussed in Chapter 14.

Special requirements for species with extensive distributions

While many of the species targeted for *in situ* conservation are restricted in distribution, if not rare, in the case of species which are widespread and of economic importance, such as major forest trees, special considerations apply when choosing which populations and areas to conserve. Sampling and conservation strategies for such species may involve including genetic core areas, important ranges of diversity, particular ecotypes or ranges of clinical variation, and outlier or marginal populations. In situations where populations of the target CWR occur in more than one area, a decision has to be made about which and how many areas should be selected for their *in situ* conservation. In the case of lychee (*Litchi chinensis*) conservation in Vietnam, it was found that a series of gene management zones was often required to ensure an adequate representation of the ecogeographic ranges needed for the selected species and populations in order to support sufficient environmental heterogeneity.

In the case of species whose populations consist of a series of isolated, widely scattered individuals – for example, in arid zones – this may require very large reserves to include a viable population. In such cases, the individual specimens may require additional protection. Rupicolous plants in inaccessible habitats and with highly niche-specific ecology, e.g. some *Brassica* wild relatives, which occur on rock faces in various parts of Europe and the Mediterranean, pose special challenges (Heywood, 2006).

Priority areas selected by the countries

Faced with financial and technical resource limitations, as well as political and socio-economic factors in certain instances, the selection of priority areas in countries was pragmatically determined, usually based on the actual presence of a priority species in an already established protected area, as well as accessibility to the area. In Bolivia, due to a moratorium imposed by the government on any activities planned within the country's protected areas, the selection of protected areas for CWR conservation was severely impacted and obviously delayed. Below is a detailed description of the protected areas and the species that were targeted for management plans by the project: wild cereals in Armenia, wild cacao in Bolivia, wild yams in Madagascar, wild cinnamon in Sri Lanka and wild almond in Uzbekistan (see Table 7.12).

Armenia

The area selected for *in situ* management is the Erebuni State Reserve. Occupying an area of approximately 89ha, the Erebuni State Reserve is Armenia's smallest protected area managed by the Reserve Park Complex of the Ministry of Nature Protection of the Republic of Armenia. It was established in 1981, in the vicinity of Yerevan, specifically to protect wild cereal species such as wheat (*Triticum araraticum*, *T. urartu*, *T. boeticum*), goatgrasses (*Aegilops* spp.), barley (*Hordeum glaucum*) and rye (*Secale vavilovii*). The reserve is also home to 292 species of vascular plants, representing 196 genera from 46 families. Participatory work carried out with local communities living in close proximity to the park has raised the profile of CWR and helped raise awareness on the need to conserve them. The reserve is located within the administrative boundaries of Yerevan city (see Chapter 9 and http://www.reservepark.mnp.am/htmls_eng/regions_1.htm).

Bolivia

Due to political delays, consultation with SERNAP (Servicio Nacional de Áreas Protegidas – the protected area authorities) commenced only in September and October 2009. SERNAP proposed working on the management plan of the Parque Nacional y Territorio Indigena Isiboro-Secure (TIPNIS) and with

Theobroma species as the target for a species management plan. Ranging in altitude from 180m to 3000m and extending for 1,372,180ha between the northern part of the Cochabamba Department and the southern part of the Beni Department, the TIPNIS (IUCN Category II – NP) is home to a high level of species and ecosystem diversity. Its range of habitats includes montane cloud forests, sub-Andean Amazonian forests, mid- to lowland evergreen rainforests and flooded savannas, each harbouring a unique flora and fauna. The protected area, established in 1965, is also an indigenous territory, property of the Chimán, Yuracaré and Moxeño tribes. SERNAP, which manages the Park, and the local organization of the indigenous people living in the Park (*Sub Central Indígena del TIPNIS*), have agreed to develop and establish a specific 'Programme for the *in situ* conservation of crop wild relatives existing within the park' and formulate a 'Management Plan for the protection of wild relatives of cocoa' to be included in the Park's management plan. The wild cacao (*Theobroma* spp.) existing inside the Park is currently threatened by habitat destruction and deforestation.

Madagascar

The area selected for *in situ* conservation of *Dioscorea maciba* and other *Dioscorea* species is Ankarafantsika National Park. *Dioscorea*, which includes over 40 species, is of high economic value as a staple food crop and several species of wild yams are now threatened due to overexploitation and are listed as critically endangered. A conservation programme has been initiated with local communities in the framework of the management plan for Ankarafantsika National Park, trying to reduce the pressure on wild species by convincing communities to grow cultivated yams. Located in the north-western part of Madagascar, the national park (IUCN Category II) was established in 1997, covers an area of 130,026km^2 and is managed by the Madagascar National Parks Association (PNM-ANGAP). See: http://www.parcs-madagascar.com/fiche-aire-protegee_en.php?Ap=15.

Sri Lanka

The area selected for *in situ* management of *Cinnamomum capparu-coronde* Blume is the Kanneliya Forest Reserve (see Chapter 9). Located in the Southern Province, near Galle, Kanneliya-Dediyagala-Nakiyadeniya (KDN) is the last large remaining rainforest in Sri Lanka, covering an area of 10,139ha. Its importance in terms of biodiversity and ecosystem services is such that it was designated as a biosphere reserve in 2004 by UNESCO. This protected area harbours many plant and animal species endemic to Sri Lanka. The Sri Lanka component of the UNEP/GEF CWR Project has worked hand in hand with the park's governing body – the Department of Forest Conservation – to modify the existing management plan for the area, which now includes a species management plan for the important endemic *Cinnamomum capparu-coronde* Blume, which is normally harvested for medicinal and commercial purposes. Awareness-raising activities have also been carried out to educate local communities on the importance of preserving such species.

Table 7.12 *Examples of CWR conserved in protected areas in Armenia, Bolivia, Madagascar, Sri Lanka and Uzbekistan*

Crop gene pool	*CWR*	*Protected area*	*Country*
Yam	*Dioscorea maciba, D. bemandry, D. antaly, D. ovinala* and *D. bemarivensis*	Ankarafantsika National Park	Madagascar
Cinnamon-tree	*Cinnamomum capparu-coronde*	Kanneliya Forest Reserve	Sri Lanka
Almond	*Amygdalus bucharica*	Chatkal Biosphere Reserve	Uzbekistan
Wheat	*Triticum araraticum, T. boeoticum, T. urartu* and *Aegilops tauschii*	Erebuni State Reserve	Armenia
Cacao	*Theobroma* spp.	Parque Nacional y Territorio Indígena Isiboro-Secure	Bolivia

Uzbekistan

Ugam-Chatkal State Natural National Park has been selected for *in situ* conservation of walnut, where this species is widely distributed (about 1500ha). The Park is located in Bostanlik region of the Tashkent district. Better forest stands with walnut (*Juglans*) are located on the Ugam range (Boguchalsay, Sidjaksay and Nauvalisay) and on the Pscem range (Aksarsay). Walnut is under better protection in the territory of Aksarsay where monitoring of the state of the walnut populations in the State Forestry Fund managed by Brichmulla Forestry has been agreed.

Ugam-Chakal State Natural National Park and Chatkal Biosphere Reserve have been chosen as areas selected for *in situ* management for barley (*Hordeum*).

Conclusions and lessons learned

In selecting species for priority conservation action, the countries used a range of criteria and a weighting mechanism. In the case of priority species, the absence of prior agreed guidelines for their selection led to considerable delays and confusion. On the other hand, it is quite clear in discussions with the five countries that the choice of areas and species was mainly influenced by the information already available on CWR conservation, as well as local knowledge of the situation in the countries concerned, and that a largely pragmatic approach was adopted. Considering that, for the purposes of the project, only a small number of priority species were selected, the choice of CWR related to important crops, especially those listed in Annex I of the ITPGRFA, and the selection of well-known protected areas in which they occurred is understandable. However, such an approach cannot be applied by the countries when the national CWR conservation strategy is implemented for all the CWR recorded.

It was also clear that there is a certain amount of confusion about the application of the global IUCN Red Listing process, its application at national level, the use of threat assessments other than those of the IUCN, the relative importance of the IUCN criteria and other threat assessment criteria.

A general conclusion that can be drawn is that it is very difficult and probably unrealistic to expect that uniform sets of criteria can be used for selecting species and selecting areas for CWR conservation. Nonetheless, it is important, especially when selecting the taxa, that as much information as possible be taken into account so that CWR representing a wide range of situations and values are chosen for conservation, subject of course to the availability of financial and technical resources.

Sources of further information

Brehm, J.M., Maxted, N., Martins-Loução, M.A. and Ford-Lloyd, B.V. (2010) 'New approaches for establishing conservation priorities for socio-economically important plant species', *Biodiversity Conservation*, vol 19, pp2715–2740.

Burgman, M.A., Keith, D.A., Rohlf, F.J. and Todd, C.R. (1999) 'Probabilistic classification rules for setting conservation priorities', *Biological Conservation*, vol 89, pp227–231.

Chape, S., Spalding, M. and Jenkins, M. (eds) (2008) *The World's Protected Areas*, Prepared by the UNEP World Conservation Centre, University of California Press, Berkeley.

CMP (2005) *Taxonomies of Direct Threats and Conservation Actions*, Conservation Measures Partnership (CMP), Washington, DC.

Dudley, N. (ed) (2008) *Guidelines for Applying Protected Area Management Categories*, IUCN, Gland, Switzerland.

Flor, A., Bettencourt, E., Arriegas, P.I. and Dias, S. (2006) 'Indicators for the CWR species' list prioritization (European crop wild relative criteria for conservation)' in B.V. Ford-Lloyd, S.R. Dias and E. Bettencourt (eds) *Genetic Erosion and Pollution Assessment Methodologies*, pp83–88, Proceedings of PGR Forum Workshop 5, Terceira Island, Autonomous Region of the Azores, Portugal, 8–11 September 2004, Published on behalf of the European Crop Wild Relative Diversity Assessment and Conservation Forum, by Bioversity International, Rome, Italy.

IUCN (1994) *IUCN Red List Categories*, International Union for Conservation of Nature (IUCN), Gland, Switzerland.

IUCN (2005) *Threats Authority File*, Version 2.1, International Union for Conservation of Nature (IUCN) Species Survival Commission, Cambridge, UK; http://www.iucn.org/about/work/programmes/species/red_list/resources/technical_documents/authority_files/threats.rtf, accessed 24 August 2009.

Lockwood, M., Worboys, G.K. and Kothari, A. (2006) *Managing Protected Areas: A Global Guide*, Earthscan, London, UK.

Maxted, N., Ford-Lloyd, B.V. and Hawkes, J.G. (eds) (1997) *Plant Genetic Conservation: The* In Situ *Approach*, Chapman and Hall, London, UK.

Notes

1. www.iucnredlist.org/apps/redlist/static/categories_criteria_3_1.
2. www.iucnredlist.org/apps/redlist/static/categories_criteria_3_1 (accessed 23 November 2010).
3. http://www.dwaf.gov.za/wfw/
4. http://www.gisp.org/
5. McNeely, J.A., Mooney, H.A., Neville, L.E., Schei, P. and Waage, J.K. (eds) (2001) *Global Strategy on Invasive Alien Species,* IUCN on behalf of the Global Invasive Species Programme, Gland, Switzerland and Cambridge, UK; http://www.gisp.org/publications/brochures/globalstrategy.pdf
6. http://www.gisinetwork.org/
7. Wittenberg, R. and Cock, M.J.W. (eds) (2001) *Invasive Alien Species: A Toolkit of Best Prevention and Management Practices*, CAB International, Wallingford, Oxon, UK; http://www.gisp.org/publications/toolkit/Toolkiteng.pdf
8. *South America Invaded,* A GISP publication (2005) written by Sue Matthews, available at: http://vle.worldbank.org/bnpp/files/TF024046BIOLOGICALINV gispSAmerica.pdf
9. Hansen and Rotella (2001)

References

Akçakaya, H.R., Butchart, S.H.M., Mace, G.M., Stuart, S.N. and Hilton-Taylor, C. (2006) 'Use and misuse of the IUCN Red List Criteria in projecting climate change impacts on biodiversity', *Global Change Biology,* vol 12, pp2037–2043

Balmford, A. (2002) 'Selecting sites for conservation', in K. Norris and D. Pain (eds) *Conserving Bird Biodiversity. General Principles and their Application,* pp74–104, Cambridge University Press, Cambridge.

Balmford, A., Carey, P., Kapos, V., Manica, A. Rodrigues, A.S.L., Scharlemann, J.P.W. and Green, R.E. (2009) 'Capturing the many dimensions of threat: Comment on Salafsky et al', *Conservation Biology*, vol 23, pp482–487

Baudoin, M. and España, R. (1997) 'Lineamientos para la elaboración de una estrategia nacional de conservación y uso sostenible de la biodiversidad', Ministerio de Desarrollo Sotenible y Medio Ambiente

Bingelli, P. (2003) 'Introduced and invasive plants', in S.M. Goodman and J.P. Benstead (eds) *The Natural History of Madagascar*, pp257–268, University of Chicago Press, Chicago, USA

Brown, K.A, Ingram, J.C., Flynn, D., Razafindrazaka, R.J and Jeannoda, V.H. (2009) 'Protected areas safeguard tree and shrub communities from degradation and invasion: A case study in eastern Madagascar', *Environmental Management,* vol 44, pp136–148

Burgman, M.A., Keith, D.A., Rohlf, F.J. and Todd, C.R. (1999) 'Probabilistic classification rules for setting conservation priorities', *Biological Conservation,* vol 89, pp227–231

Cavers, S., Navarro, C. and Lowe, A.J. (2004) 'Targeting genetic resource conservation in widespread species: A case study of *Cedrela odorata* L.', *Forest Ecology and Management,* vol 197, pp285–294

CMP (2005) *Taxonomies of Direct Threats and Conservation Actions*, Conservation Measures Partnership (CMP), Washington, DC

Coart, E., Van Glabeke, S., De Loose, M., Larsen, A.S. and Roldán-Ruiz, I. (2006) 'Chloroplast diversity in the genus *Malus*: New insights into the relationship between the European wild apple (*Malus sylvestris* (L.) Mill.) and the domesticated apple (*Malus domestica* Borkh.)', *Molecular Ecology,* vol 15, no 8, pp2171–2182

Dulloo, M.E., Labokas, J., Iriondo, J.M., Maxted, N., Lane, A., Laguna, E., Jarvis, A. and Kell, S.P. (2008) 'Genetic reserve location and design', in J.M. Iriondo, N. Maxted and M.E. Dulloo (eds) *Conserving Plant Genetic Diversity in Protected Areas,* pp23–64, CAB International

ECODIT (2009) 'Biodiversity analysis update for Armenia final report: Prosperity, livelihoods and conserving Ecosystems (PLACE), IQC Task order #4', Prepared by: Armenia Biodiversity Update Team, Assembled by ECODIT, Inc. Arlington, VA, USA

Ervin, J., Gidda, S.B., Salem, S. and Mohr, J. (2008) 'The programme of work on protected areas – A view of global implementation', *Parks,* vol 17, pp4–11

Fernández, M. (2009) 'Distribución de plantas invasoras en caminos cercanos a la ciudad de La Paz', Tesis de licenciatura en Biología, Universidad Mayor de San Andrés, La Paz, Bolivia, p50

Flor, A., Bettencourt, E., Arriegas, P.I. and Dias, S. (2006) 'Indicators for the CWR species' list prioritization (European crop wild relative criteria for conservation)' in B.V. Ford-Lloyd, S.R. Dias and E. Bettencourt (eds) *Genetic Erosion and Pollution Assessment Methodologies,* pp83–88, Proceedings of PGR Forum Workshop 5, Terceira Island, Autonomous Region of the Azores, Portugal, 8–11 September 2004, Published on behalf of the European Crop Wild Relative Diversity Assessment and Conservation Forum by Bioversity International, Rome, Italy

Foden, W., Mace, G., Vié, J.-C., Angulo, A., Butchart, S., DeVantier, L., Dublin, H., Gutsche, A., Stuart, S. and Turak, E. (2008) 'Species susceptibility to climate change impacts', in J.-C. Vié, C. Hilton-Taylor and S.N. Stuart (eds) *The 2008 Review of The IUCN Red List of Threatened Species,* International Union for Conservation of Nature (IUCN), Gland, Swtizerland

Ford-Lloyd, B., Kell, S.P. and Maxted, N. (2008) 'Establishing conservation priorities for crop wild relatives', in N. Maxted, B.V. Ford-Lloyd, S.P. Kell, J.M. Iriondo, M.E. Dulloo and J. Turok (eds) *Crop Wild Relative Conservation and Use,* pp110–119, CAB International, Wallingford, UK

Gardenfors, U., Rodriguez, J.P., Hyslop, C., Mace, G.M., Molur, S. and Poss, S. (1999) 'Draft guidelines for the application of IUCN Red List criteria at regional and national levels', *Species,* vol 31/32, pp58–70

Hansen, A.J. and Rotella, J.J. (2001) 'Nature reserves and land use: Implications of the "place" principle', in V.H. Dale and R.A. Hauber (eds) *Applying Ecological Principles to Land Management,* Springer, Berlin, Germany

Heywood, V. (2006) 'On the rocks', *Geneflow '06,* Bioversity International, p38

Heywood, V.H. and Dulloo, M.E. (2005) In Situ *Conservation of Wild Plant Species – A Critical Global Review of Good Practices,* IPGRI Technical Bulletin, no 11, FAO and IPGRI, International Plant Genetic Resources Institute (IPGRI), Rome, Italy

Iriondo, J.M., Maxted, N. and Dulloo, M.E. (eds) (2008) *Conserving Plant Diversity in Protected Areas,* CAB International, Wallingford, UK

IUCN (1994) *IUCN Red List Categories,* International Union for Conservation of Nature (IUCN), Gland, Switzerland

IUCN (1996) *The 1996 IUCN Red List of Threatened Animals,* International Union for Conservation of Nature (IUCN), Gland, Switzerland

IUCN (2000) 'Background to IUCN's system for classifying threatened species', CITES

Inf. ACPC.1.4.(Document CWG1-3.4), International Union for Conservation of Natures (IUCN), http://www.cites.org/eng/com/aC/joint2/ACPC1-Inf4.pdf, accessed 24 August 2009

IUCN (2005a) *Threats Authority File,* Version 2.1, International Union for Conservation of Nature (IUCN) Species Survival Commission, Cambridge, UK, http://www.iucn.org/about/work/programmes/species/red_list/resources/technical_documents/authority_files/threats.rtf, accessed 24 August 2009

IUCN (2005b) *Conservation Actions Authority File,* Version 1.0, International Union for Conservation of Nature (IUCN), http://www.iucn.org/about/work/programmes/species/red_list/resources/technical_documents/authority_files/consactions.rtf, accessed 24 August 2009

IUCN (2008) *Species Susceptibility to Climate Change Impacts,* International Union for Conservation of Nature (IUCN), http://cmsdata.iucn.org/downloads/climate_change_and_species.pdf, accessed 24 August 2009

Juniper, B. and Mabberley, D. (2006) *The Story of the Apple,* Timber Press, Portland, OR, USA

Kjær, E., Amaral, W., Yanchuk, A. and Graudal, L. (2004) 'Chapter 2: Strategies for conservation of forest genetic resources', in *Forest Genetic Resources Conservation and Management,* vol 1, *Overview, Concepts and Some Systematic Approaches,* FAO/FLD/IPGRI, International Plant Genetic Resources Institute, Rome, Italy

Maxted, N. and Kell, S.P. (2009) *Establishment of a Global Network for the* In Situ *Conservation of Crop Wild Relatives: Status and Needs,* FAO Commission on Genetic Resources for Food and Agriculture, Rome, Italy

Maxted, N., Ford-Lloyd, B.V. and Hawkes, J.G. (1997) 'Complementary conservation strategies', in N. Maxted, B.V. Ford-Lloyd and J.G. Hawkes (eds) *Plant Genetic Conservation: The* In Situ *Approach,* Chapman and Hall, London, UK

MDSP (2001) *Estrategia Nacional de Conservación y Uso Sostenible de la Biodiversidad,* Ministerio de Desarrollo Sostenible y Planificación (MDSP), La Paz, Bolivia

Pressey, R.L., Humphries, C.J., Margules, C.R., Vane-Wright, R.E. and Williams, P.H. (1993) 'Beyond opportunism: Key principles for systematic reserve selection', *Trends in Ecology and Evolution,* vol 8, pp124–128

Pressey R., Possingham, H. and Day, J. (1997) 'Effectiveness of alternative heuristic algorithms for identifying indicative minimum requirements for conservation reserves', *Biological Conservation,* vol 80, pp207–219

Rico, A. (2009) 'Informe Final Técnico y Financiero Donaciones para la Digitalización de Datos Red Temática de Especies Invasoras del Proyecto: "Establecimiento en Bolivia de Bases de Datos sobre Especies Exóticas Invasoras, como parte de la Red Interamericana de Información en Biodiversidad, –IABIN"', La Paz, Bolivia

Salafsky, N., Salzer, D., Stattersfield, A.J., Hilton-Taylor, C., Neugarten, R., Butchart, S.H.M., Collen, B., Cox, N., Master, L.L., O'Connor, S. and Wilkie, D. (2008) 'A standard lexicon for biodiversity conservation: Unified classifications of threats and actions', *Conservation Biology,* vol 22, no 4, pp897–911

Salafsky, N., Butchart, D.H.M., Salzer, D., Stattersfield, A.J., Neugarten, R., Hilton-Taylor, C., Collen, B., Master, L.L., O'Connor, S. and Wilkie, D. (2009) 'Pragmatism and Practice in Classifying Threats: Reply to Balmford et al', *Conservation Biology,* vol 23, pp488–493

Saterson, K.A. (1995) 'Foreword' in N.C. Johnson, *Biodiversity in the Balance: Approaches to Setting Geographic Conservation Priorities,* Biodiversity Support Program, Washington, DC

Stolton, S., Maxted, N., Ford-Lloyd, B., Kell, S.P. and Dudley, N. (2006) *Food Stores: Using Protected Areas to Secure Crop Genetic Diversity,* World Wide Fund for Nature (WWF) Arguments for Protection Series, WWF International, Gland, Switzerland

Thi Hoa, T., Dinh, L.T., Thi Ngoc Hue, N., Van Ly, N. and Ngoc Hai Ninh, D. (2005) '*In situ* conservation of native lychee and their wild relatives and participatory market analysis and development – The case of Vietnam', in N. Chomchalow and N. Sukhvibul (eds) *Proc. 2nd International Symposium on Lychee, Longan, Rambutan & Other Sapindaceae Plants. Acta Horticulturae,* vol 665, pp125–140

WWF (2004) *How Effective are Protected Areas?* Preliminary analysis of forest protected areas by WWF – the largest ever global assessment of protected area management effectiveness. Report prepared for the Seventh Conference of the Parties of the Convention on Biological Diversity, February 2004, World Wide Fund for Nature (WWF), Gland, Switzerland

Chapter 8

Establishing an Information Baseline: Ecogeographic Surveying

> *Before sensible conservation decisions can be made, a basic understanding of the taxonomy, genetic diversity, geographic distribution, ecological adaptation and ethnobotany of a plant group, as well as of the geography, ecology, climate and human communities of the target region, is essential* (Guarino et al, 2005).

Aims and purpose

Before any conservation action on a target taxa can be undertaken, sufficient information about the taxa must be gathered in order to make informed decisions and establish appropriate priorities for the development of a practical conservation strategy. Box 8.2 outlines the different kinds of information about the taxa that should be gathered. This information can be obtained from the literature, herbarium specimens, genebanks, botanic gardens, arboreta and meteorological stations, as well as from field surveys, so as to establish a knowledge baseline.

> *An ecogeographic study is a process of gathering and synthesizing ecological, geographic and taxonomic information. The results ... can be used to help formulate conservation strategies and collecting priorities* (Maxted et al, 1995).

The process of gathering this information is sometimes referred to as an *ecogeographical survey or study* (IBPGR, 1985; Maxted et al, 1995; Dulloo et al, 2008) and is a key first step in the development of any conservation strategy, whether *in situ* or *ex situ*. An ecogeographical survey aims to determine: (i) the distributions of particular taxa in particular regions and ecosystems; (ii) the patterns of infraspecific diversity; and (iii) the relationships between survival and frequency of variants and associated ecological conditions. The term ecogeographic survey

Box 8.1 Examples of ecogeographic surveys

Coffea (Dulloo et al, 1999; Maxted et al, 1999): an herbarium-based ecogeographic survey was made, supplemented by detailed field surveys of wild *Coffea* species in the Mascarene Islands. The geographical and ecological distribution of the different *Coffea* species in the Mascarene Islands, principally in Mauritius, was determined. Genetic diversity hotspots were mapped and an assessment was made of the IUCN conservation status of native *Coffea* species.

Vicia (Maxted, 1995; Bennett and Maxted, 1997).

Corchurus (Edmonds, 1990).

Medicago (Bennett et al, 2006).

Phaseolus (Nabhan, 1990).

Lens (Ferguson and Robertson, 1996).

Leucaena (Hughes, 1998).

Annual legumes (Ehrman and Cocks, 1990).

South American *Solanum* (Smith and Peralta, 2002).

Trifolium (Bennett and Bullitta, 2003): an ecogeographical analysis of six species of *Trifolium* from Sardinia, with the aim of designing future collection missions and for the designation of important *in situ* reserves in Sardinia.

African *Vigna* (Maxted et al, 2004).

applies to various processes of gathering and collating information on the taxonomy, geographical distribution, ecological characteristics, genetic diversity and ethnobiology of the target species, as well as the geography, climate and the human setting of the regions under study (Guarino et al, 2002).[1] Ecogeographic information can be used to locate significant genetic material and representative populations can be monitored to guide the selection of representative samples for conservation and utilization (IBPGR, 1985). Although originally designed and applied in the context of conservation of gene pools of wild species such as CWR, the ecogeographic survey approach can be modified so as to apply to crops (Guarino et al, 2005).

A full ecogeographic survey requires considerable resources to carry out and may take several years to complete, especially in the case of wide-ranging species. While highly desirable, especially for CWR of major importance, this will seldom be possible and much more concise studies are often undertaken. Examples of ecogeographic surveys are given in Box 8.1.

The Bioversity International series 'Systematic and Ecogeographic Studies on Crop Genepools' covers some of the most important CWR and is available for download at: http://www.bioversityinternational.org/publications.

Ecogeographic surveys carried out during the UNEP/GEF CWR Project

In the course of the CWR Project, ecogeographical studies were made of the following species:

Armenia

Desktop studies were made of 99 species, of which 79 were the subject of field studies (Table 8.1).

Bolivia

Researchers from national partner institutions participating in the CWR Project in Bolivia, gathered ecogeographic data through field trips to the areas of distribution of species in different regions. During the period 2006 to 2009, researchers collected field data on 149 (out of 201) species identified in 2005, covering 14 genera (*Anacardium*, *Ananas* and *Pseudoananas*, *Annona*, *Arachis*, *Bactris*, *Chenopodium*, *Cyphomandra*, *Ipomoea*, *Manihot*, *Phaseolus*, *Rubus*, *Solanum*, *Theobroma* and one *Vasconcellea* segregated from *Carica*). The 14 genera were prioritized from an original group of 52 genera previously identified, based on criteria such as: potential use and importance of economic, social and cultural state of knowledge, including taxa in the International Treaty on Plant Genetic Resources for Food and Agriculture. The species are listed in Annex I.

The researchers also collected specimens that were then incorporated into the collections of the Herbaria Bolivia (BOLV, USZ and LPB) and accessions that

Table 8.1 *List of species surveyed ecogeographically in Armenia*

*Triticum araraticum, T. boeoticum, T. urartu, *Aegilops crassa, A. tauschii, A. cylindrica, A. triunccialis, A. biunccialis, A. triaristata, A. columnaris, Ambylopyrum muticum, Hordeum spontaneum, H. glaucum, H. murinum, H. geniculatum, H. marinum, H. violaceum, H. bulbosum, *H. hrasdanicum, Secale vavilovii, S. montanum, *Cicer anatolicum, *Lens ervoides, L. orientalis, *Pisum arvense, P. elatius, Vavilovia formosa, Vicia villosa, *V. ervilia, V. cappadoixcica, Lathyrus pratensis, L. tuberosus, Onobrychis transcaucasic, O. altissima, O. hajastana, O. cadmea, O. oxytropoides, Medicago sativa, M. lupulina, *Trifolium sebastianii, T. hybridum, T. pratense, T. repens, Beta macrorhiza, B. corolliflora, B. lomatogona, *Spinacia tetrandra, Asparagus officinalis, A. verticillatus, A. persicus, Rumex acetosa, R. crispus, R. tuberosus, *R. scutatus, R. obtusifolius, Chaerophyllum aureum, C. bulbosum, Daucus carota, Falcaria vulgaris, Heracleum trachyloma, Allium atroviolaceum, A. rotundum, A. victorialis, Cucumis melo, Malva neglecta, Lactuca serriola, Malus orientalis, Pyrus caucasica, P. syriaca, P. takhtadzhianii, P. salicifolia, P. zangezura, P. tamamschjanae, P. medvedevii, P. pseudosyriaca, Sorbus hajastana, S. aucuparia, S. takhtajanii, S. subfusca, S. roopiana, S. persica, Crataegus orientalis, Crataegus pontica, Ficus carica, Armeniaca vulgaris, Amygdalus nairica, A. fenzliana, Cerasus avium, Prunus spinosa, P. divaricata, Diospyros lotus, Rubus idaeus, R. cartalinicus, R. armeniacus, *Ribes armenum, *R. biebersteinii, Punica granatum, Cornus mas, Juglans regia.*

* Species that were the subject of desktop studies only

were added to local genebanks. Such efforts support an increased knowledge about CWR and help to ensure that key information is available for decision-making regarding research, production, public planning, conservation and use of CWR, and the design of policies and standards related to research, conservation and use of biodiversity.

Sri Lanka

Ecogeographic surveys of the following species were made: *Oryza nivara*, *Vigna aridicola*, *V. trilobata*, *V. stipulacea*, *V. dalzelliana*, *V. marina*, *V. radiata* var. *sublobata*, *Musa acuminata*, *M. balbisiana*, *Piper chuvya*, *P. longum*, *P. siriboa*, *P. walkeri*, *P. trineuron*, *P. zeylanicum*, *Cinnamomum dubium*, *C. ovalifolium*, *C. litseaefolium*, *C. capparu-coronde*, *C. citriodorum*, *C. sinharajaense* and *C. rivulorum*.

Uzbekistan

The following species were surveyed:

Malus sieversii (apple), *Allium pscemense* (onion), *Amygdalus communis*, *A. bucharica*, *A. spinosissima*, *A. petunnikovii* (almond), *Pistacia vera* (pistachio), *Juglans regia* (walnut), *Hordeum spontaneum* and *H. bulbosum (barley).*

It should be noted that these surveys carried out as part of the UNEP/GEF CWR Project are probably the largest set of ecogeographic assessments ever undertaken and represent a major contribution to the practice.

The components of the knowledge baseline

The knowledge baseline component of an ecogeographic survey brings together a wide range of information about the target species, its distribution, habitat, uses and its presence in protected areas and the availability of germplasm collections (see Box 8.2). The amount of detail will depend largely on how well the species is known, how common it is, its economic uses and where it occurs. There is no 'correct' data set in this regard and a great deal of pragmatism must be applied in practice.

The main stages involved in an ecogeographic survey are given in Box 8.3. The Crop Genebank Knowledge Base has a useful training module on ecogeographic surveys, outlined in Box 8.3. It is available for download at: http://cropgenebank.sgrp.cgiar.org/index.php?option=com_content&view=article&id=378&Itemid=538&lang=english (accessed 27 October 2010).

Gathering and collation of desktop *in situ* information

Much of the desktop information will be available from a country's national CWR or plant genetic resources strategy, if these exist, but the information will still need to be compiled. The national biodiversity strategies and action plans and various

Box 8.2 Elements needed for knowledge baseline

- Bringing together information on the main wild species of economic use in the country or region on:
  - the correct identity;
  - distribution;
  - reproductive biology;
  - breeding system;
  - demography; and
  - conservation status.
- Gathering information on:
  - how they are used, including local traditional knowledge;
  - the nature and extent of trade in these species;
  - the extent to which (if relevant) they are harvested from the wild and the consequences of this on the viability of wild populations; and
  - their cultivation and propagation.
- Establishing which species occur in protected areas, and to what extent.
- Gathering information on the availability of germplasm and authenticated stock for cultivation.
- Ecogeographic conspectus for each species.

Source: Heywood and Dulloo, 2005

Box 8.3 Phases of an ecogeographic study or survey

Phase I – Project design

- project commission;
- identification of taxon expertise;
- selection of target taxon taxonomy;
- delimitation of target region;
- identification of taxon collections;
- designing and building of ecogeographic database structure.

Phase II – Data collection and analysis

- listing of germplasm conserved;
- survey of taxonomic, ecological and geographical data sources;
- collection of ecogeographical data;
- data verification;
- analysis of taxonomic, ecological and ecogeographical data.

Phase III – Product generation

- data synthesis;
- ecogeographical database, conspectus and report;
- identification of conservation priorities.

Source: Maxted et al, 1995; Maxted and Kell, 1998

national reports submitted to the Convention on Biological Diversity (CBD) will also contain valuable information, as will country reports for the first *State of the World's Plant Genetic Resources for Food and Agriculture* (FAO, 1998) and the second Report approved at the Twelfth Regular Session of the Commission on Genetic Resources for Food and Agriculture in October 2009.

Data may be gathered from a range of sources (see also Chapter 6):

- literature, including floras, monographs, checklists and phytosociological studies;
- herbaria;
- botanic gardens and arboreta;
- passport data from genebanks;
- national or local meteorological service data sets (for annual and monthly rainfall, monthly minimum and maximum temperature);
- National Soil Survey and data sets;
- international, regional and national biodiversity databases and information systems.

Taxonomic information

Although it might appear obvious, correct identification of the taxa being surveyed or selected for conservation is essential. This is much more difficult than it seems, as the level of accuracy of identification of plant taxa in scientific literature is very variable and often quite low. Though fundamental when conducting research, often the scientific identification of the taxa is not checked for accuracy. Numerous cases where plants of reference have been misidentified can be cited; the consequences can be serious and very costly.

Reasons for the difficulties experienced in ensuring correct taxonomic identification include the fact that taxonomy and classification are highly specialized subjects and, with certain exceptions such as student Floras and Faunas, and simplified guides for amateurs, the formal products of taxonomy have traditionally not been user-friendly. Floras, monographs, revisions, checklists can be very daunting to the non-specialist as they are highly technical and often cater to the needs of taxonomists rather than the interests of less specialized users. Some identification guides are not written clearly and leave out fundamental information, such as an illustration of the species. Concern for the needs of users, including taxonomists, is a relatively recent development and components such as keys are missing from many classic works. Even when such components are included, they are often highly technical and difficult for an inexperienced user to understand (Heywood, 2004).

A particular case where taxonomic tools such as Floras are critical for conservation is in the preparation of lists of endangered species (Red Data Lists or Handbooks). Floras are a prime resource for the preparation of Red Data Lists and are relied upon, along with herbarium specimens, as a source of data for this purpose, particularly in developing countries. Floras are interpreted to estimate and infer the distribution ranges of the taxa concerned and their degree of rarity

(Golding and Smith, 2001). Unfortunately, Floras were not designed for this purpose, and the extraction and proper interpretation of data can be quite difficult without the assistance of a professional taxonomist.

It must also be stressed that despite the unique role of species as the basic unit in both biological classification and biological diversity, there is no universal agreement on how to define a species. The actual named species we handle in biodiversity studies are comparable only by designation, not in terms of their degree of evolutionary, genetic, ecological or morphological differentiation. In the majority of cases, it is likely that a conventional taxonomic species concept, i.e. one based primarily on morphological differentiation (see Bisby, 1995: Box 2.1–4), will be employed for identifying target species. In practical terms, the standard Flora(s) of the country should be used for species identification and the nomenclature adopted by the Flora(s) should be followed unless it is possible to determine the correct name (if different) through other sources. If a recent revision of the genus or group of species is available that should be used.

In addition, it must be recognized that species concepts differ from group to group and there are often national or regional differences in the way in which the species category is employed (Gentry, 1990; Heywood, 1991), which makes comparisons difficult. Species may be interpreted in some Floras in a wide sense, including species that are regarded as separate ones in other Floras. Likewise, some Floras will treat a particular taxon as a species, while others will treat the same taxon as a subspecies or even as a variety. In fact, infra-specific variants such as subspecies, ecotypes or chemical races or individual populations, rather than species, may be the focus of attention in agrobiodiversity (Yanchuk, 1997). There is a widespread tendency in much work on biodiversity and conservation (e.g. in Red Lists) to treat most species as though they were uniform, whereas many do, in fact, contain a great deal of variation that has been recognized taxonomically or genecologically. It will clearly make a difference when planning conservation actions if distinctive variants are recognized, as their behaviour and underlying genetic differentiation will vary from one to another and require appropriate treatment. This is especially true for CWR where particular alleles in a species' population may be the focus of interest.

While it is likely that, in the case of well-known rare and endangered wild species, few problems of identification will arise, for widespread species occurring in more than one country care should be taken, as the same species may be listed under different names in the Flora, depending on the country. In the absence of any agreed nomenclature, specialist taxonomic advice should be sought.

The same considerations apply at the generic level: an example concerns the genera *Triticum* and *Aegilops*, which have been commonly treated as separate while some taxonomists include *Aegilops* in *Triticum*. This is a matter of taxonomic opinion and neither interpretation is 'correct'. The consequence of these discrepancies is that the same CWR taxon may occur in the taxonomic literature under a range of different names or synonyms.

The problem of synonymy, whereby the same taxon (species, genus etc.) occurs in the literature and herbarium under more than one name, can be

Box 8.4 Practical hints for dealing with taxonomy and names

Remember that if a species has a different name in a Flora or in a herbarium specimen from the one you recognize or are used to, it does not necessarily mean that it is a separate species – it may just be a synonym of that species.

Remember that the names given to species in the literature (scientific papers in journals, inventories, phytosociological or ecological surveys, etc.) may be incorrect and need checking.

If you cannot find a species in a particular Flora or handbook, consider whether it may be 'masquerading' under a different name (synonym) or in a different genus.

If you are unable to identify a specimen, prepare a herbarium sample to take to a taxonomist for identification. Make sure the sample has flowers and fruits, if possible.

If in doubt, consult with a taxonomist for assistance or advice.

intractable for the non-specialist. A plant may have more than one name because:

- it has been described independently more than once by different taxonomists;
- a taxon, such as a species, is later shown to be the same as other earlier published species; or
- a taxon, such as a species, is treated by different taxonomists at different ranks, such as subspecies, or variety, or is placed in different genera by different specialists.

It is important, therefore, that those using taxonomic literature in compiling ecogeographical surveys be aware of these pitfalls.

Sources of taxonomic information

The taxonomic literature is enormous, stretching back centuries and is daunting for the non-professional user. Chapter 2 of the *Global Biodiversity Assessment* (Heywood, 1995) on the characterization of biodiversity (Bisby, 1995) is a valuable source of information. As noted in Chapter 6, in recent years, much taxonomic information has been stored electronically in databases and information systems. Electronic databases and electronic floras are increasingly being developed and should be consulted when available. They range from major international enterprises such as GRIN Taxonomy for Plants (http://www.ars-grin.gov/cgi-bin/npgs/html/index.pl), TROPICOS (http://www.tropicos.org/) and Species 2000 (http://www.sp2000.org/), to national, local and specialized databases.

Taxonomic and other information about biodiversity (natural history collections, library materials, databases, etc.) is not distributed evenly around the globe. The Global Biodiversity Information Facility (GBIF) estimates that

Box 8.5 What is the Global Biodiversity Information Facility (GBIF)?

GBIF enables free and open access to biodiversity data online. It is an international government-initiated and -funded initiative focused on making biodiversity data available to all and anyone, for scientific research, conservation and sustainable development.

GBIF provides three core services and products:

- An information infrastructure – an internet-based index of a globally distributed network of interoperable databases that contain primary biodiversity data – information on museum specimens, field observations of plants and animals in nature and results from experiments – so that data holders across the world can access and share them.
- Community-developed tools, standards and protocols – the tools data providers need to format and share their data.
- Capacity building – the training, access to international experts and mentoring programmes that national and regional institutions need to become part of a decentralized network of biodiversity information facilities.

Source: About GBIF http://www.gbif.org/index.php?id=269

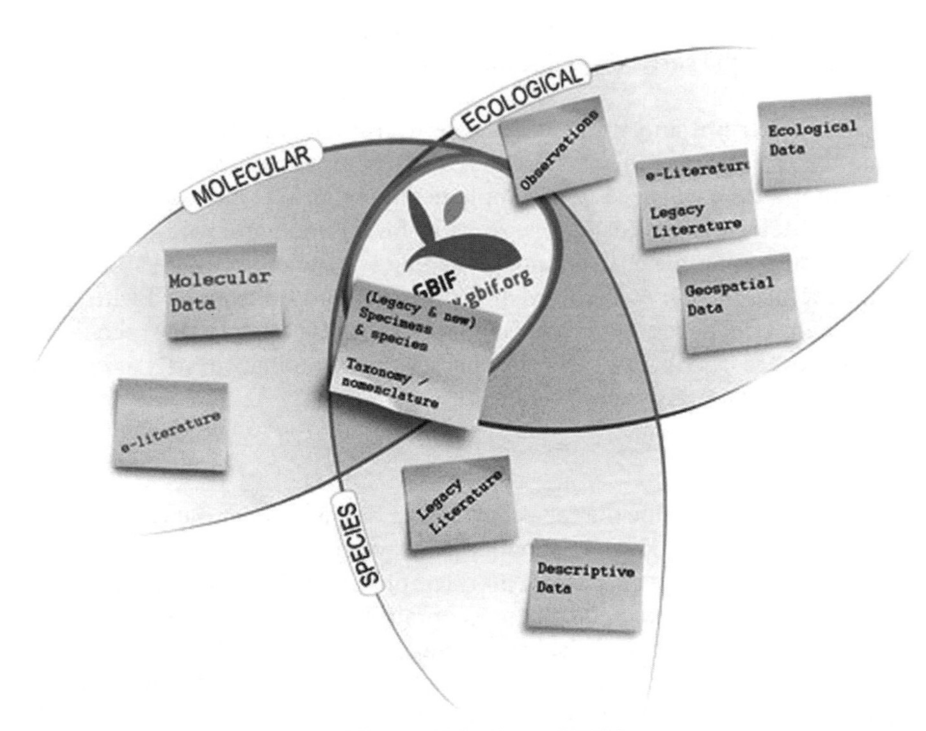

Figure 8.1 *About GBIF*

three-quarters or more of biodiversity data is stored in the developed world. However, most of the data that may be needed cannot be transferred because they are not digitized and/or the capacity to handle digital information is lacking. In order to address this issue and facilitate access to information on biodiversity, the GBIF (see Box 8.5) was initiated. GBIF is a global network of data providers building biodiversity information infrastructure and promoting the growth of biodiversity information content on the internet by working with partner initiatives and coordinating activities worldwide. It aims to be the first port of call for those seeking information on biodiversity.

Herbarium specimens are also a useful source of information (Pearce and Bytebier, 2002) and have been used in many ecogeographical surveys to help determine the distribution of taxa. Maxted (1995), for example, consulted material in 18 major international herbaria in his survey of *Vicia* subgenus *Vicia*. A study on American wild potatoes (Bamberg et al, 2003), included a survey of available herbarium material to help determine the location and distribution of the species and potential collection sites; information was also obtained from local botanists.

Herbarium label data are often insufficient or incomplete, or even difficult to interpret or decipher; the geographical location may be incomplete and the localities given cannot be traced. Likewise, ecological data are often poorly recorded, if at all, and this is especially true for older specimens. It should also be remembered that there is no guarantee that material in herbaria is correctly identified and, even if it is, it will not necessarily bear the correct name based on current research. If any doubts about the correct identification exist, professional assistance should be sought.

While herbarium and floristic data are useful sources of ecogeographical information, in the case of taxa that have not been extensively collected, the desktop information will need to be supplemented by field exploration. Field data are, in fact, desirable in most cases, so as to be able to gather information on ecology, demography, genetic variation, breeding system and so on.

The use of *common names* to identify taxa should be exercised with great caution. Many taxa have several common names that are often locally specific but not unique over larger areas. Common names are often inaccurately associated with scientific names (Kanashiro et al, 2002).

Distribution data

It is important to determine the full geographical distribution of the CWR species being targeted. Distributional information, like taxonomic data, may be obtained from a variety of sources: Floras and monographs; geobotanical, phytosociological and vegetation studies, which often contain lists of species recorded from particular areas; herbarium labels; biodiversity databases; etc. Again, it is important to remember that CWR species may occur in the literature and on herbarium specimen labels under a range of different synonyms. Moreover, they may be polymorphic and contain one or more named and distinct subspecies or varieties.

Box 8.6 Geographic information system (GIS)

Put simply, a GIS is a collection of computer hardware and software tools used to enter, edit, store, manipulate and display spatial (geographically referenced) data. The data input can be from maps, aerial photos, satellites, surveys and other sources, and can be presented in the form of maps, reports and plans.

Typically, a GIS is used for manipulating maps with linked databases. These maps may be represented as several different layers where each layer holds data about a particular kind of feature. Each feature is linked to a position on the graphical image of a map. Layers of data are organized in a particular manner for study and statistical analysis. GIS organizes geographic data into a series of thematic layers and tables.

Georeferencing is the process of converting text descriptions of locations to those which can be read by a computer, and which can be used by software such as GIS. The BioGeomancer Project (http://www.biogeomancer.org/understanding.html) provides tools to improve results for organizations with large amounts of data to georeference by: automating the georeferencing of bulk data; learning from existing georeferences; accessing map and place-name gazetteers; generating computer-readable geographic locations and error descriptions according to accepted standards; and providing tools for validating results.

BioGeomancer is a worldwide collaboration of natural history and geospatial data experts. The primary goal of the project is to maximize the quality and quantity of biodiversity data that can be mapped in support of scientific research, planning, conservation and management. The project promotes discussion, manages geospatial data and data standards, and develops software tools in support of this mission.

The BioGeomancer consortium is developing an online workbench, web services and desktop applications that will provide georeferencing for collectors, curators and users of natural history specimens, including software tools to allow natural language processing of archival data records collected in many different formats.

Various methods and tools have been developed for the prediction of the geographic distribution of species. A recent study (Elith et al, 2006) compares the performance of 16 methods such as GARP, Domain, Bioclim and Maxent, on over 226 species from six regions of the world (see also Lobo, 2008). These methods require the use of a geographic information system (GIS) (Box 8.6), and commercial software packages such as ESRI's ArcGIS (ArcInfo, ArcEditor, ArcView), MapInfo, ERDAS ER Mapper and IDRISI Taiga GIS can be used for this purpose. In addition, some GIS software has been specially developed for conducting work with genetic resources, such as FloraMap,[2] which was developed and widely used at the International Centre for Tropical Agriculture (CIAT), although, it is now rather outdated and has been discontinued in favour of Maxent. Another package is DIVA-GIS, developed by the International Potato Centre (CIP) in collaboration with the International Plant Genetic Resources Institute (IPGRI) (now Bioversity International), and with support from the System-wide Genetic Resources Programme (SGRP). The software is available

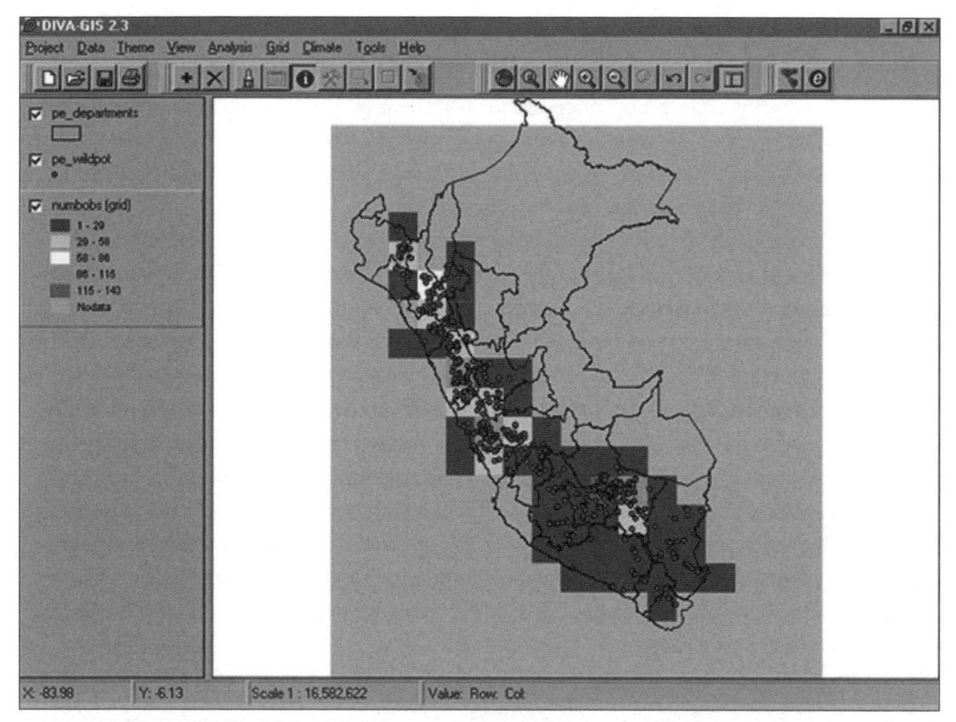

Figure 8.2 *DIVA-GIS – The Create Grid and Output Options windows, together with a main map window showing wild potato species richness in Peru*

Source: Hijmans et al, 2001

free of charge from the CIP website (https://research.cip.cgiar.org/confluence/display/divagis/Downloads) and from http://diva-gis.org/ (Hijmans et al, 2001) (Figure 8.2). It is important to carefully consider the choice of GIS software and the hardware used to run such programs, as mistakes can be costly. Peterson (2001) has developed models of species' ecological niches using an artificial-intelligence algorithm projected onto geography to predict species' distributions.

The Botanical Research and Herbarium Management System (BRAHMS) (http://dps.plants.ox.ac.uk/bol/) is a valuable resource for information on collection management, taxonomic work, botanical surveys and diversity analysis. A PowerPoint presentation on mapping the distribution of five species of Passiflora in Andean countries is available (see http://www.floramap-ciat.org/ing/poster-ppt.htm).

The Gap Analysis[3] team led by Andy Jarvis at CGIAR/IRRI/CIAT aims to develop an approach that will allow collectors (and other people related with *ex situ* and *in situ* conservation) to determine in which areas around the world traits and taxa are still unrepresented among target genebank collections managed by CGIAR-supported centres.

Bioclimatic niche-modelling techniques, which are used for projections of species distributions in climate change studies are discussed in Chapter 14.

Genetic variation

Genetic variation is at the heart of CWR conservation. It occurs at various levels in the populations of species and, in the case of CWR, particular alleles could provide the basis of valuable traits for breeding programmes. In order to be able to capture the desired amount of genetic variation in the conservation of a CWR, this will normally require a detailed understanding of the structure and partitioning of the genetic variation that occurs in a species and its populations. This will largely determine the location of the reserves and the design of the conservation strategy and management plan. Moreover, the more genetic variation is captured, the more likely is the species to continue to evolve and generate new variation favouring its long-term persistence and survival; the species will also stand a better chance of adapting to face the challenges of climate change.

Previously, genetic variation was commonly inferred from morphological differentiation; however, in recent years, biochemical and molecular techniques have been developed, such as isozyme analysis and DNA-based techniques such as sequencing, AFLP, RFLP, polymerase chain reaction (PCR) techniques, RAPD and microsatellite or SSR. The advantage of using molecular markers for studying genetic diversity is that they are not influenced by environmental factors and reflect genetic similarity without previous knowledge of pedigree information (Kuleung et al, 2006). The various molecular markers perform differently and each has its own characteristics. As such, there is no single molecular approach suitable for all purposes and more than one type should be used to ascertain which is most appropriate for a particular species or for the different issues involved in CWR conservation and management.

A comparison of the different molecular screening approaches is given in Table 8.2, but it should be noted that this is a rapidly evolving field and the assistance of specialists should be sought when planning to undertake screening. Further details of these techniques and their application can be obtained from texts such as Barnes and Breen (2009), de Vicente and Fulton (2004), de Vicente et al (2004), and from the review by Karp (2002).

An example of the genetic diversity of the CWR *Malus sieversii* is given in Box 8.7 and the assessment of genetically significant variation in *Coffea* is given in Box 8.8. In the case of *Malus orientalis* (Volk et al, 2009), genotypic (seven microsatellite markers) and disease resistance data were collected for 776 trees from Armenia, Georgia, Turkey and Russia. A total of 106 alleles were identified in the trees from Georgia and Armenia and the average gene diversity ranged from 0.47 to 0.85 per locus; it was found that the genetic differentiation among sampling locations was greater than that between the two countries.

While it is often stated that as much as possible of the genetic variation in a species should be captured for genetic conservation (e.g. Hawkes, 1987), such a laudable aim should not be pursued if it is at the cost of allowing many other species to become extinct. How much effort should be devoted to the genetic sampling of a particular CWR will depend on the priority given to that species, the finances and resources available, and how easy or difficult it is to measure the

Table 8.2 *Common genetic marker technologies and their main characteristics*

	Abundance	Level of polymorphism	Locus specificity	Co-dominance of alleles	Reproducibility	Labour intensity
Allozymes	low	low	yes	yes	high	low
RFLP	high	medium	yes	yes	high	high
Mini-satellites	medium	high	no/yes	no/yes	high	high
PCR-sequencing	low	low	yes	yes	high	high
RAPD	high	medium	no	no	low	low
Micro-satellites	high	high	yes	yes	high	low
ISSR	medium-high	medium	no	no	medium-high	low
SSCP	low	low	yes	yes	medium	low-medium
CAPS	low	low-medium	yes	yes	high	low-medium
SCAR	low	medium	yes	yes/no	high	low
AFLP	high	medium	no	no/yes	high	medium

	Technical demands	Operational costs	Development costs	Quantity of DNA required	Amenability to automation
Allozymes	low	low	low	-	no
RFLP	high	high	medium-high	high	no
Mini-satellites	high	high	medium-high	high	no
PCR-sequencing	high	high	high	low	yes
RAPD	low	low	low	low	yes
Microsatellites	low-medium	low-medium	high	low	yes
ISSR	low-medium	low-medium	low	low	yes
SSCP	medium	low-medium	high	low	no
CAPS	low-medium	low-medium	high	low	yes
SCAR	low	low	high	low	yes
AFLP	medium	medium	low	medium	yes

Source: Centre for Genetic Resources, the Netherlands at Wageningen University. http://www.cgn.wur.nl/UK/CGN+Plant+Genetic+Resources/Research/Molecular+markers/-+Overview+marker+technology/; accessed 20 December 2009

variation. Of course, even if it is possible to make a detailed survey of the genetic variation in a CWR, it does not necessarily mean that it will be possible or feasible to include all of this variation in genetic reserves, but it will help select which populations should be conserved.

On the other hand, it has to be accepted that for many species, perhaps the majority, detailed data on genetic information is unlikely to become available in the foreseeable future, simply because of the costs and labour involved. As Gole et al (2002) note in connection with the conservation of the *Coffea arabica* gene pool, knowledge of the distribution and genetic structure of its populations 'is one

Box 8.7 Genetic structure of the apple CWR *Malus sieversii* population from Xinjiang, China, revealed by SSR markers

A total of 109 *Malus sieversii* accessions from four geographical populations located in: Kuerdening in Mohe town, Gongliu County; Jiaowutuohai, Xinyuan County, Daxigou in Huocheng County of Ily State; and Baerluke Mountain in Yumin County of Tacheng State, Xinjiang Uygur Autonomous Region of China, were studied by simple sequence repeat (SSR) markers. The purpose of the study was to determine the genetic structure and diversity in these ecogeographical populations with eight pair SSR primers of apple. The results indicated an average of 16 bands were detected in the four populations. The percentage of polymorphic bands in Gongliu population (89.06 per cent) was the highest in the four populations. The average Nei's gene diversity index was 0.257 for all the loci. In total, 128 polymorphic loci were detected and the percentage of polymorphic loci (P) was 100 per cent, 88.28 per cent, 84.83 per cent, 87.50 per cent and 87.12 per cent, respectively, at the species level and Gongliu, Xinyuan, Huocheng and Yumin population levels. The Nei's gene diversity index (H = 0.2619) and Shannon's information index (I = 0.4082) in the species level were higher than in the population level. The Nei's gene diversity index and Shannon's information index in the four populations were: Gongliu > Huocheng > Xinyuan > Yumin. Gongliu population and Xinyuan population were the highest in genetic identity and the closest in genetic distance. Gene flow between the populations was 7.265, based on the genetic differentiation coefficient (GST = 0.064). The UPGMA cluster analysis indicated that the genetic relationships between the Gongliu and Xinyuan population were the closest, and the Yumin population had the greatest difference with the other three populations. The UPGMA cluster analysis indicated that the four geographical populations were relatively independent populations. Concurrently, there was also mild gene exchange between the populations. On the basis of the study of population genetic structure and the highest genetic diversity, the Gongliu population should be considered as high priority for the *in situ* conservation of *Malus sieversii* populations.

Source: Zhang et al, 2007

of the major challenges for coffee research in Ethiopia since it is expensive and needs up-to-date laboratory equipment and highly skilled personnel, which Ethiopia cannot afford'. Faced with such a situation, recourse will have to be made to proxy information (Dulloo et al, 2008; see also Box 8.8) such as the use of morphological differentiation to reflect underlying genetic differences and genecological zonation which assumes that genetic variation will be reflected in the patterns of ecological variation (Theilade et al, 2000).

How many individuals, how many populations?

How many individuals and populations of a target species should be conserved so as to remain viable are among the most difficult questions in the conservation

Box 8.8 Assessment of genetic variation in *Coffea*

The patterns of genetic variation within and among 14 populations of three wild *Coffea* species endemic to Mauritius were studied using RAPD molecular techniques as a tool to assist in gap analysis of actively conserved biodiversity. Sites were principally sampled from Mauritius with a view to determine the genetic relationships within and between sites, as well as evaluating the effectiveness of the protected areas system in Mauritius in conserving genetic diversity of *Coffea* found on the island. Two other populations of *Coffea mauritiana* from the neighbouring island of La Réunion were also sampled. Cluster analysis of RAPD data confirmed the taxonomic classification of these taxa in three clusters corresponding to the species *C. macrocarpa*, *C. mauritiana* and *C. myrtifolia* and, in addition, showed the distinctiveness of Montagne des Creoles accessions as a separate entity. The results showed that there is as much variation within as among populations (Wright F coefficient = 0.522). Of the 85 polymorphic bands, 25 were unique to one of the above four clusters and 60 (75 per cent) were variable among the four clusters. Almost all individuals of the same population grouped together. The total genetic diversity across all the accessions studied is 0.216. The population genetic parameters, when calculated for the different clusters, show that there is more variation within the clusters than among them. The gene diversity indices (Hj) within each cluster, 'macrocarpa', 'mauritiana', 'MDC' and 'myrtifolia' were 0.168, 0.169, 0.159 and 0.117, respectively. Within the 'mauritiana' cluster, there was a clear distinction between the *C. mauritiana* accessions from Mauritius and La Réunion. Further, the 'mauritiana' cluster contained two samples from the Mondrain population, previously classified as *C. macrocarpa*. In the 'macrocarpa' cluster, the *C. macrocarpa* populations divided into two main groups. Bassin Blanc and the different morphotype in the Mondrain population formed a distinct group, while the rest of the *C. macrocarpa* populations clustered together in the second group. In the 'myrtifolia' cluster, there is a clear demarcation between the western and the eastern populations of *C. myrtifolia* consistent with geographical distribution of the populations.

Source: Dulloo, 1998

biology of species. As Heywood and Dulloo (2005) note, 'The number of individuals needed to maintain genetic diversity within populations has been the subject of considerable work and a great body of literature exists on topics such as population viability analysis (PVA), minimum viable population size (MVP), minimum effective population size and, in the case of metapopulations, the minimum viable metapopulation size (MVM) and minimum available suitable habitat (MASH) (Hanski et al, 1996)'. The minimum available habitat is a relatively new concept which has great potential in restoration, sampling for alleles or heterozygosity (see Box 8.9). Likewise, the question of how many populations should be included in a reserve or network of reserves so as to include the maximum representation of the genetic variation of the CWR must be addressed and will depend on the distribution of the species and populations, and how that variation is partitioned between the different populations, which may require

Box 8.9 Population and metapopulation viability concepts

Population viability analysis (PVA) is the methodology of estimating the probability that a population of a specified size will persist for a specified length of time. It is a comprehensive analysis of the many environmental and demographic factors affecting survival of a (usually small) population (Morris and Doak, 2002).

The minimum viable population (MVP), a concept introduced by Soulé (1986) to population biology, is the smallest population size that will persist for some specified length of time with a specified probability.

The minimum amount of suitable habitat (MASH) is the number (as a rule of thumb, 15–20) of well-connected patches needed for the long-term survival of a metapopulation (Hanski et al, 1996; Hanski, 1999).

The minimum viable metapopulation size (MVM) is an estimate of the minimum number of interacting local populations necessary for long-term survival of a metapopulation (Hanski et al, 1996).

Source: Heywood and Dulloo, 2005

considerable effort to ascertain (see Dulloo et al, 2008, pp31–32 for a review and discussion on this topic). However, as a rule of thumb, a minimum of five populations per genetic reserve is recommended for *in situ* conservation (Dulloo et al, 2008; Brown and Briggs, 1991). In many cases it may not be possible for practical, political or economic reasons to attempt a comprehensive coverage of the genetic variation.

Ecological information

Ascertaining the ecological conditions under which the selected species grows is one of the main concerns of an ecogeographic survey. Although some information may be derived from the literature and herbarium specimen label data, in most cases field exploration is essential. There are no agreed criteria for collecting ecological information but those commonly recommended are:

- habitat types – although there is no generally accepted global set of habitat types, many countries have produced their own classifications for use in official documents; the European Union Directive on the Conservation of Natural Habitats and of Wild Fauna and Flora (Habitats Directive) lists 218 habitat types in its Annex 1 (see Evans, 2006, for a list and discussion of the issues involved);
- condition of the habitat;
- disturbance regimes;
- threats to the habitat;
- topography;
- altitudinal range;

- soil types;
- slope and aspect;
- land use and/or agricultural practice.

For some species, a phytosociological characterization may be available or may be developed through field-work.

Reference should also be made to the list of descriptors that has been developed by Bioversity International to provide a standard format for the gathering, storage, retrieval and exchange of farmers' knowledge of plants (Bioversity and The Christensen Fund, 2009). For applying participatory approaches to data gathering, see Chapter 5 and Hamilton and Hamilton (2006) and Cunningham (2001).

Methodologies for field surveys

The amount of field-work that can be carried out will depend on the particular target species and local circumstances. Basically, at each site, latitude, longitude, and altitude should be determined by GPS, and location descriptors (geographical region, road or settlement name, proximity to prominent land marks) and site physical characteristics (habitat type, slope, aspect, and precise location of target species plants at the site, if found) should be recorded. Details of how to prepare for field-work are given by Hawkes et al (2000), although it should be noted that the recommendations are for *ex situ* approaches but can often be applied to surveying for *in situ* conservation. Training will need to be provided (see Chapter 15) although few centres or universities offer appropriate courses.

Data analysis and products

The data gathered in ecogeographical surveys may be analysed in various ways such as discriminant analysis or principal component analysis. For the visualization, analysis and management of spatial data, GIS-based packages such as ArcInfo, WorldMap or DIVA may be used.

One of the main products of an ecogeographic survey is the 'ecogeographic conspectus', which is a formal summary of the available taxonomic, geographic and ecological information of the target taxon, gathered from the herbarium and field surveys (Maxted et al, 1995). The conspectus is arranged by species and includes the following information: the accepted taxon name, authors, dates of publication, synonyms, morphological description, distribution, phenology, altitude, ecology and conservation notes. For example, Dulloo et al (1999) published an ecogeographic survey of the genus *Coffea* in the Mascarenes, which includes an ecogeographic conceptus (Box 8.10).

Box 8.10 Example of an ecogeographic conspectus

C. mauritiana Lam., Encycl. 1:550 (1783); DV Prodr. 4: 499 (1830); Bojer, H.M.: Baker, F.M.S.: 152; Cordem., F.R.: 506; R.E. Vaughan. Maur. Inst. Bull. 1:44 (1937); A. Chevalier, Rev. Bot. Appl. 18: 830 (1938); Rivals, Et. Veg. Nat. Réunion: 174 (1960).

Synonyms: *C. sylvestris* Willd. ex. Roemer et Schultes, Syst. Vég. 5: 201 (1819). Type La Réunion. *C. nossikumbaensis* A.Chev., Rev. Bot. Appl. 18: 830 (1938). Type Nossi Kumba. *C. campaniensis* Leroy. Journ. Agr. Trop. Bot. Appl. 9: 530 (1962) Type Mauritius. *Geniostoma reticulatum* Cordem., F.R.: 464 Type La Réunion.

Morphological description: Shrub or small tree, reaching about 6m in height, with verticillate branches. Leaves glabrous, leathery, obovate to elliptical, acuminate, cuneiform and decurrent, 4–10cm long by 2–6cm wide with 6–8 pairs of secondary veins. Petiole 3–10mm in length. Stipule deltoid, 2–8mm long. Inflorescence auxiliary and upright. Fruit ovoid to oblong, 18–20mm long, yellowish green becoming purple at maturity.

Distribution: Endemic to Mauritius and Réunion. In Mauritius, *C. mauritiana* is restricted to Plaine Champagne, Mt Cocotte, Pétrin and Les Mares. The species has historically been recorded in three other localities, namely: Le Pouce Mountain, Nouvelle Découverte and Mon Gout. This species is more widespread on Réunion.

Phenology: Bud, August to November; flowers, November to December; fruits, April to August.

Altitude: 270–1500m. In La Réunion, *C. mauritiana* has a broad range of altitude, occurring at 270m at Mare Longue to c.1500m a.s.l. at Bebour. In Mauritius, the species altitudinal range is very narrow (700–760m).

Ecology: Mid to high altitude wet montane rainforest. In Mauritius, *C. mauritiana* is very localized and occurs on the upland plateau in the super-humid zones (rainfall varies between 2500 and 5000mm per annum (Vaughan and Wiehe, 1937) at Mt Cocotte and Plaine Champagne. Plaine Champagne, situated on an area of ground water laterite consisting of highly ferruginous slabs of cuirasses (Parish and Feillafe, 1965), sustains an open canopy of dwarf thickets of native species rarely exceeding more than 5m in height. The area has a rich floristic composition principally composed of *Sideroxylon cinereum* and *S. puberulum* (Sapotaceae), *Aphloia theiformis* (Flacourtiaceae), *Olea lancea* (Oleaceae), *Gaertnera* spp. (Rubiaceae), *Nuxia verticillata* (Loganiaceae), *Antirhea borbonica* (Rubiaceae) and *Syzygium glomeratum* (Myrtaceae). Because of the high rainfall of the area, the ground is covered with a thick cushion of bryophytes with many epiphytic and ground ferns and orchids. The habitat is highly invaded with *Psidium cattleianum* (Myrtaceae), which is the dominant species in the area.

The habitat at Mt Cocotte has been described as a cloud or mossy forest (Vaughan and Wiehe, 1937; Lorence, 1978). It is characterized by very high rainfall often exceeding 5000mm and is often enveloped in clouds and nocturnal mists (Vaughan and Wiehe, 1937). The vegetation community at Mt Cocotte is a relict of the original native vegetation of the area and is composed of such species as *Nuxia verticillata* (Loganiaceae), *Tambourissa* spp., *Monimia ovalifolia* (Monimiaceae), *Syzygium mammillatum*, *Eugenia* spp.

(Myrtaceae) and *Casearia mauritiana* (Flacourtiaceae). The vegetation is poorly stratified. The whole area is now very degraded, with high infestation of alien plants such as *Psidium cattleianum* (Myrtaceae), *Homalanthus populifolius* (Euphorbiaceae) and *Rubus alceifolius* (Rosaceae).

Conservation notes. IUCN Status: Mauritius CR (B 1,2); Réunion VU (C 2a). The IUCN conservation status for *C. mauritiana*, in Mauritius, is here classified as Critically Endangered (CR), under criteria B 1,2. The area of occupancy is less than $1 km^2$ and it is considered that there is only one major population at P. Champagne. The other sites (Mt Cocotte, Les Mares and Pétrin) all have very scattered individuals and do not form any population as such. The site is heavily invaded with alien plants, principally *Psidium cattleianum* (Chinese guava) and there is no sign of regeneration of *C. mauritiana*. The population is estimated to be between 350 and 400 plants at this site, contained within an area of about four to five hectares. In addition, there is a high influx of visitors into this area for the picking of Chinese guava fruits; this is a favourite pastime for many Mauritians, which can be damaging for the threatened flora of the island. At the other sites, particularly at Les Mares and Pétrin, only few specimens are known. At Les Mares, there is only one plant growing under high tension wire at the side of a road. Most of this area has been converted into exotic plantations of forestry species such as *Pinus elliottii*, and *Eucalyptus* spp. At Mt Cocotte, there is only a small sterile population of *C. mauritiana* population (15 individuals). These are located within a conservation management area, an intensely managed forest plot where alien species are excluded (Dulloo et al, 1996), and unfortunately is not regenerating. Over the past three years, two of the plants at this site have died.

In Réunion, *C. mauritiana* is more common than in Mauritius. During the course of this survey, only a few sites were visited and the species was found to be occasional in these areas. Consequently, it is difficult to assess the overall conservation status for the whole island. However, discussions with field-workers at the University of La Réunion suggest that the IUCN status for *C. mauritiana* may be considered as Vulnerable (T. Pailler, personal communication).

Source: modified from Dulloo et al, 1999

Results from each country

Armenia

Desktop study

The first step involved collating available information on the taxonomy, occurrence and distribution, biological features, conservation status and uses of the 104 target CWR species. This was done by searching the literature and by examining passport data from herbaria at the Institute of Botany of the National Academy of Sciences, Plant Genetic Resource Laboratory of Armenian State Agrarian University and the Department of Botany of Yerevan State University, as well as seed bank (*ex situ*) collections records at Armenian State Agrarian University. Literature sources consulted include: Takhtajan, *Flora of Armenia*; Grossheim,

Box 8.11 Analytical tools used for the assessment of the CWR status and monitoring in each country

Armenia: DIVA-GIS and other GIS software were used in Red Listing and monitoring.

Bolivia: DIVA-GIS, ArcView and ArcGIS were used in determining the collection sites of species of 13 genera in the different departments of Bolivia, within and outside protected areas, and within and outside the community lands. During 2007 and 2008, Bioclim, Domain and Maxent prediction models were used to determine the potential distribution of the CWR and Maxent to determine the effect of climate change on distribution in priority selected species. GisWeb has also been developed as a tool for the visualization of maps of different types using the services of Google Maps. Maps include major and minor rivers and coverage of CWR in the national CWR portal. GisWeb offers satellite images of maps that can be zoomed in on to display further detail (Bellot and Cortez, 2010; Bellot and Justiniano, 2010).

Madagascar: Data analysis was done using Domain, FloraMap, ArcGIS and other GIS software. The Information Management Committee (IMC) is testing DIVA-GIS for data analysis.

Sri Lanka: DIVA-GIS was used to map current distribution and FloraMap to predict potential distribution.

Uzbekistan: DIVA-GIS and MapSource were used to generate species-distribution maps.

Flora of the Caucasus; Red Data Book of Armenia; Gabrielian and Zohary (2004), 'Wild relatives of food crops native to Armenia and Nakhichevan'; Czerepanov, *Vascular Plants of Russia and Adjacent States*; the Germplasm Resources Information Network (GRIN)/United States Department of Agriculture (USDA) database; and other relevant sources. Relevant experts were consulted when necessary.

Field-work

Extensive field surveys were conducted in the administrative regions (marzes) of Armenia and Yerevan city (Table 8.3) during two consecutive years, 2006 and 2007, from late spring till autumn. Surveys were conducted, where possible, during the flowering or fruit-bearing stage, when identification of species is easy. Slight adjustments were made for individual species and altitudes in different regions. For example, sites located at relatively high altitudes (1500–2000m) were visited later in the season (July–August) compared with lowland areas.

The team conducting field surveys comprised experts, including taxonomists, from the Institute of Botany, Armenian State Agrarian University and the local CWR team. Although the majority of field surveys were organized (and funded) by the UNEP/GEF CWR Project, a few field trips were also supported by other projects underway at the Institute of Botany and Armenian State Agrarian University (Table 8.3).

Table 8.3 *Ecogeographic surveys and administrative regions (marzes) visited in Armenia*

Date	*Administrative regions (marzes)*	*Organized by*	*Expedition*
01.06.2006	Ararat marz	CWR-Armenia and Armenian State Agrarian University	Omargo_1_2006
03.06.2006	Yerevan city and Kotayk marz	CWR-Armenia	Erebuni_1_2006
12.06.2006	Shirak and Aragatsotn marzes	CWR-Armenia	Talin_1_2006
20.06.2006	Kotayk marz	CWR-Armenia	Abovian_1_2006
06.07.2006	Yerevan city, Kotayk, Ararat, Aragatsotn and Tavush marzes	CWR-Armenia and Institute of Botany	O_6_2006
15.07.2006	Ararat, Vayots Dzor and Gegharkunik marzes	CWR-Armenia	Eghegnadzor_1_2006
02.08.2006	Vayots dzor, Kotayk, Lori and Tavush marzes	CWR-Armenia and Institute of Botany	O_5_2006
03.08.2006	Ararat marz	CWR-Armenia	Khosrov_1_2006
10.08.2006	Syunik marz	CWR-Armenia and Institute of Botany	O_4_2006
17.08.2006	Tavush, Lori and Aragatsotn marzes	CWR-Armenia	Dilijan_1_2006
20.08.2006	Ararat marz	CWR-Armenia and Institute of Botany	O_3_2006
27.08.2006	Aragatsotn marz	CWR-Armenia and Institute of Botany	O_2_2006
29.09.2006	Aragatsotn and Kotayk marzes	CWR-Armenia	AknaLich_1_2006
08.10.2006	Syunik marz	CWR-Armenia	ShikahoghZ_1_2006
01.06.2007	Aragatsotn and Kotayk marzes	CWR-Armenia and Institute of Botany	O3_Ivan_2007
10.06.2007	Vayots dzor and Syunik marzes	CWR-Armenia and Institute of Botany	O2_Ivan_2007
16.06.2007	Aragatsotn, Shirak and Lori marzes	CWR-Armenia	Stepanavan_1_2007
04.07.2007	Kotayk marz	CWR-Armenia	Erebuni_2_2007
14.07.2007	Ararat, Vayots Dzor and Syunik marzes	CWR-Armenia	Syunik_1_2007
21.07.2007	Kotayk, Syunik, Vayots Dzor and Lori marzes	CWR-Armenia and Institute of Botany	O_7_2007
24.07.2007	Yerevan city and Kotayk marz	CWR-Armenia	Erebuni_1_2007
28.07.2007	Gegharkunik marz	CWR-Armenia	Sevan _1_2007
29.07.2007	Tavush marz	CWR-Armenia and Institute of Botany	O1_Ivan_2007
28.08.2007	Yerevan city and Kotayk marz	CWR-Armenia	Garni_1_2007
30.08.2007	Aragatsotn marz	CWR-Armenia	Bjurakan_1_2007
07.09.2007	Syunik marz	CWR-Armenia and Institute of Botany	O4_Ivan_2007
23.09.2007	Tavush and Gegharkunik marzes	CWR-Armenia	Shamshadin_1_2007

The data collected during field surveys included:

- latitude, longitude and altitude (collected using a GPS);
- description of location, including administrative unit and nearest settlement;
- soil characteristics;
- conservation status of the area;
- average density (number of plants per unit area);
- approximate area occupied by each subpopulation, plant community;
- phenology of the populations (time of leaf break, flowering, etc.);
- current and potential threats to the populations.

Special questionnaires were developed to collect the data, which were entered into a database. If the species could not be properly identified, a specimen was taken for determination at the herbarium. Where possible, seeds (collected as heads) were collected for *ex situ* conservation at Armenian State Agrarian University seed bank as a complementary measure. The collection was done in such a way as to capture maximum genetic diversity of the population and not to endanger the natural population, following the IUCN technical guidelines on the management of *ex situ* populations (IUCN, 2002). The data collected were entered into a database (Microsoft Access).

Summary results of desktop research

It is important to note that the original list of 104 species was reduced to 99: it was decided to exclude two species from the list, as their presence in Armenia was debatable (*Aegilops umbellulata* and *Cicer minutum*); *Crysopsis sebastianii* was excluded as it is a synonym of *Trifolium sebastianii*, already included in the project; and *Vitis vinifera* was excluded from the list since an extensive project on the species was already funded by another international agency.

The information collected was used to draft the preliminary distribution of the species, as well as to plan the timetable and routes for field studies.

Summary results of field surveys

Field studies covered almost all of Armenia and all administrative regions (marzes) except for Armavir marz. In total, 571 populations representing 79 species were studied in the field and their details recorded. The remaining 20 species were not found in the field for different reasons: one species (*Aegilops crassa*) is assumed to be extinct in the wild in Armenia; others are rare and were not found in the field either because of time constraints or because passport data were not detailed enough and did not allow field-work to be planned to cover their locations.

As identified during field surveys, major threats to the populations included uncontrolled grazing and hay harvesting, urbanization (especially for the populations extending to Yerevan city), land privatization accompanied by construction of buildings and agricultural activities, road construction, mining activities in southern Armenia, climate change (especially increasing aridity) and the wild harvesting of early leafy vegetables and wild fruits/berries. Results for selected species are presented in Tables 8.4–8.7.

Bolivia

Types of data recorded

From early 2006 until mid-2009, researchers from project partner institutions undertook field assessments and gathered ecogeographic data, having previously identified areas with potential distribution maps. Data was collected using field sheets prepared by each partner institution, taking into account the fields/descriptors of the database for the National Information System of Crop

Table 8.4 *Triticum araraticum*

	Erebuni State Reserve (Monitoring data)	*Armenia (Field survey data)*
Population size:	1,832,000	65,900,000
Area occupied:	20.9ha	3200ha
Threats:	illegal grazing, chemical deposition	agricultural expansion, land use
General trend:	stable	decline

Table 8.5 *Triticum boeoticum*

	Erebuni State Reserve (Monitoring data)	*Armenia (Field survey data)*
Population size:	42,354,000	6,853,000,000
Area occupied:	52.3ha	14,400ha
Threats:	illegal grazing, chemical deposition	agricultural expansion, land use
General trend:	stable	decline

Table 8.6 *Triticum urartu*

	Erebuni State Reserve (Monitoring data)	*Armenia (Field survey data)*
Population size:	837,000	837,000
Area occupied:	5.2ha	5.2ha
Threats:	illegal grazing, chemical deposition	illegal grazing, chemical deposition
General trend:	decline	decline

Table 8.7 *Aegilops tauschii*

	Erebuni State Reserve (Monitoring data)	*Armenia (Field survey data)*
Population size:	3,400,000	5,647,000,000
Area occupied:	15ha	62,400ha
Threats:	illegal grazing, chemical deposition	agricultural expansion, land use
General trend:	stable	decline

Wild Relatives (NISCWR), which are organized into seven groups (taxon, site, contact, resource, accessions, specimen and population). These data are included in the NISCWR database, available to the general public through the National and International Portals: www.cropwildrelatives.org and www.cwrbolivia.gob.bo.

From 2006 to 2009, researchers from the partner institutions systematized desktop information and field data gathered from a total of 201 species of 14 genera: *Anacardium*, *Ananas* and *Pseudoananas*, *Annona*, *Arachis*, *Bactris*, *Capsicum*, *Chenopodium*, *Cyphomandra*, *Euterpe*, *Ipomoea*, *Manihot*, *Phaseolus*, *Rubus*, *Solanum*, *Theobroma* and *Vasconcellea* (*Carica*). The data recorded for the species of 14 genera according to the seven groups of descriptors/fields of the NISCWR are reflected in Table 8.8.

Population data for priority CWR taxa surveyed are given in Tables 8.9 and 8.10.

Sri Lanka

Ecogeographic surveys were conducted throughout the country, except for the Northern Province. A total of 1121 GPS locations were assigned to wild relatives of priority crops from the field survey, passport data, herbarium specimens and literature survey. Total GPS locations are given in the accompanying map (Figure 8.3) and presented separately for each genus.

Assemblage geographic coordinates were entered in the FloraMap, Garmin map sources and DIVA-GIS software programs for the preparation of distribution maps and predictive maps to identify remaining areas to be surveyed and to identify gaps in the surveys.

Two types of surveys were conducted during the project period. A literature survey was conducted to collect basic information on CWR while an actual field survey was launched to determine the present situation of past known locations and to find out new locations of the CWR. The field survey was conducted in different parts of the island as indicated in Table 8.11 and Figure 8.3. Ecogeographic descriptors were prepared using available and collected information of priority CWR for all possible species.

Ecogeographic surveying was carried out for priority wild species of Sri Lanka from August 2005 to December 2007, in targeted areas. Habitat and taxonomic data were recorded in a field data record form and herbarium specimens were prepared. Photographs were taken to highlight habitats and specific characteristics of the plants. A global positioning system (GPS) with map datum WGS 84 was used to mark the locations where wild species were found. Since most of the herbarium specimens lacked geographic coordinates, they were manually examined and approximate geographic coordinates were assigned using a coordinates book published on a website on the internet. Geographic coordinates obtained from the field survey, passport data and herbarium specimens were entered into the Garmin map sources and DIVA-GIS software for the preparation of distribution maps for each species. The distribution of the wild

Table 8.8 *Kinds of data recorded for each genus from desktop revision and field assessments during the period 2006–2009, in Bolivia*

Data type	*Genera and the species for which data was recorded*															
	Anacard-ium	*Ananas Pseudo-ananas*	*Annona*	*Arachis*	*Bactris*	*Capsicum*	*Cheno-podium*	*Cypho-mandra*	*Euterpe*	*Ipomoea*	*Manihot*	*Phaseolus*	*Rubus*	*Solanum*	*Theo-broma*	*Vascon-cellea*
1. Taxon																
1.1 Taxonomy	✗	✗	✗	✗	✗	✗	✗	✗	✗	✗	✗	✗	✗	✗	✗	✗
1.2 Common names	✗	✗	✗	✗			✗	✗	✗	✗	✗	✗	✗	✗	✗	✗
1.3 Biological data	✗	✗	✗	✗	✗		✗	✗	✗	✗	✗	✗	✗	✗	✗	✗
1.4 Uses	✗		✗	✗			✗	✗	✗	✗	✗	✗	✗	✗	✗	✗
1.5 Conservation actions			✗	✗			✗	✗		✗	✗			✗	✗	
1.6 Legislation affecting the taxon																
1.7 Category of threat (Red List)	✗		✗	✗			✗	✗			✗	✗	✗	✗	✗	✗
1.8 Distribution	✗	✗	✗	✗	✗	✗	✗	✗	✗	✗	✗	✗	✗	✗	✗	✗
2. Site																
2.1 Administrative unit	✗	✗	✗	✗	✗		✗	✗	✗	✗	✗	✗	✗	✗	✗	✗
2.2 Locality	✗	✗	✗	✗	✗		✗	✗	✗	✗	✗	✗	✗	✗	✗	✗
2.3 Climate				✗			✗	✗	✗	✗	✗	✗	✗	✗	✗	✗
2.4 Habitat	✗	✗	✗	✗	✗		✗	✗	✗	✗	✗	✗	✗	✗	✗	✗
2.5 Relief and Topography							✗							✗		
2.6 Soil							✗							✗		
2.7 Administrative status																
2.7 Soil use							✗							✗		
3. Contact																
3.1 Category, institution, person, address, comments	✗						✗							✗	✗	
4. Source																
4.1 URL, Title, Author, Theme, Key words, Date, Editor. Name of Editor, Place of Publication	✗						✗							✗	✗	
5. Accession																
5.1 Number				✗			✗					✗		✗		
5.2 Date of acquisition				✗			✗					✗		✗		
5.3 Owner				✗			✗					✗		✗		
5.4 Taxon				✗			✗					✗		✗		
5.5 Source				✗			✗					✗		✗		

Table 8.8 *continued*

Data type	*Genera and the species for which data was recorded*															
	Anacard-ium	*Ananas Pseudo ananas*	*Annona*	*Arachis*	*Bactris*	*Capsicum*	*Cheno-podium*	*Cypho-mandra*	*Euterpe*	*Ipomoea*	*Manihot*	*Phaseolus*	*Rubus*	*Solanum*	*Theo-broma*	*Vascon-cellea*
5.6 Distribution				✗			✗					✗		✗		
5.7 Name of accession				✗			✗					✗		✗		
5.8 Donation																
5.9 Biology				✗			✗					✗		✗		
5.10 Other duplicate number				✗			✗					✗		✗		
5.11 Source																
5.12 Distribution				✗			✗					✗		✗		
6. Specimen																
6.1 Number of specimen	✗	✗	✗	✗	✗	✗	✗	✗	✗	✗	✗	✗	✗	✗	✗	✗
6.2 Date of acquisition	✗	✗	✗	✗	✗		✗	✗	✗	✗	✗	✗	✗	✗	✗	✗
6.3 Entity	✗	✗	✗	✗	✗		✗	✗	✗	✗	✗	✗	✗	✗	✗	✗
6.4 Taxonomy	✗	✗	✗	✗	✗		✗	✗	✗	✗	✗	✗	✗	✗	✗	✗
6.5 Source (action, location)																
6.6 Sample	✗	✗	✗	✗	✗		✗	✗	✗	✗	✗	✗	✗	✗	✗	✗
6.7 Donation (number of specimen, entity, date of donation, donation agreement, material transfer agreement, comments)																
6.8 Biology	✗	✗	✗	✗	✗		✗	✗	✗	✗	✗	✗	✗	✗	✗	✗
6.9 Action (dates, actors, collection, ID of collection)	✗	✗	✗	✗	✗		✗	✗	✗	✗	✗	✗	✗	✗	✗	✗
6.10 Sample (Identification storage, voucher, distribution)																
7. Population																
7.1 ID	✗			✗												
7.2 Date	✗			✗												
7.3 Actor	✗			✗												
7.4 Methodology	✗			✗												
7.5 Characterization (abundance, density cover, threats)	✗		✗	✗									✗			

Source: Beatriz Zapata Ferrufino, CWR Project, Bolivia

Table 8.9 *Number of populations for 14 wild species of* Arachis *genus studied by biogeographic zones of Bolivia through thesis work (2007–2009), with field assessments and desk data within the framework of the CWR Project*

Genus	Province Biogeographic	Species	Populations recorded	Populations studied	% Populations studied
Arachis	Chaco boreal – western sector	*Arachis batizocoi*	23	5	13.0
		Arachis duranensis	51	5	7.8
	Cerrado – Chiquitano sector	*Arachis cardenasii*	51	20	39.2
		*Arachis cruziana**	18	10	55.5
		*Arachis chiquitana**	4	2	50.0
		Arachis glandulifera	45	11	24.4
		*Arachis herzogii**	16	11	68.8
		*Arachis kempff-mercadoi**	45	5	11.1
		*Arachis krapovickasii**	6	2	33.3
		Arachis magna	26	8	30.7
		Arachis sp.*	5	5	100.0
	Beni – Lllanos de Moxos sector	*Arachis benensis**	5	2	40.0
		*Arachis trinitensis**	4	3	75.0
		*Arachis willamsii**	7	4	57.0
TOTAL			306	90	29%

* species is endemic
Source: Ramos Canaviri, 2009

Table 8.10 *Number of localities in Bolivia visited by year for ecogeographic surveys*

Genus 2006–2009	Number of localities visited by year 2006	2007	2008	2009	Total localities
Anacardium	0	2	0	0	2
Ananas-Pseudoananas	5	5	2	0	12
Annona	32	53	30	0	115
Arachis	0	108	0	0	108
Bactris	0	0	0	0	0
Cyphomandra	0	12	1	0	13
Chenopodium	0	0	12	0	12
Euterpe	1	0	0	0	1
Ipomoea	10	27	31	0	68
Manihot	13	46	36	0	95
Phaseolus	0	0	22	0	22
Theobroma	20	16	21	0	57
Rubus	0	58	9	0	67
Solanum	6	20	9	5	40
Vasconcellea	0	26	11	0	37
Total number of localities	87	373	184	5	
Total number of localities in the period 2006–2009					649

Source: VMABCCGDF–Bioversity International, 2010. Informes Técnicos de Fase 2006–2008 de las instituciones socias del Proyecto CPS & Inventario de Especímenes colectados por las instituciones socias del Proyecto CPS en el periodo 2006–2009)

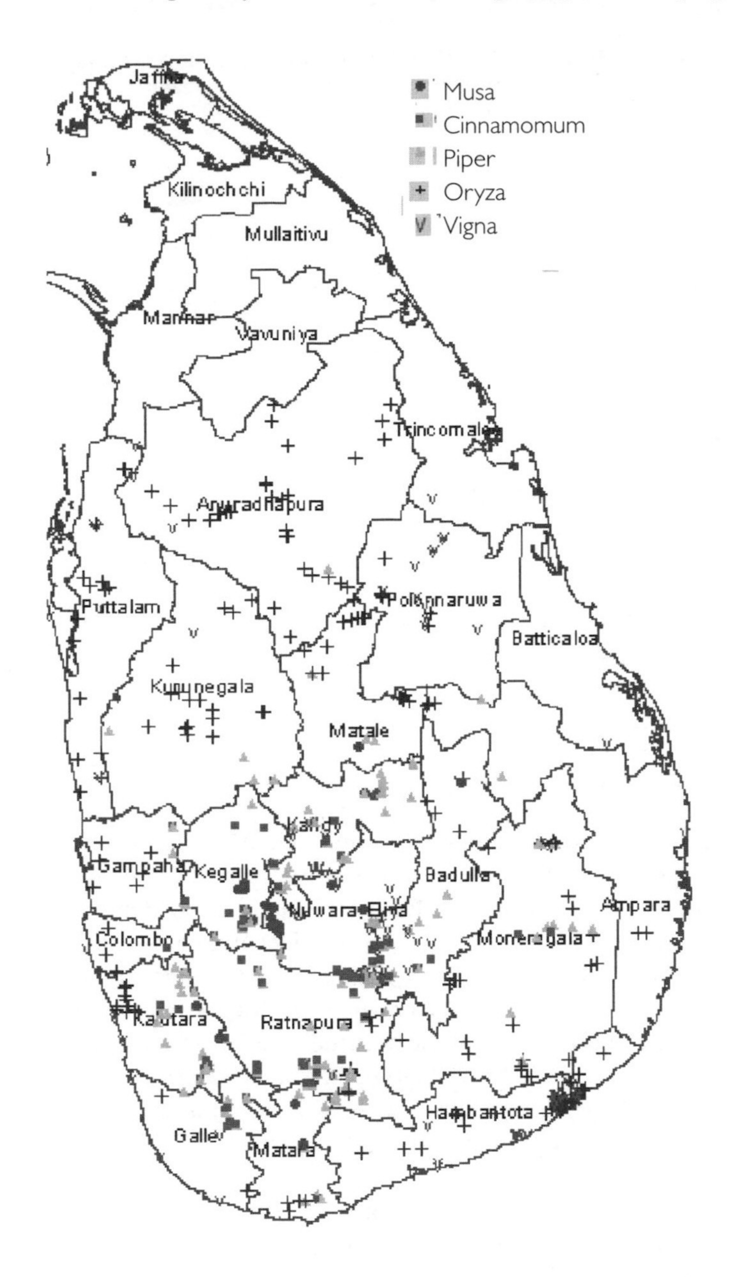

Figure 8.3 *Map of ecogeographic survey locations for priority CWR in Sri Lanka*

CWR of the priority crops occurring in Sri Lanka is depicted in the maps. GPS data were analysed by FloraMap distribution modelling and DIVA-GIS Bioclim modelling to map the probable areas in which wild species could be expected. The Red Listing category (conservation status) for each species was assigned on the basis of the ecogeographic survey data.

Table 8.11 *GPS locations summary for Sri Lanka*

Genus	Field surveys	PP+HS+Lit*	Total
Oryza	111	180	291
Musa	30	3	33
Vigna	129	56	185
Piper	241	100	341
Cinnamomum	182	89	271
Totals	693	428	1121

* PP=Passport data, HS=Herbarium specimen label data, Lit=Literature data

Uzbekistan

Summary of results of desktop research

Ecogeographic surveys were carried out for six priority species of CWR (see Figures 8.4–8.9):

- *Malus* – apple;
- *Amygdalus* – almond;
- *Juglans* – walnut;
- *Pistachia* – pistachio;
- *Allium* – onion;
- *Hordeum* – barley.

Field surveys were conducted by the Scientific Plant Production Centre 'Botanica'; R. Shreder Scientific Research Institute of Gardening, Viticulture and Winemaking; Uzbek Scientific Research Institute of Plant Industry; and the Republican Scientific Production Centre of Decorative Gardening and Forestry.

Surveys were conducted over four years at different stages of vegetation development in order to cover the current areas of distribution. A single methodology developed by project experts at the beginning of the project was employed. Before field surveys were undertaken, literature data and herbarium material were studied.

Field surveys were conducted by establishing pilot plots in various populations of priority species. The following data were studied during the surveys:

- composition of the plant communities in which the CWR populations occur;
- conservation status of the populations;
- threats to the populations;
- growth habit of the priority species;
- physical and geographical conditions of the area where pilot plots were established;
- longitude, latitude;
- local name of the plants;
- biometrical data;
- soil conditions;
- level of soil erosion.

Arachis rigonii, *Bolivia*

Source: G. Seijo, IBONE

Rubus megalococcus, *Bolivia*

Source: Saul Job Altamirano Azurduy, CBG-BOLV

Wild yam, Dioscorea seriflora, *Madagascar*

Source: V. Jeannoda, University of Antananarivo

Ensete perrieri, *Madagascar*

Source: Danny Hunter, Bioversity International

Coccinia grandis, *Sri Lanka*

Source: A. Wijesekara, CWR Project, Sri Lanka

Wild pepper, *Sri Lanka*

Source: E. Dulloo, Bioversity International

Plate 6: *CWR species in Bolivia, Madagascar, Sri Lanka*

Aegilops columnaris, *Armenia*

Source: A. Melikyan, Armenian Agrarian University

Drying garlands of Rumex, *an important traditional wild food in Armenia*

Source: A. Melikyan, Armenian Agrarian University

Pyrus caucasica, *Armenia*

Source: Ivan Gabrielyan, Institute of Botany

Prof U. Pratov observes the variation of wild apple from western regions in Tien Shan

Source: A.Yuldashev, SPC 'Botanica' ASc RUz

Allium pskemense, *Uzbekistan*

Source: U. Pratov

Amygdalus bucharica *faces severe threats by deforestation and erosion, Uzbekistan*

Source: M.Yu. Djavakyntc, Uzbek Scientific Research Institute of Gardening, Viticulture and Winemaking

Plate 5: *CWR species in Armenia and Uzbekistan*

Preparing herbarium samples on an ecogeographic survey in Madagascar
Source: V. Jeannoda, University of Antananarivo

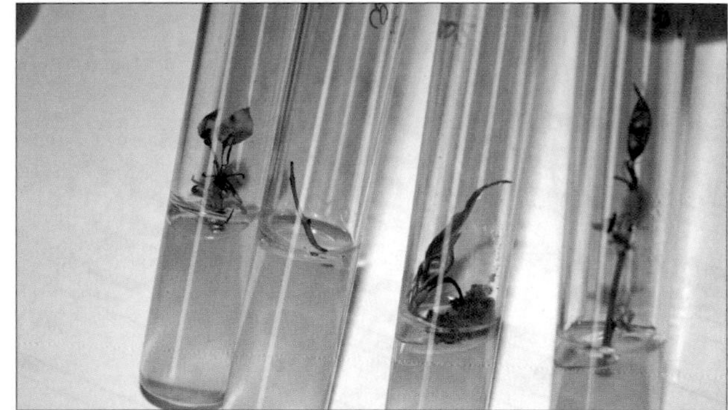

In vitro *techniques are used as a complementary strategy for the conservation of wild pear germplasm in Armenia*
Source: D. Hunter, Bioversity International

Wild relatives of potato will be important in helping Peruvian farmers adapt to climate change in the Andes
Source: E. Fox, Bioversity International

Monitoring populations of wild relatives of wheat in the Erebuni State Reserve, Armenia
Source: D. Hunter, Bioversity International

Plate 4: *Eco-geographic surveying, complementary conservation, adapting to global change and monitoring CWR species and populations*

Kanneliya-Dediyagala-Nakiyadeniya (KDN) Biosphere Reserve, Sri Lanka

Source: A. Wijesekara, CWR Project, Sri Lanka

Territorio Indígena Isiboro-Secure (TIPNIS), Bolivia

Source: Fundación TIPNIS

Plate 3: *Protected areas*

Erebuni State Reserve, Armenia
Source: I. Gabrielyan, Institute of Botany

Ugam Chatkal Nature Reserve, Uzbekistan
Source: S. Djataev

Ankarafantsika National Park, Madagascar
Source: D. Andrianasolo

Plate 2: *Protected areas*

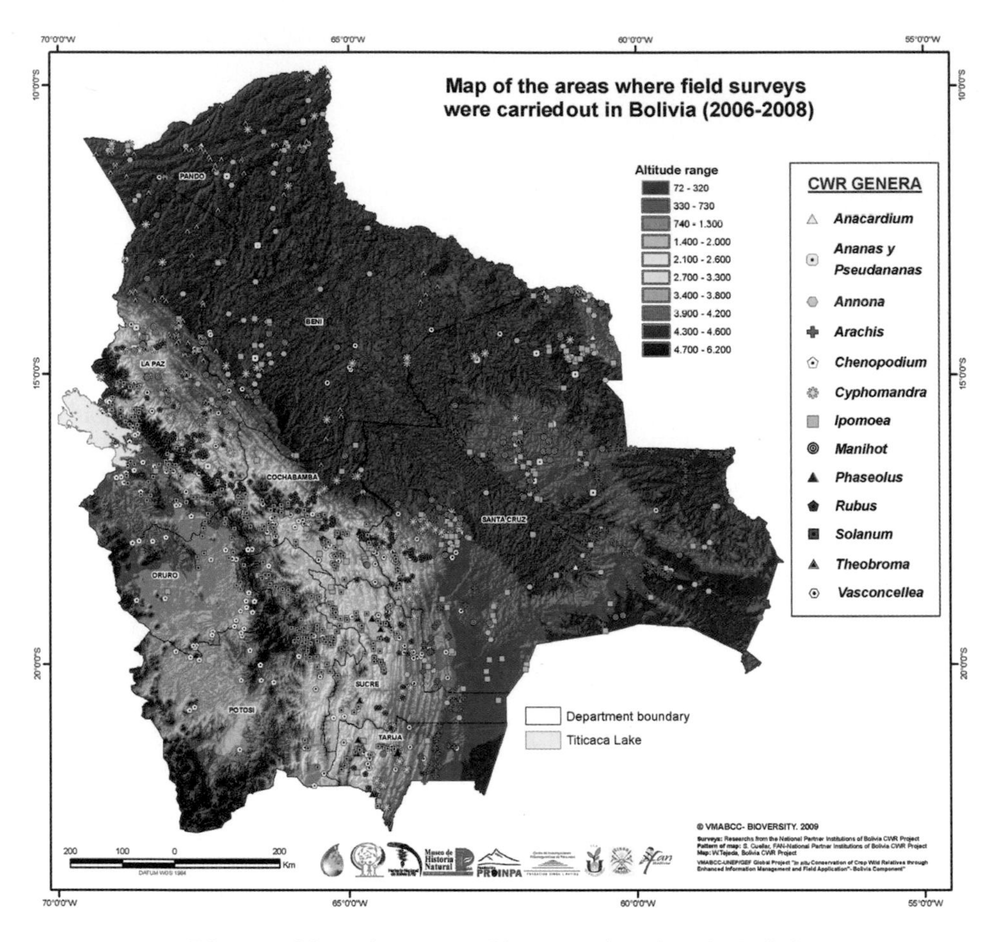

Plate 1: *Map of ecogeographic survey locations in Bolivia*

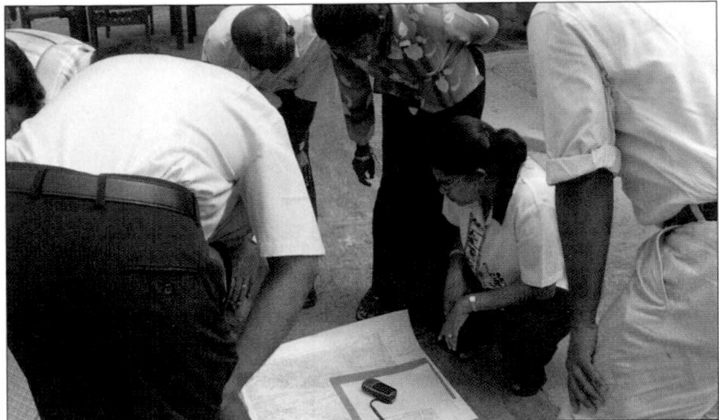

Conservation practitioners receive GIS training at the University of Peradeniya, Sri Lanka

Source: A. Wijesekara, CWR Project, Sri Lanka

Community consultation with indigenous people in the Territorio Indigena Isiboro-Secure (TIPNIS) in Bolivia

Source: Fundación TIPNIS

Discussing the implications of wild yam conservation for local communities bordering Ankarafantsika National Park, Madagascar

Source: D. Hunter, Bioversity International

Presenting research findings on wild yams to communities bordering Ankarafantsika National Park, Madagascar – an important element of the participatory process

Source: D. Hunter, Bioversity International

Plate 7: *Capacity building and participatory work*

Signboards display the importance of endemic wild relatives within the Sri Lankan Department of Agriculture's CWR Information Park

Source: A. Wijesekara, CWR Project, Sri Lanka

Visitors to the Tashkent Botanic Gardens in Uzbekistan can observe demonstration plots of wild onions. Botanic gardens have an important role in raising awareness of crop wild relatives

Source: G. Reimova

Plate 8: *Raising awareness*

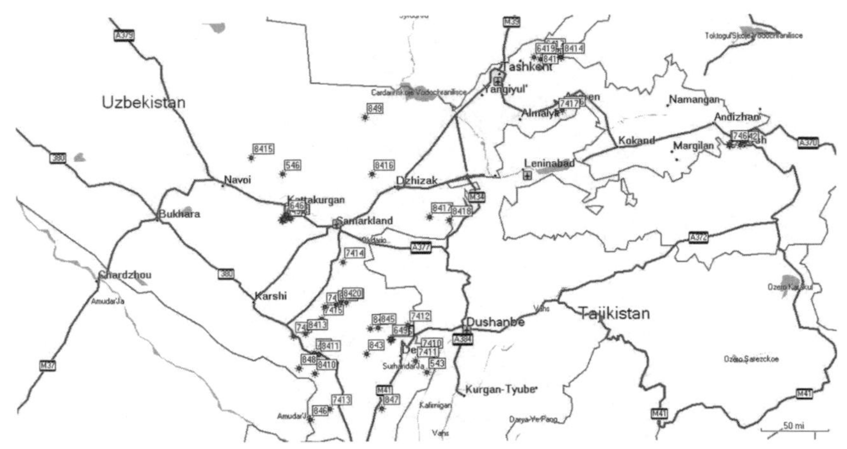

Figure 8.4 *Distribution of wild pistachio in Uzbekistan*

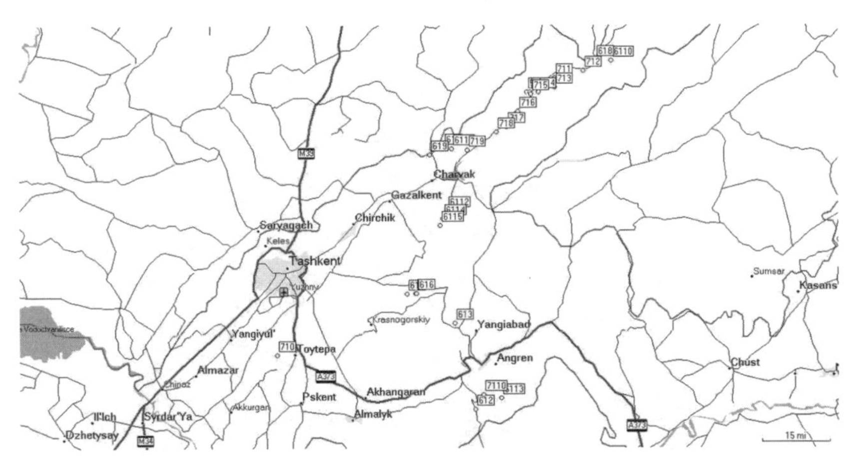

Figure 8.5 *Distribution of wild onion in Uzbekistan*

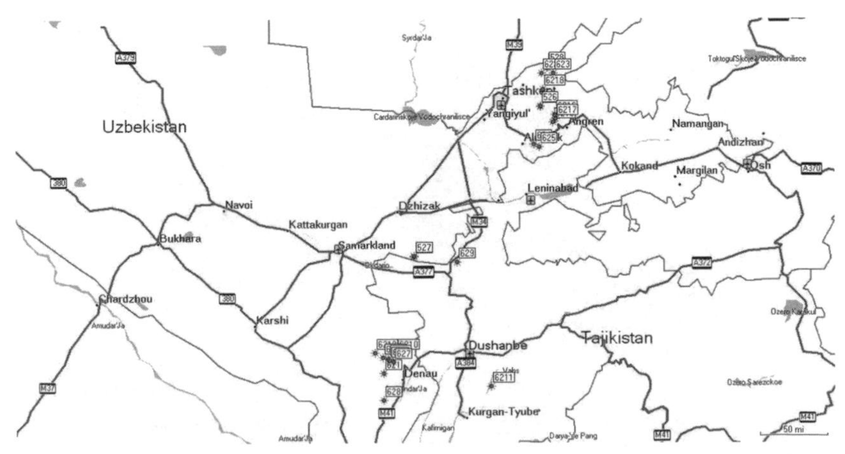

Figure 8.6 *Distribution of wild almond in Uzbekistan*

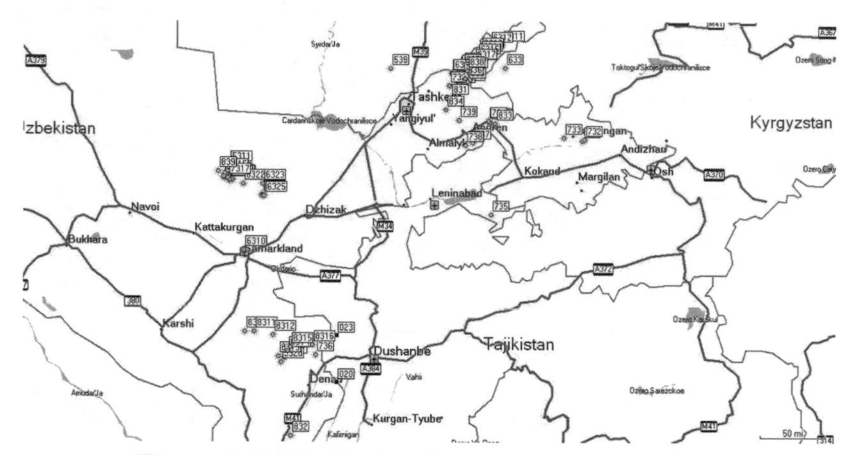

Figure 8.7 *Distribution of wild walnut in Uzbekistan*

Figure 8.8 *Distribution of wild apple tree in Uzbekistan*

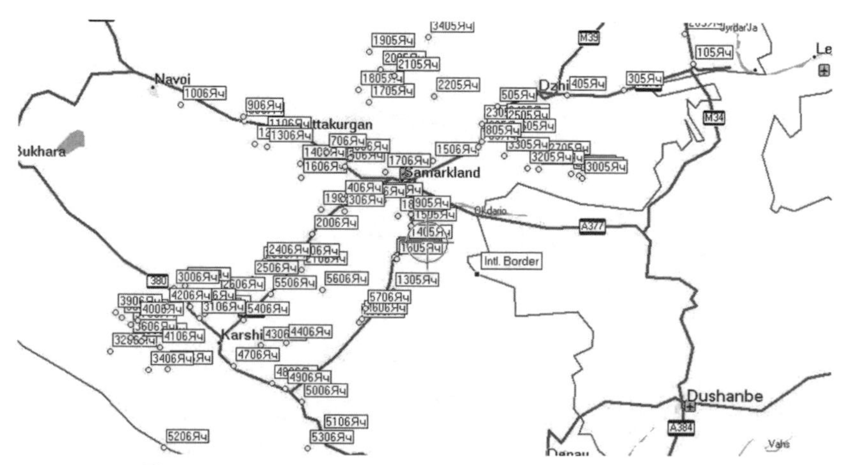

Figure 8.9 *Distribution of wild barley in Uzbekistan*

In total, 30 sets of data were used to define the current status and extent of distribution of the target species. All data were included in the database (in Russian and English). The database is being prepared for online use and access. The main threats for surveyed species were identified as: cattle grazing, uncontrolled collection of the fruit and using trees as timber.

Technical difficulties experienced

In Armenia, it is important to note that the distribution information extracted from herbarium passport data was treated with much caution. Some records were quite old and difficulties were encountered in handling outdated information, particularly old names of settlements and administrative subunits. In addition, only a few location descriptions were accompanied by coordinate readings. So mapping the possible distribution sites was a complex process. It started with an expert botanist putting dots on the map based on the information provided in the passport data, using his or her judgement on the possible collection site, after which the maps were georeferenced by a GIS expert.

In Bolivia, the following technical difficulties were noted:

- While the Global Project Coordination Unit at Bioversity had provided documents with methodologies for conducting ecogeographic surveys, there was no common understanding among the five project countries as to the meaning and scope of 'Geographical Survey'.
- Initially, because the descriptors (fields) of the CWR-Genetic Resources Information System (CWR-GRIS), developed by Bioversity for the national information systems of the CWR Project countries, were not agreed at an early stage between the five countries and Bioversity, field data were gathered using non-standard survey data sheets according to descriptors produced by the Bolivian national partner institutions. The herbaria emphasized the importance of collecting specimens and the genebanks emphasized the importance of collecting data of accessions.
- The relationship between components 2 and 3 (information system and *in situ* conservation actions) of the CWR Project were not clearly set out, leading to a failure to link the data required for database NISCWR and those needed for conservation and monitoring.
- It was not possible to gather all necessary data or conduct the required field assessments due to the different life cycles of the species, distance and travel costs. To gather population data, for example, more field trips and funds were needed; therefore, only data for two prioritized species were used. These data were collected through thesis work financed by the project.

In Sri Lanka, the following technical problems were experienced in undertaking ecogeographic surveys of CWR:

Season – Some species of CWR are annual and seasonal so the survey teams needed to visit the field at the correct time to find the species. In addition, funding was not received at the correct time, which also affected the survey work. For example, some populations of *O. rufipogon* are best seen in the field from December to March. Unavailability of project funds at the beginning of the year caused the surveying of *O. rufipogon* sites to be delayed until January and February.

Single or few plants in a population – Intrapopulation variation cannot be obtained when the number of individuals in a population is limited. For instance, only a single or few individuals were observed for most of the *Cinnamomum* species; this may cause problems in determining their estimated future rate of survival in the wild.

Preparation of herbarium specimens – Different species of CWR have different flowering times. Therefore, several visits had to be made to the same location to obtain herbarium specimens for correct identification. Additionally, some plants are very tall and unavailability of lower branches is also a problem in collecting herbarium specimens.

Distribution in specific areas – The ecogeographic surveys were conducted on a multi-species basis to save time and resources. Hence, the survey team concentrated on areas/region where the maximum number of species was expected. However, some species were distributed in very specific regions or areas and additional survey trips had to be made to those specific areas; only a few locations were found on these visits.

Unavailability of past survey information – Past survey information has not been well documented.The DIVA-GIS uses a UGS84 system while the maps used in the project were from the Sri Lanka Meteorology Department, which use a different coordinate system. There was limited capacity to convert these formats to make the systems compatible. As such, surveyed gaps were not clear, meaning that one must conduct an entire survey of a given species.

Incorrectly identified and old herbarium specimens – It was found that some herbarium specimens were incorrectly identified by authors and could therefore not be found when visits were made to areas where the specimens had been collected. Another problem was the unavailability of newly collected specimens in the National Herbarium for most CWR. Old herbarium specimens are difficult to handle, particularly flowering parts that are damaged.

Weak satellite signal – CWR are not always located in open areas. It was difficult to receive a satellite signal inside the forest areas when the upper forest canopy was thick. In this situation, GPS coordinates were taken at the nearest location with good satellite reception.

Lack of awareness – Protected areas of Sri Lanka are rich in wild relatives of food crops and wild food plants. Before conducting intensive surveys in these areas, the authorities of protected areas must have a good understanding of the importance of CWR in protected areas.

Identification of threat factors – Within a short period, it is difficult to identify the factors that threaten wild populations. Therefore, there should be several visits to the same location to identify the real threats to the populations; local knowledge will be important to gather further information.

Preparation of maps – GPS coordinates could not be plotted on the Sri Lankan Survey Department digital maps provided. Distribution prediction maps were prepared using available DIVA-GIS and FloraMap software. However, these software packages have only world climate data files that were too broad and were not specific for the localities surveyed.

The main problems encountered in Uzbekistan were:

- the wide distribution of priority species – not all distribution areas of apple and almond were surveyed during the four years of the project;
- the distribution areas of almond and pistachio are very remote;
- the number of ecogeographic surveys undertaken did not allow a full description of the phenology of populations;
- some territories where surveys were planned are in border regions with other countries and are not accessible.

Further sources of information

Bioversity International hosts a series of training modules on its website, including one on ecogeographical surveys: http://training/training_materials/ecogeographic_surveys.html

http://cropgenebank.sgrp.cgiar.org/images/flash/ecogeographic_surveys/index.htm

Brown, A.H.D. and Briggs, J.D. (1991) 'Sampling strategies for genetic variation in *ex situ* collections of endangered plant species', in D.A. Falk and K.E. Holsinger (eds) *Genetics and Conservation of Rare Plants,* pp99–119, Oxford University Press, New York

Dulloo, M.E., Maxted, N., Newbury, H., Florens, D. and Ford-Lloyd, B.V. (1999) 'Ecogeographic survey of the genus *Coffea* in the Mascarene Islands' *Botanical Journal of the Linnean Society,* vol 131, pp263–284

Dulloo, M.E., Labokas, J., Iriondo, J.M., Maxted, N., Lane, A., Laguna, E., Jarvis, A. and Kell, S.P. (2008) 'Genetic reserve location and design', in J.M. Iriondo, N. Maxted and M.E. Dulloo (eds), *Conserving Plant Genetic Diversity in Protected Areas,* pp23–64, CAB International, Wallingford

Maxted, N., van Slageren, M.W. and Rihan, J.R. (1995) 'Ecogeographic surveys', in L. Guarino, V. Ramanatha Rao and R. Reid (eds), *Collecting Plant Genetic Diversity: Technical Guidelines,* pp255–287, CAB International, Wallingford

Notes

1. Bennett (1997) uses the term ecogeography in this context which he defines as the collection and synthesis of ecological, geographical and taxonomic data.
2. CIAT (http://www.floramap-ciat.org/) and Jarvis et al (2005)
3. http://gisweb.ciat.cgiar.org/GapAnalysis/

References

Bamberg, J.B., del Rio, A.H., Huaman, Z., Vega, S., Martin, M., Salas, A., Pavek, J., Kiru, S., Fernandez, C. and Spooner, D.M. (2003) 'A decade of collecting and research on wild potatoes of the southwest USA', *American Journal of Potato Research,* vol 80, pp159–172

Barnes, M.R. and Breen, G. (eds) (2009) *Genetic Variation,* Springer-Verlag, New York, NY

Bellot, Y. and Cortez, M. (2010) *Manual de Mantenimiento,* Version 0.2 Sistema Nacional de de Información sobre Parientes Silvestres de Cultivos, SNIPSC – Unidad de Enlace, Fundacion Amigos de la Naturaleza – Proyecto UNEP/GEF 'Conservación *in situ* de parientes silvestres de especies cultivadas a traves del manejo de información y su aplicación en campo'

Bellot, Y. and Justiniano, R. (2010) *Manual de Usuario,* Version 0.2, Sistema Nacional de de Información sobre Parientes Silvestres de Cultivos, SNIPSC – Unidad de Enlace, Fundacion Amigos de la Naturaleza – Proyecto UNEP/GEF 'Conservación *in situ* de parientes silvestres de especies cultivadas a traves del manejo de información y su aplicación en campo'

Bennett, S.J. (1997) 'Ecogeographic assessment of Mediterranean environments for targeting legume collections', *International Grasslands Organization,* http://www.internationalgrasslands.org/publications/pdfs/1997/1_01_023.PDF

Bennett, S.J. and Bullita, S. (2003) 'Ecogeographical analysis of the distribution of six *Trifolium* species in Sardinia', *Biodiversity and Conservation,* vol 12, pp1455–1466

Bennett, S.J. and Maxted, N. (1997) 'An ecogeographic analysis of the *Vicia narbonensis* complex', *Genetic Resources and Crop Evolution,* vol 44, pp411–428

Bennett, S.J., Broughton, D.A. and Maxted, N. (2006) 'Ecogeographical analysis of the perennial *Medicago*', *CRC Salinity Bulletin,* vol 1, pp1–62

Bioversity and The Christensen Fund (2009) *Descriptors for Farmers' Knowledge of Plants,* Bioversity International, Rome, Italy and The Christensen Fund, Palo Alto, California, USA

Bisby, F.A. (1995) 'Chapter 2: Characterization of biodiversity', in V.H. Heywood (ed), *Global Biodiversity Assessment,* pp21–106, Cambridge University Press, Cambridge

Brown, A.H.D. and Briggs, J.D. (1991) 'Sampling strategies for genetic variation in *ex situ* collections of endangered plant species', in D.A. Falk and K.E. Holsinger (eds) *Genetics and Conservation of Rare Plants,* pp99–119, Oxford University Press, New York

Cunningham, A.B. (2001) *Applied Ethnobotany: Wild Plant Use and Conservation,* Earthscan, London, UK

Dulloo, M.E. (1998) 'Diversity and conservation of wild *Coffea* germplasm in the Mascarene Islands', PhD thesis, University of Birmingham, UK

Dulloo, M.E, Kell, S.P. and Jones, C.G. (1996) 'Impact and control of invasive alien species on small islands', *International Forestry Review,* vol 4, no 4, pp277–291

Dulloo, M.E., Maxted, N., Newbury, H., Florens, D. and Ford-Lloyd, B.V. (1999) 'Ecogeographic survey of the genus *Coffea* in the Mascarene Islands', *Botanical Journal of the Linnean Society,* vol, 131, pp263–284

Dulloo, M.E., Labokas, J., Iriondo, J.M., Maxted, N., Lane, A., Laguna, E., Jarvis, A. and Kell, S.P. (2008) 'Genetic reserve location and design', in J.M. Iriondo, N. Maxted and M.E. Dulloo (eds), *Conserving Plant Genetic Diversity in Protected Area,* pp23–64, CAB International, Wallingford, UK

Edmonds, J.M. (1990) *Herbarium Survey of African* Corchorus *L. Species,* Systematic and Ecogeographic Studies on Crop Genepools 4, IBPGR, Rome, Italy

Ehrman, T. and Cocks, P.S. (1990) 'Ecogeography of annual legumes in Syria: Distribution patterns', *Journal of Applied Ecology,* vol 27, pp578–591

Elith, J., Graham, C.H., Anderson, R.P., Dudik, M., Ferrier, S., Guisan, A., Hijmans, R.J., Huettmann, F., Leathwick, J.R., Lehmann, A., Li, J., Lohmann, L.G., Loiselle, B.A., Manion, G., Moritz, G., Nakamura, M., Nakazawa, Y., Overton, J. McC., Peterson, A.T., Phillips, S.J., Richardson, K., Scachetti-Pereira, R., Schapire, R.E., Soberón, J., Williams, S.,. Wisz, M.S. and Zimmermann, N.E. (2006) 'Novel methods improve prediction of species' distributions from occurrence data', *Ecography,* vol 29, pp129–151

Evans, D. (2006) 'The habitats of the European Union Habitats Directive' *Proceedings of the Royal Irish Academy,* vol 106B, pp167–173

FAO (1998) *The State of the World's Plant Genetic Resources for Food and Agriculture,* Food and Agriculture Organization of the United Nations (FAO), Rome, Italy

Ferguson, M. and Robertson, L.D. (1996) 'Genetic diversity and taxonomic relationships within the genus *Lens* as revealed by allozyme polymorphism', *Euphytica,* vol 91, pp163–172

Gabrielian, E. and Zohary, D. (2004) 'Wild relatives of food crops native to Armenia and Nakhichevan', *Flora Mediterranea,* vol 14, pp 5–80

Gentry, A.H. (1990) 'Herbarium taxonomy versus field knowledge', *Flora Malesiana Bulletin, s*pecial vol 1, pp31–35

Golding, J.S. and Smith, P.P. (2001*)* 'A 13-point flora strategy to meet conservation challenges', *Taxon,* vol 50, pp475–477

Gole, T.W.G., Denich, M., Teketay, D. and Vlek, P.L.G. (2002) 'Human impacts on the *Coffea arabica* genepool in Ethiopia and the need for its *in-situ* conservation', Chapter 22, in V. Ramanatha Rao, A.D.H. Brown, and M.T. Jakson (eds), *Managing Plant Genetic Diversity,* IPGRI, Rome, Italy

Guarino, L., Jarvis, A., Hijmans, R.J. and Maxted, N. (2002) 'Geographic information systems (GIS) and the conservation and use of plant genetic resources', in J. Engels, V. Ramanatha Rao, A.H.D. Brown and M/T. Jackson (eds) *Managing Plant Genetic Diversity,* pp387–404, CAB International, Wallingford, UK

Guarino, L., Maxted, N. and Chiwona, E.A. (2005) *Ecogeography of Crops,* IPGRI Technical Bulletin No. 9, International Plant Genetic Resources Institute, Rome, Italy

Hamilton, A. and Hamilton, P. (2006) *Plant Conservation: An Ecosystem Approach,* Earthscan, London, UK

Hanski, I. (1999) *Metapopulation Ecology,* Oxford University Press, Oxford

Hanski , I., Moilanen, A. and Gyllenberg, M. (1996) 'Minimum viable metapopulation size,' *American Naturalist,* vol 147, pp527–541

Hawkes, J.G. (1987) 'A strategy for seed banking in botanic gardens', in D. Bramwell, O. Hamman, V. Heywood and H. Synge (eds) *Botanic Garden and the World Conservation Strategy,* pp131–149, Academic Press, London, UK

Hawkes J.G., Maxted N. and Ford-Lloyd, B.V. (2000) *The* Ex Situ *Conservation of Plant Genetic Resources,* Kluwer Academic Publishers, Dordrecht, The Netherlands

Heywood V.H. (1991) 'Developing a strategy for germplasm conservation in Botanic Gardens', in V.H. Heywood and P.S. Wyse Jackson (eds) *Tropical Botanic Gardens – Their Role in Conservation and Development,* pp11–23, Academic Press, London

Heywood, V.H. (ed) (1995) *Global Biodiversity Assessment,* Cambridge University Press, Cambridge

Heywood, V. (2004) 'Meeting the demands for taxonomic information from users in conservation and genetic resources', *Phytologia Balcanica,* vol 93, pp425–434

Heywood, V.H. and Dulloo, M.E. (2005) In Situ *Conservation of Wild Plant Species – A Critical Global Review of Good Practices,* IPGRI Technical Bulletin, no 11, FAO and IPGRI, IPGRI, Rome, Italy

Hijmans, R.J., Guarino, L., Cruz, M. and Rojas, E. (2001) 'Computer tools for spatial analysis of plant genetic resources data:1 DIVA-GIS', *Plant Genetic Resources Newsletter,* vol 127, pp15–19

Hughes, C. (1998) *The Genus* Leucaena: *A Plant Genetic Resources Manual,* Tropical Forestry Papers 34, Oxford Forestry Institute, Oxford, UK

IBPGR (1985) *Ecogeographical Surveying and* In Situ *Conservation of Crop Relatives,* International Board for Plant Genetic Resources (IBPGR), Rome, Italy

IUCN (2002) *IUCN Technical Guidelines on the Management of* Ex-Situ *Populations for Conservation,* as approved at the 14th Meeting of the Programme Committee of IUCN Council, Gland, Switzerland on 10 December 2002, International Union for Conservation of Nature, http://intranet.iucn.org/webfiles/doc/SSC/SSCwebsite/Policy_statements/IUCN_Technical_Guidelines_on_the_Management_of_Ex_situ_populations_for_Conservation.pdf

Jarvis, A., Williams, K., Williams, D., Guarino, L., Caballero, P.J. and Mottram, G. (2005) 'Use of GIS in optimizing a collecting mission for a rare wild pepper (*Capsicum flexuosum* Sendtn.) in Paraguay', *Genetic Resources and Crop Evolution,* vol 52, no 6, pp671–682.

Kanashiro, M., Thompson, I.S., Yared, J.A.G., Loveless, M.D., Coventry, P., Martinsda-Silva, R.C.V., Degen, B. and Amaral, W. (2002) 'Improving conservation values of managed forests: The Dendrogene Project in the Brazilian Amazon', *Unasylva,* vol 53, no 209, pp25–33

Karp, A. (2002) 'The new genetic era: Will it help us in managing genetic diversity?', in J.M.M. Engels, V.R. Rao, A.H.D. Brown and M.T. Jackson (eds) *Managing Plant Genetic Diversity,* pp43–56, CAB Publishing, Wallingford, UK

Kuleung, C., Baenziger, P.S., Kachman, S.D. and Dweikat, I. (2006) 'Evaluating the genetic diversity of Triticale with wheat and rye SSR markers', *Crop Science,* vol 46, pp1692–1700

Lobo, J.M. (2008) 'More complex distribution models or more representative data?', *Biodiversity Informatics,* vol 5, pp14–19

Lorence, D.H. (1978) 'The pteridophytes of Mauritius (Indian Ocean): Ecology and distribution', *Botanical Journal of the Linnean Society,* vol 76, pp207–247

Maxted, N. (1995) 'An herbarium based ecogeographic study of *Vicia* subgenus *Vicia*', Systematic and Ecogeographic Studies on Crop Genepools 8, IPGRI (nternational Plant Genetic Resources Institute), Rome, Italy

Maxted, N. and Kell, S. (1998) 'Ecogeographic techniques and *in situ* conservation: A case study for the legume genus *Vicia* in Turkey', in N. Zencirci, Z. Kaya, Y. Anikster and W.T. Adams (eds) *Proceedings of an International Symposium on* In Situ

Conservation of Plant Diversity 4–8 November, 1996, pp323–344, Central Research Institute for Field Crops, Ankara, Turkey

Maxted, N., van Slageren, M.W. and Rihan, J.R. (1995) 'Ecogeographic surveys', in L. Guarino, V. Ramanatha Rao and R. Reid (eds), *Collecting Plant Genetic Diversity: Technical Guidelines,* pp255–287, CAB International, Wallingford, UK

Maxted, N., Dulloo, M.E. and Eastwood, A. (1999) 'A model for genetic reserve conservation: A case study for *Coffea* in the Mascarene Islands', *Botanica Lithuanica* Supplementum, vol 2, pp61–78

Maxted, N., Mabuza-Diamini, P., Moss, H., Padulosi, S., Jarvis, A. and Guarino, L. (2004) *Systematic and Ecogeographic Studies on Crop Genepools 11: An Ecogeographic Study African* Vigna. International Plant Genetic Resources Institute (IPGRI), Rome, Italy

Morris, W.F. and Doak, D.F. (2002) *Quantitative Conservation Biology: Theory and Practice of Population Viability Analysis,* Sinauer Associates, Sunderland, MA, USA

Nabhan, G.P. (1990) *Systematic and Ecogeographic Studies on Crop Genepools* 5: *Wild* Phaseolus *Ecogeography in the Sierra Madre Occidentalis, Mexico: Areographic techniques for targeting and conserving species diversity,* International Board for Plant Genetic Resources (IBPGR), Rome, Italy

Parish, D.H. and Feillafe, S.M. (1965) *Notes on the 1:100,000 Soil Map of Mauritius,* Mauritius Sugar Industry Research Institute (MSIRI) Occasional Paper 22, Mauritius

Pearce, T. and Bytebier, B. (2002) 'The role of a herbarium and its database in supporting plant conservation,' in M. Maunder, C. Clubbe, C. Hankamer and M. Groves (eds) *Plant Conservation in the Tropics: Perspectives and Practice,* pp49–67, Royal Botanic Gardens, Kew, UK

Peterson, A.T. (2001) 'Predicting species' geographic distributions based on ecological niche modelling', *The Condor,* vol 103, no 3, pp599–605

Ramos Canaviri, C.L. (2009) 'Estudio poblacional de especies silvestres del género *Arachis* (Maní) y estrategias para su conservación *in situ* en Bolivia', Tesis de Licenciatura, UMSS, Cochabamba, Bolivia

Smith, S.D. and Peralta, I.E. (2002) 'Ecogeographic surveys as tools for analyzing potential reproductive isolating mechanisms: An example using *Solanum juglandifolium* Dunal, *S. ochranthum* Dunal, *S. lycopersicoides* Dunal, and *S. sitiens* I. M. Johnston', *Taxon,* vol 51, pp341–349

Soulé, M.E. (1986) *Conservation Biology: The Science of Scarcity and Diversity,* Sinauer Associates, Sunderland, MA, USA

Theilade, I., Graudal, L. and Kjær, E. (2000) 'Conservation of the genetic resources of *Pinus merkusii* in Thailand', DFSC Technical Note 58, Danida Forest Seed Centre (DFSC), Humlebaek, Denmark

Vaughan, R.E. and Wiehe, P.O. (1937) 'Studies on the vegetation of Mauritius I: A preliminary survey of the plant communities', *Journal of Ecology,* vol 25, pp289–243

de Vicente M.C. and Fulton, T. (2004) *Using Molecular Marker Technology Effectively in Plant Diversity Studies, Vol 1: Learning Module,* CD-ROM, International Plant Genetic Resources Institute (IPGRI), Rome, Italy and Institute for Genomic Diversity, Cornell University, Ithaca, NY, USA

de Vicente, M.C., Lopez, C. and Fulton, T. (2004) *Genetic Diversity Analysis with Molecular Marker Data, Vol 2: Learning Module,* CD-ROM, International Plant Genetic Resources Institute (IPGRI), Rome, Italy, Universidad Nacional Agraria 'La molina', Peru and Institute for Genomic Diversity, Cornell University, Ithaca, NY, USA

Volk, G.M., Ruichards, C.M., Henk, A.D., Reilley, A.A., Reeves, P.A., Forsline, P.E. and Ardwinckle, H.S. (2009) 'Capturing the diversity of wild *Malus orientalis* from

Georgia, Armenia, Russia, and Turkey', *Journal of the American Society for Horticulture Science,* vol 134, pp453–459

Yanchuk, A.D. (1997) 'Conservation issues and priorities for the conifer genetic resources of British Columbia, Canada', *Forest Genetic Resources,* vol 25, pp2–9

Zhang, C., Chen, X., He, R., Liu, X., Feng, R. and Yuan, Z. (2007) 'Genetic structure of *Malus sieversii* population from Xinjiang, China, revealed by SSR markers', *Journal of Genetics and Genomics,* vol 34, pp947–955

Part III

Conservation Actions

This part covers the management actions at both the reserve level and the species/population level needed to maintain the target CWR and to control, mitigate or eliminate any threats to them. It also outlines the monitoring procedures that may be needed to assess the effectiveness of these actions.

Chapter 9

Protected Areas and CWR Conservation

> *In general, the idea that the conservation of agrobiodiversity is a potentially valuable function of a protected area is as yet little recognized. … Indeed, a study by WWF found that the degree of protection in places with the highest levels of crop genetic diversity is significantly lower than the global average; and even where protected areas did overlap with areas important for crop genetic diversity (i.e. landraces and crop wild relatives) little attention was given to these values in the management of the area* (Amend et al, 2008).

The role of protected areas in CWR conservation

A system of protected areas constitutes the basic underpinning of the conservation strategy of most countries. On the other hand, the predicted impacts of accelerated climate change are beginning to bring our reliance on such an approach as our main tool for *in situ* conservation of biodiversity into clear focus (Spalding and Chape, 2008). Questions are being raised about the effectiveness of protected areas as a long-term strategy in conserving biodiversity and several surveys have been undertaken to assess this (e.g. WWF, 2004). This is discussed in more detail in Chapter 14.

> *It is apparent that nature conservation has become one of the most important human endeavours on the planet, and the area under protection now exceeds the total area of permanent crops and arable land* (Chape et al, 2008).

Protected areas cover at least 114,000 sites and occupy more than 19 million km^2, representing 12.9 per cent of the earth's land surface. Data for the five project countries are given in Table 9.1.

Table 9.1 *Areas protected, by country (2005)*

Country	*Land area (km²)*	*Total protected area (km²)*	*Total number of sites*
Armenia	29,800	2991	28
Bolivia	1,098,580	230,509	50
Madagascar	587,040	18,458	60
Sri Lanka	65,610	14,877	264
Uzbekistan	447,400	20,503	24

Some of them have a long history while others are of recent creation. In Sri Lanka, for example, wildlife sanctuaries were set up in the 3rd century BC by King Devanampiya Tissa in the area around Mihintale, apparently the first in the world.

Protected area (PA) is a general term used to cover a wide variety of situations. The definition adopted by the International Union for Conservation of Nature (IUCN) has recently been revised as 'a clearly defined geographical space, recognised, dedicated and managed, through legal or other effective means, to achieve the long-term conservation of nature with associated ecosystem services and cultural values' (Dudley, 2009). A similar definition is given by the Convention on Biological Diversity (CBD): 'A geographically defined area which is designated or regulated and managed to achieve specific conservation objectives'. Protected areas vary enormously in size, ranging from tens of thousands of hectares to many which are relatively small (1000–10,000ha), often representing remaining fragments that, although valuable, may be inadequate for maintaining large-scale processes. There is also a great diversity of types of area in terms of their conservation objectives, the degree of human activity permitted and the extent of involvement of stakeholders. There are also evident gaps in coverage of existing protected area networks and urgent priorities for the expansion of the global protected area system include the Andes, Madagascar and Sri Lanka (Chape et al, 2008: Chapter 2).

The relevant targets of the Global Strategy for Plant Conservation (GSPC) are given in Box 9.1; although, it should be noted that these are currently (April 2010) under review. A summary of Madagascar's progress towards the GSPC Targets 4 and 5 is given in Box 9.2.

The different categories of protected areas recognized by IUCN are widely used. It should be noted, however, that they have recently been redefined as indicated in Box 9.3. A set of guidelines on how to apply these categories has been published by Dudley (2009). It is recommended that in applying the categories system, the first step is to determine whether or not the site meets the IUCN definition (see above) and the second step is to decide on the most suitable category.

Categories 1 and 2 are likely to be the most appropriate for CWR conservation, but CWR occur in all types of protected areas even though the suitability of

Box 9.1 Targets 4 and 5 of the GSPC referring to protected areas

Target 4: At least 10 per cent of each of the world's ecological regions effectively conserved

The target implies: 1) increasing the representation of different ecological regions in protected areas; and 2) increasing the effectiveness of protected areas. Effective conservation is understood to mean that the area is managed so as to achieve a favourable conservation status for plant species and communities. Favourable conservation status is not defined.

Target 5: Protection of 50 per cent of the most important areas for plant conservation

Important plant areas are defined by criteria such as endemism, species richness and/or uniqueness of habitats and take into account the provision of ecosystem services.
The failure to agree on a set of defining criteria makes implementation at a national level difficult to apply or assess.

Box 9.2 Progress in implementing Targets 4 and 5 of GSPC in Madagascar

Target 4: Protected areas represented only 3 per cent of the total area of the country, but during the World Parks Congress in Durban in 2003, Madagascar pledged to extend the protected areas to cover 6 million hectares (10 per cent of the country's area) by 2010. This is known as the Durban vision. In 2009, all the future new protected areas (NAP) have been identified and half of them (2 million hectares) have already been given a temporary protection status. The creation of the rest of the protected areas is ongoing. All the existing and future protected areas will be part of what is called the System of Protected Areas of Madagascar or SAPM.

Target 5: In the frame of a project conducted by GSPM (Madagascar Plant Specialists Group*) and Botanic Gardens Conservation International (BGCI) in 2008 to 2009, which aimed at the conservation of wild plants for food and medicine, all protected areas were assessed according to the important plant areas (IPA) qualification processes, e.g. Plantlife criteria (threatened species presence, floristic richness, and presence of threatened habitats) or Priority Area for Plant Conservation (PAPC) criteria. The assessment revealed that 40 out of the 52 current PAs managed by Madagascar National Parks are IPAs, while 26 out of the 35 NAPs are also IPAs. In addition, all the Madagascar Key Biodiversity Areas (20) identified by Conservation International and other sites assessed through the PAPC process are also IPAs.

* GSPM is a member of IUCN Species Survival Committee and is, above all, responsible for the validation of the status of the species submitted to the IUCN Red List.

many of them for genetic conservation is limited. The problems of adapting existing protected areas for targeted CWR conservation are discussed in the section, 'Protected area management'.

It should be noted that, in practice, many, if not most, countries use different or additional categories and definitions. For example, in the UNEP/GEF CWR Project countries: specially protected nature areas in Armenia can have the status of state reserve, national park, reservation and nature monument; in Sri Lanka, there are basically eight types of national protected areas, depending on their objective: strict nature reserves, national parks, nature reserves, jungle corridors, refuges, marine reserves, buffer zones and sanctuaries.

It should also be noted that national parks occur in all six categories and as Dudley (2008) points out, 'the fact that a government has called, or wants to call, an area a national park does not mean that it has to be managed according to the guidelines under category II. Instead, the most suitable management system should be identified and applied; the name is a matter for governments and other stakeholders to decide'.

A small number of protected areas are specifically tailored for the genetic conservation of target species such as genetic reserves, gene management zones, *in situ* gene conservation forests, gene parks and genetic resources management units (see Heywood and Dulloo, 2005: 2.2.5; Iriondo et al, 2008). Thomson and Theilade (2001) suggest that a case can be made for designation of *in situ* gene conservation areas as a special category of protected area on the basis that:

- they have conservation of within-species genetic variation as the major objective;
- the gene pools of concern are primarily of economic species; and
- provision is made for the use of the gene pool by researchers, tree breeders and for *ex situ* conservation purposes.

Sacred groves, forests, sites

An important type of traditional nature conservation, practised as part of the religion-based conservation ethos of ancient people in many parts of the world, is the protection of small areas of forest as sacred groves or forests or of particular tree specimens as sacred trees. A characteristic of such traditional ecosystem approaches is that they require a belief system that includes a number of prescriptions, such as taboos, that regulate human behaviour and lead to a restrained use of the resource. Such sacred sites (including sacred natural sites and landscapes) that fit into national and international definitions of protected areas can potentially be recognized as legitimate components of protected area systems and can be attributed to any of the six IUCN protected area categories. If the site's management objectives meet the IUCN definition of a protected area and the requirements of a particular category and if the faith group so desires, particular sacred natural sites can be formally included in national protected area systems. Examples of sacred sites in Sri Lanka are Yala National Park (Category Ia), which is significant to Buddhists and Hindus and requires high levels of protection for

Box 9.3 The IUCN protected area management categories

CATEGORY Ia: Strict nature reserve – strictly protected areas set aside to protect biodiversity and also possibly geological/geomorphological features, where human visitation, use and impacts are strictly controlled and limited to ensure protection of the conservation values. Such protected areas can serve as indispensable reference areas for scientific research and monitoring. Their primary objective is to conserve regionally, nationally or globally outstanding ecosystems, species (occurrences or aggregations) and/or geodiversity features: these attributes will have been formed mostly or entirely by non-human forces and will be degraded or destroyed when subjected to all but very light human impact.

CATEGORY Ib: Wilderness area – protected areas that are usually large unmodified or slightly modified areas, retaining their natural character and influence, without permanent or significant human habitation, which are protected and managed so as to preserve their natural condition. Their primary objective is to protect the long-term ecological integrity of natural areas that are undisturbed by significant human activity, free of modern infrastructure and where natural forces and processes predominate, so that current and future generations have the opportunity to experience such areas.

CATEGORY II: National park – protected areas that are large natural or near-natural areas set aside to protect large-scale ecological processes, along with the complement of species and ecosystems characteristic of the area, which also provide a foundation for environmentally and culturally compatible spiritual, scientific, educational, recreational and visitor opportunities. Their primary objective is to protect natural biodiversity along with its underlying ecological structure and supporting environmental processes and to promote education and recreation.

CATEGORY III: Natural monument or feature – protected areas that are set aside to protect a specific natural monument, which can be a landform, sea mount, submarine cavern, geological feature such as a cave or even a living feature such as an ancient grove. They are generally quite small protected areas and often have high visitor value. Their primary objective is to protect specific outstanding natural features and their associated biodiversity and habitats.

CATEGORY IV: Habitat/species management area – protected areas that aim to protect particular species or habitats and management reflects this priority. Many category IV protected areas will need regular, active interventions to address the requirements of particular species or to maintain habitats, but this is not a requirement of the category. Their primary objective is to maintain, conserve and restore species and habitats.

CATEGORY V: Protected landscape/seascape – a protected area where the interaction of people and nature over time has produced an area of distinct character with significant ecological, biological, cultural and scenic value: and where safeguarding the integrity of this interaction is vital to protecting and sustaining the area and its associated nature conservation and other values. Their primary objective is to protect and sustain important landscapes/seascapes and the associated nature conservation and other values created by interactions with humans through traditional management practices.

CATEGORY VI: Protected area with sustainable use of natural resources – protected areas that conserve ecosystems and habitats, together with associated cultural values and traditional natural resource management systems. They are generally large, with most of the area in a natural condition, where a proportion is under sustainable natural resource management and where low-level non-industrial use of natural resources compatible with nature conservation is seen as one of the main aims of the area. Their primary objective is to protect natural ecosystems and use natural resources sustainably, when conservation and sustainable use can be mutually beneficial.

Based on Dudley, 2008

Box 9.4 Ankodida, a community-managed protected area and sacred forest in Madagascar

Ankodida is a newly established, community-managed Category V protected area in south-eastern Madagascar, which protects a sacred forest, the former home of a pre-colonial Tandroy king. The forest also shelters spirits that play an important role in the spiritual life of the Tandroy tribe and provides the bulk of household income for local populations, thereby making it of great cultural, spiritual and material importance. Six of the protected area's seven zones are composed of traditional village territories managed under devolved management contracts and, in addition, there is a priority conservation zone covering the sacred forest managed by local communities according to traditional regulations. Management of Ankodida is focused on the reinforcement of management through the legal empowerment of its traditional guardians. Ankodida houses two critically endangered *Aloe* species, the endangered palm *Ravenea xerophila* and 30 to 40 per cent of the world's population of the triangle palm, *Dypsis decaryi*.

Source: Gardner et al, 2008

faith reasons; and Peak Wilderness Park, (Sri Pada-Adams Peak), a sacred natural site for Islam, Buddhism, Hinduism and Christianity, attracting many pilgrims of all these faiths. Such sacred sites or forests may be of interest for *in situ* conservation of any target species that occur within them as they provide a degree of protection and are a focus of community interest. An example from Madagascar is given in Box 9.4. An overview and examples of cultural and spiritual values of protected landscapes is given by Mallarach (2008).

Protected area ownership and governance

Enormous variation exists in the ways in which protected areas are owned and governed. They may be managed by government, community or co-managed or they may be private. In many countries, public protected areas are supplemented by extensive private reserves or other forms of protection. In the US, for example, The Nature Conservancy currently owns and manages approximately 15 million acres of the national territory and globally protects more than 116 million acres of the most ecologically important places in the US and 28 other countries.

The main types of governance of protected areas are given in Table 9.2. Any of these can be associated with any management objective.

In the UNEP/GEF CWR Project countries, for example, the Chatkal State Biosphere Reserve, Uzbekistan, was established in 1947 and has had a varied history, changing in size several times (in 1952, 1960, 1993 and 1996) and in status, being designated as a United Nations Educational, Scientific and Cultural Organization (UNESCO) Biosphere Reserve in 1978 and part of the State Committee for Nature Protection. Since 2001, the reserve has been a separate

Table 9.2 *Modes of protected area governance*

Mode	*Type*
Government	National
	State or province
	Local
	Delegated (to another government agency)
	Delegated (to statutory authority)
	Delegated (to local government or community group)
Co-management	Collaborative
	Joint
Private	Individual
	Not-for-profit organization
	Commercial organization
Community	Indigenous
	Local

Source: Chape et al, 2008

legal entity as part of Ugam-Chatkal State Nature National Park, reporting to the Khokim (Governor) of Tashkent Oblast.

In Armenia, the Erebuni Reserve was established in 1981 in the vicinity of Yerevan, specifically to protect wild cereal species – *Triticum araraticum*, *T. urartu*, *T. boeticum*, four species of *Aegilops*, *Hordeum glaucum* and *Secale vavilovii*. It is the

Figure 9.1 *The Chatkal State Reserve, Uzbekistan*

Table 9.3 *Governance of protected areas in Madagascar*

Type of governance	*A: PA managed by government*			*B: PA under participative management (co-management)*			*C: Private PA*			*D: PA of community patrimony*	
Category of PA according to IUCN	*National or federal ministry or national agency*	*Local or municipal ministry or agency*	*Delegated management by the government (i.e. NGO)*	*Transboundary management*	*Collaborative management*	*Joint management*	*Declared and managed by owners as individuals*	*Managed by non-profit organization such as universities, NGOs*	*Managed by profit-making organizations such as tourism firms*	*Declared and managed by indigenous communities*	*Declared and managed by local communities*
I. Natural integral Reserve	✗		✗		✗						
II. National park	✗	✗	✗		✗	✗					
III. Natural monument	✗	✗	✗		✗	✗	✗	✗	✗		✗
IV. Special reserve	✗	✗	✗		✗	✗	✗	✗	✗		
V. Protected landscape or seascape					✗	✗					✗
VI. Natural resources PA	✗	✗	✗		✗	✗					✗

smallest reserve in Armenia (89ha) and is the only reserve which is not a 'state non-commercial organization' (SNCO) with a charter approved by the government and which does not have its own management system but remains under the jurisdiction of the 'Reserve Park Complex' of the Ministry of Nature Protection of the Republic of Armenia.

Protected areas constitute the Madagascar National Parks (MNP) formerly known as ANGAP (*Association Nationale pour la Gestion des Aires Protégées*) network and are managed by the MNP itself or by NGOs. MNP is the national association that managed all the protected areas before the creation of other categories of PA within the framework of the Durban vision (see Box 9.2).

The Kanneliya-Dediyagala-Nakiyadeniya (KDN) Biosphere Reserve in Sri Lanka is managed by the Forest Department along with other biosphere reserves and national heritage and wilderness areas and conservation forests, whereas 60 per cent of Sri Lanka's protected areas are under the jurisdiction of the Department of Wildlife Conservation. There are 78 villages surrounding the reserve, and 50 per cent of the households are below the poverty line and depend on the forest for timber and non-timber forest products, such as medicinal plants, fuelwood, poles and posts for subsistence rather than trade. Their needs have been taken into account by the Forest Department in the Management Plan for the forest.

Good governance

IUCN has identified the following principles of good governance, any of which can be associated with any management objective (Dudley, 2008):

- **Legitimacy and voice** – social dialogue and collective agreements on protected area management objectives and strategies on the basis of freedom of association and speech with no discrimination related to gender, ethnicity, lifestyles, cultural values or other characteristics.
- **Subsidiarity** – attributing management authority and responsibility to the institutions closest to the resources at stake.
- **Fairness** – sharing equitably the costs and benefits of establishing and managing protected areas and providing a recourse to impartial judgement in case of related conflict.
- **Do no harm** – making sure that the costs of establishing and managing protected areas do not create or aggravate poverty and vulnerability.
- **Direction** – fostering and maintaining an inspiring and consistent long-term vision for the protected area and its conservation objectives.
- **Performance** – effectively conserving biodiversity while responding to the concerns of stakeholders and making a wise use of resources.
- **Accountability** – having clearly demarcated lines of responsibility and ensuring adequate reporting and answerability from all stakeholders about the fulfilment of their responsibilities.
- **Transparency** – ensuring that all relevant information is available to all stakeholders.

Box 9.5 Activities that may be involved in establishing and maintaining a network of protected areas

- preparation of information and publicity material;
- scientific studies to identify and designate sites – survey including inventory, mapping, condition assessment;
- administration of selection process;
- consultation, public meetings, liaison with landowners, complaints;
- pilot projects;
- pre-designation phase;
- preparation and review of management plans, strategies and schemes;
- establishment and running costs of management bodies;
- provision of staff (wardens, project officers), buildings and equipment;
- consultation – public meetings, liaison with landowners;
- costs for statutory and case work (environmental impact assessments, legal interpretation, etc.);
- management planning and administration;
- conservation management measures – e.g. maintenance of habitat or status of species;
- management schemes and agreements with owners and managers of land or water;
- fire prevention and control;
- research monitoring and survey;
- provision of information and publicity material;
- training and education;
- visitor management;
- 'ongoing' management actions and incentives;
- restoration or improvement of habitat or status of species;
- compensation for rights forgone, loss of land value, etc.;
- land purchase, including consolidation;
- infrastructure for public access, interpretation works, observatories and kiosks, etc.;
- habitat type survey and GIS data.

Source: Natura, 2000; http://ec.europa.eu/environment/nature/natura2000/index_en.htm

- **Human rights** – respecting human rights in the context of protected area governance, including the rights of future generations.

The setting up and maintenance of a protected area covers a wide range of activities (Box 9.5) and involves many kinds of professional and stakeholders.

Protected area management

Although protected area management is the responsibility of those in charge of the area, it is important that those engaged in targeted *in situ* species conservation are aware of the main issues involved when cooperating with protected area managers or negotiating with them over management interventions for target species. It would not be appropriate in this manual to enter into details of protected area management, which is a vast and highly complex topic and is beyond the remit of this manual, so the reader is referred to the IUCN *Guidelines for Management Planning of Protected Areas* (Thomas and Middelton, 2003), which will provide information on the key management planning processes in protected areas and on developing management plans; the *Management Guidelines for IUCN Category V Protected Areas: Protected Landscapes/Seascapes* (Phillips, 2002) will also be a useful resource.

According to Thomas and Middleton (2003), the most commonly found contents of a management plan include:

- executive summary;
- introduction (e.g. purpose and scope of plan, reason for designation of protected area and authority for plan);
- description of the protected area;
- evaluation of the protected area;
- analysis of issues and problems;
- vision and objectives;
- zoning plan (if appropriate);
- management actions (list of agreed actions, identifying schedule of work, responsibilities, priorities, costs and other required resources);
- monitoring and review.

The quality and effectiveness of the management of protected areas varies considerably and can be a cause of major concern. Various tools and guidelines have been developed to assess management effectiveness (Chape et al, 2008). Management challenges include land encroachment, illegal logging or permitted destructive logging practices, unsustainable agricultural practices in buffer zones and lack of proper management mechanisms and institutional capacity.

Adapting protected area management plans to cover the conservation needs of CWR

Many of the populations of the target species selected for *in situ* conservation will be found to grow in one or more protected areas and consequently benefit from some degree of protection (but see below). As already noted, most protected areas do not include genetic management as one of their management objectives. The management needs of the populations of the CWR target species are quite

Box 9.6 Management responses to deal with threats facing protected areas

- **regeneration**, which involves the recovery of natural integrity following disturbance or degradation, with minimum human intervention;
- **restoration**, which requires returning existing habitats to a known past state or to an approximation of the natural condition by repairing degradation, by removing introduced species, or by reinstatement;
- **reinstatement**, which means reintroduction to a place of one or more species or elements of habitat or geodiversity that are known to have existed there naturally at a previous time, but that can no longer be found at that place;
- **enhancement**, which involves introduction to a place of additional individuals of one or more organisms, species or elements of habitat or geodiversity that naturally exist there.
- **preservation**, which means maintaining the biodiversity and/or an ecosystem of a place at the existing stage of succession, or maintaining existing geodiversity;
- **modification**, which involves altering a place to suit proposed uses that are compatible with the natural significance of the place;
- **protection**, which requires taking care of a place by maintenance and by managing impacts to ensure that natural significance is retained;
- **maintenance**, which involves continuous protective care of the biological diversity and geodiversity of a place.

Source: ACIUCN, 2002; Chape et al, 2008

specific and separate from the management of the protected area itself which is why the concept of genetic reserves was introduced (see Chapter 3). Many management actions are responses to threats and unwanted changes to the area (see Box 9.6). Area-based management interventions include nutrient control, erosion control, burning, control of invasive species, habitat disturbance and grazing control (Maxted et al, 2008).

It is important to become informed of the management interventions that are practised in the candidate protected area as these may affect the decision as to whether to select that particular area for the conservation of CWR; there may well be potential management conflicts. For example, nature reserve design and management practices that focus on the landscape level, community level or species level may conflict with one another. If the management goal is to perpetuate natural fluctuations in landscape structure, then certain species dependent on landscape structure may fluctuate as well. Maintaining stable populations of these species may entail landscape manipulations that lower the value of the reserve for perpetuating landscape processes and structures. In the majority of cases, the management plan of the protected area in which CWR are found to occur will not include specific prescriptions that will favour the conservation of individual target species.

Box 9.7 Steps to enhance the conservation role of protected areas for forest genetic resources

- collate information on tree species present in PA;
- make a comprehensive botanical inventory;
- identify priority forest and tree genetic resources;
- for each priority species determine whether there is a need for special protective and management measures;
- develop overall and individual species management plans;
- conduct focused research on target species;
- implement species management plans;
- monitoring and detailed survey of priority species;
- review management plan(s).

Source: Thomson and Theilade, 2001

It may be possible, in some cases, to enhance the capacity of a protected area to protect target species, subject to the degree of flexibility of the management plans for the area and the willingness of the protected area manager to undertake the required actions.

In the case of forest genetic resources, the sequence of stages that may be followed so as to achieve this improved conservation capacity is given by Thomson and Theilade (2001) and apply equally well to other target species, including CWR (Box 9.7). The principles of genetic conservation in tropical forest management are analysed in detail by Kemp et al (1993).

It is often assumed that once the protected area in which the target CWR occurs has been selected and the management needs of the target species have been decided, it will simply be a matter of persuading the protected area manager to amend the area's management plan accordingly. This is by no means certain and often PA managers are resistant to such proposed changes for a variety of reasons.[1] Managers tend to be generalists and are interested in matters that relate to the current concerns and issues in their park. The distribution of genetic variation among the populations of a target species is unlikely to have much management relevance unless the area was set up with the needs of the target species specifically in mind. Consequently, in conserving CWR the project team will need to review the effectiveness of the protected areas in which it is planned to undertake conservation of targeted CWR, examine their current management policy and governance, and engage with protected area managers to assess what changes are needed to favour the maintenance of viable populations of these CWR and negotiate for the introduction of specific management interventions to achieve this. Of course, it may not prove possible to come to a satisfactory arrangement and a dedicated genetic reserve may have to be established if the circumstances and resources permit.

In order to undertake targeted management of CWR within protected areas, an assessment will need to be made of the changes to existing PA management plans that are required to favour the maintenance of healthy populations of CWR (i.e. targeted management) and allow implementation of specific management interventions to ensure the survival of the populations of the target species. Then negotiations will have to be entered into with protected area management to allow such interventions to take place.

In the very small number of cases where reserves have been established primarily to conserve the genetic resources of CWR, as in the case of the Erebuni Reserve, Armenia, the management plan of the reserve and management needs of the CWR may coincide to some degree; although, this depends on the size of the area and the number of populations of the target CWR being considered for management. The management plan of the protected area is primarily concerned with maintaining the integrity, functioning and health of the area as a whole, while

Box 9.8 Species management actions in Erebuni Reserve Management Plan

Action	*Methodology*	*Timescale*
Collecting biodiversity data	Field surveys to collect herbarium specimens, living material or any other data regarding plants and animals of the reserve, including information on their distribution.	2008–2009
Creating updated maps of the distribution of CWR of the reserve	Field surveys to identify biological characteristics of the species of interest and collect data on their distribution.	2008–2009
Estimating resources of CWR	Field surveys to collect resource data of crop wild relatives.	2008–2009
Creating maps of the flora of the reserve	Field surveys for collecting specimens and distribution data with subsequent identification of collected material in the lab.	2008–2012
Creating maps of the fauna of the reserve	Field surveys for collecting specimens and distribution data with subsequent identification of collected material in the lab.	2008–2012
Creating a database to store information about the reserve	Developing a database to store information regarding current state, scientific, economical and social values, qualitative and quantitative characteristics of biodiversity components.	2010–2012

Source: Erebuni State Reserve Management Plan, 2007 – developed by the Institute of Botany, Armenian Agrarian University, Yerevan State University and Jrvegh Reserve Park Complex; http://cwr.am/index.php?menu=output

the species management plan is directed at the maintenance and survival of viable populations of CWR.

The Erebuni State Reserve is one of the very few specifically established for the conservation of wild relatives of cereal crops. It is characterized by the presence of wild wheats (*Triticum*), Vavilov's rye (*Secale vavilovii*), wild barley (*Hordeum*), *Amblyopyrum muticum*, goatgrasses (*Aegilops*), along with their rich inter-specific diversity. The main issue here is whether it is possible to conserve more than one CWR in the reserve without, in effect, creating a separate management regime for each of the species. It will be interesting to see how this works out in practice. A fully detailed management plan for the Erebuni Reserve has, in fact, been prepared and its action plan includes both habitat and species management actions (see Box 9.8).

Box 9.9 Central Asia Transboundary Biodiversity Project

The World Bank is currently developing a transboundary project in the West Tien Shan Mountains of Central Asia. The Central Asia Transboundary Biodiversity Project currently comprises four discontinuous protected areas in a three-country transborder region of the Kyrgyz Republic, Kazakhstan and Uzbekistan as follows:

- Aksu-Djabagly Reserve: Kazakhstan (IUCN Category Ia, 8575ha) (juniper forests, steppe and meadows);
- Sary Chelek Reserve: Kyrgyz Republic (IUCN Category Ia, 2390ha) (juniper forests with walnut, spruce, fir apple);
- Besh Aral: Kyrgyz Republic (IUCN Category Ia, 6329ha) (juniper forests, steppe and meadows);
- Chatkal Reserve: Uzbekistan (IUCN Category Ia, 3570ha) (juniper and tugai forests, steppe and meadows).

Discussions are underway on an interstate agreement for a West Tien Shan transboundary conservation area. Additional improvements to the network are underway, focusing on improving coverage of representative habitats and connectedness.

The West Tian Shan mountains house unique stands of walnut (*Juglans regia*) forest, wild ancestors of cultivated fruit-bearing species such as apple, pear, pistachio and almond, as well as medicinal plants and many endemic plant species.

Support is being provided to the four key protected areas through a mix of investments in capacity building (including training, transport, communications and infrastructure), community awareness and education, and research and monitoring. The project has established new technical standards for protected area management and methods for involving local communities. A small grants programme provides financial and technical assistance to buffer zone communities and community-based organizations to finance demand-driven activities in sustainable agriculture, alternative livelihoods and alternative energy systems.

Source: http://www.tbpa.net/case_07.htm

Transboundary protected areas

Some CWR have populations that occur in adjacent reserves in more than one country or administrative district in a country. Such areas are known as **transboundary protected areas (TBAs)**. They are defined by IUCN as:

> *an area of land and/or sea that straddles one or more borders between states, sub-national units such as provinces and regions, autonomous areas and/or areas beyond the limit of national sovereignty or jurisdiction, whose constituent parts are especially dedicated to the protection and maintenance of biological diversity, and of natural and associated cultural resources, and managed cooperatively through legal or other effective means.*

In such areas, the interests and concerns of the different countries or administrations may be taken into account through their representation on their steering or management committees. The level of cooperation varies widely and a set of good practice guidelines has been proposed by the IUCN World Commission on Protected Areas (WCPA).[2] An example of a TBA that contains CWR is the Central Asia Transboundary Biodiversity Project (Box 9.9).

A project to establish protected areas to conserve biodiversity in the Javakhq border region of Armenia with Georgia and Turkey is planned to link up with a similar project in Georgia in order to form a transboundary cooperative arrangement (Box 9.10).

Enhancing protected areas for conserving forestry genetic resources

In some cases it will be possible to enhance the capacity of protected areas to protect target species, provided the management plans for the areas permit this. In the case of forest genetic resources, the sequence of stages that may be followed to achieve this improved conservation capacity is presented in a review by Thomson and Theilade (2001) (see also Box 9.11):

- broadening participation in design of protected area management plans and expanding the range of issues addressed by those plans;
- elaborating the management objectives to include the full scope of conservation of biological diversity and genetic resources;
- improving management and monitoring of protected areas;
- enhancing the ecological and social value of protected areas through land purchase and zoning outside the protected area;
- identifying, securing and developing new sources of financing for protection and management; and
- providing financial incentives for conservation on adjacent private lands.

Box 9.10 Establishment of protected areas in Armenia's Javakhq (Ashotsk) border region

A project has been developed by WWF-Germany, WWF-Armenia and the WWF Caucasus Programme Office. In September 2007, KfW (the German Development Bank) and the Armenian Ministry of Nature Protection granted WWF the task of implementing the project in close collaboration with the Ministry. The project aimed to conserve the unique biodiversity of the Javakheti-Shirak plateau in Armenia along the border to Georgia and Turkey and, at the same time, enhance sustainable rural development in the northern Shirak region through establishment of the Lake Arpi National Park and implementation of a support zone programme, targeting around 15 villages. The project will explore new development opportunities in the region linked to summer and winter tourism, alternative energy production and climate change, but it will also explore how more traditional land-use activities can fit into a more dynamic future perspective. The overall budget of the project is €2.2 million. The project will also promote the area internationally. To ensure a connection to the local and regional agenda, a project implementation unit (PIU) has been established in the town of Gyumri – capital of the Shirak region. A regional advisory council, with representatives from four Ministries, the Shirak region, more than 15 communities, and other national and international stakeholders, serves as a reference body for the planning and implementation of the new national park and the support zone developments.

The Javakheti-Shirak plateau in Armenia is part of a large high mountain plateau of volcanic origin with mountain steppes, subalpine grasslands as well as lakes and wetlands. Due to its uniqueness in the Caucasus, the plateau was selected as a priority conservation area in an ecoregional conservation plan for the Caucasus launched at a Ministerial Conference in March 2006, with the governments of Armenia, Azerbaijan, Georgia and Germany.

The Javakheti-Shirak ecosystem is recognized as a globally important area for birds, reptiles and plants, of which several are listed as endangered in the IUCN Red Data Book. Preserving this unique ecosystem calls for a coordinated approach to nature conservation and management across national boundaries, accompanied by sustainable development measures for local people.

WWF has also been asked to implement a similar project on the Georgian side of the Javakheti-Shirak region, which creates interesting opportunities for synergy, learning and cooperation across the border. A transboundary cooperation board, with representatives from both countries, will be asked to facilitate the collaboration, supported by the Transboundary Joint Secretariat for the Southern Caucasus, with offices in all three countries, including Armenia.

Source: http://www.panda.org/who_we_are/wwf_offices/armenia/newsroom/?123460/
Lake-Arpi-National-Park-Bringing-welfare-to-people-and-nature-in-the-northern-Shirak-Region-Armenia

Box 9.11 Main steps in planning a programme to conserve the genetic resources of a particular tree species

- Set overall priorities, i.e. identification of genetic resources at the species level based on their present or potential socioeconomic value and their conservation status.
- Determine or infer the genetic structure of the priority species at the landscape level.
- Assess the conservation status of the target species and their populations.
- Identify specific conservation requirements or priorities, typically at the population level for single species and at the ecosystem level for groups of species, i.e. identify geographical distribution and number of populations to be conserved.
- Identify the specific populations to be included in the network of *in situ* conservation stands.
- Choose conservation strategies or identify conservation measures.
- Organize and plan specific conservation activities.
- Provide management guidelines.

Source: Graudal et al, 2004

Further sources of information

Chape, S., Spalding, M. and Jenkins, M. (eds) (2008) *The World's Protected Areas,* Prepared by the UNEP World Conservation Centre, University of California Press, Berkeley.

Iriondo, J.M., Maxted, N. and Dulloo, M.E. (eds) (2008), *Conserving Plant Diversity in Protected Areas,* CAB International, Wallingford, UK.

Maxted, N., Kell, S., Ford-Lloyd, B. and Stolton, S. (2010) 'Food stores: Protected areas conserving crop wild relatives and securing future food stocks', in S. Stolton and N. Dudley (eds) *Arguments for Protected Areas: Multiple Benefits for Conservation and Use,* Earthscan, London.

Stolton, S., Maxted, N., Ford-Lloyd, B., Kell, S.P. and Dudley, N. (2006) *Food Stores: Using Protected Areas to Secure Crop Genetic Diversity,* WWF Arguments for Protection series, WWF, Gland, Switzerland.

Notes

1. Maxted and Kell (2009) are overoptimistic in their claim that it is relatively easy to amend the existing site management plan to facilitate genetic conservation of CWR species. While the changes required may, in some instances, be minor, getting them approved and implemented may prove difficult, if at all possible.
2. Sandwith et al, 2001.

References

ACIUCN (2002) *Australian Natural Heritage Charter* (2nd edition) Australian Heritage Commission, in association with the Australian Committee for the International Union for the Conservation of Nature (ACIUCN), Canberra

Amend, T., Brown, J., Kothari, A., Phillips, A. and Stolton, S. (eds) (2008) *Protected Landscapes and Agrobiodiversity Values,* Vol 1 in the series Protected Landscapes and Seascapes, International Union for Conservation of Nature (IUCN) and Deutsche Gesellschaft für Technische Zusammenarbeit (GTZ), Kasparek Verlag, Heidelberg, Germany, p139

Chape, S., Spalding, M. and Jenkins, M. (eds) (2008) *The World's Protected Areas,* Prepared by the UNEP World Conservation Centre, University of California Press, Berkeley, CA, USA

Dudley N. (ed) (2008) *Guidelines for Applying Protected Area Management Categories,* International Union for conservation of Nature (IUCN), Gland, Switzerland.

Dudley, N. (ed) (2009) *Guidelines for Applying Protected Area Management Categories,* International Union for Conservation of Nature (IUCN), Gland, Switzerland

Gardner, C.J., Ferguson, B., Rebara, F. and Ratsifandrihama, A.N. (2008) 'Integrating traditional values and management regimes into Madagascar's expanded protected area system: The case of Ankodida', in J.-M. Mallarach (ed) *Protected Landscapes and Cultural and Spiritual Values,* Vol 2 in the series Values of Protected Landscapes and Seascapes, IUCN, GTZ and Obra Social de Caixa Catalunya, Kasparek Verlag, Heidelberg, Germany

Graudal, L., Yanchulk, A. and Kjaer, E. (2004) 'Chapter 3: National planning', in FAO, FLD, IPGRI, *Forest Genetic Resources Conservation and Management, Vol 1, Overview, Concepts and Some Systematic Approaches,* International Plant Genetic Resources Institute (IPGRI), Rome, Italy

Heywood, V.H. and Dulloo, M.E. (2005) In Situ *Conservation of Wild Plant Species – A Critical Global Review of Good Practices,* IPGRI Technical Bulletin, no 11, FAO and IPGRI, International Plant Genetic Resources Institute (IPGRI), Rome, Italy

Iriondo, J.M., Maxted, N. and Dulloo, M.E. (eds) (2008) *Conserving Plant Diversity in Protected Areas,* CAB International, Wallingford, UK

Kemp, R.H., Namkoong, G. and Wadsworth, F.M. (1993) *Conservation of Genetic Resources in Tropical Forest Management: Principles and Concepts,* Food and Agriculture Organization of the United Nations (FAO) Forestry Papers 107, FAO, Rome, Italy

Mallarach, J.-M. (ed) (2008) *Protected Landscapes and Cultural and Spiritual Values,* Vol 2 in the series Values of Protected Landscapes and Seascapes, IUCN, GTZ and Obra Social de Caixa Catalunya, Kasparek Verlag, Heidelberg, Germany

Maxted, N. and Kell, S.P. (2009) *Establishment of a Global Network for the* In Situ *Conservation of Crop Wild Relatives: Status and Needs,* FAO Commission on Genetic Resources for Food and Agriculture, Rome, Italy

Maxted, N., Dulloo, M.E., Ford-Lloyd, B.V., Iriondo, J. and Jarvis, A. (2008) 'Gap analysis: A tool for complementary genetic conservation assessment', *Diversity and Distributions,* vol 14, no 6, pp1018–1030

Phillips, A. (2002) *Management Guidelines for IUCN Category V Protected Areas: Protected Landscapes/Seascapes,* no 9, IUCN, Gland, Switzerland and Cambridge, UK

Sandwith, T., Shine, C., Hamilton, L. and Sheppard, D. (2001) *Transboundary Protected Areas for Peace and Cooperation,* IUCN, Gland, Switzerland and Cambridge, UK

Spalding, S. and Chape, M. (2008) in S. Chape, M. Spalding and M. Jenkins (eds) *The World's Protected Areas*, Prepared by the UNEP World Conservation Centre, University of California Press, Berkeley, CA, USA

Thomas, L. and Middleton, J. (2003) *Guidelines for Management Planning of Protected Areas,* International Union for Conservation of Nature (IUCN), Gland, Switzerland and Cambridge, UK

Thomson, L. and Theilade, I. (2001) 'Protected areas and their role in conservation of forest genetic resources', in FAO, DFSC and IPGRI (eds) *Forest Genetic Resources Conservation and Management, vol 2, Managed Natural Forests and Protected Areas (*In Situ), International Plant Genetic Resources Institute, (IPGRI), Rome, Italy

WWF (2004) *How Effective are Protected Areas?* Preliminary analysis of forest protected areas by WWF – the largest ever global assessment of protected area management effectiveness. Report prepared for the Seventh Conference of the Parties of the Convention on Biological Diversity, February 2004, World Wide Fund for Nature (WWF), Gland, Switzerland.

Chapter 10

Species and Population Management/Recovery Plans

For most wild species the best that we can hope for is to establish and monitor their presence in some form of protected area where, provided the area itself is not under threat, and subject to the dynamics of the system and the extent of human pressures, some degree of protection may be afforded. We are a long way from achieving even this. Moreover, the fact is that most species currently (and for the foreseeable future) occur outside currently protected areas (Heywood, 2005).

Introduction: The aims and purpose of species management or recovery plans

The actions taken to ensure the maintenance of viable populations are at the core of targeted *in situ* conservation of species and are referred to as species management, action, conservation or recovery plans, depending on the degree of intervention required, which will, in turn, reflect the conservation status of the species concerned. Many conservationists (e.g. Sutherland, 2000) regard species management as a confession of failure – that is, failure to provide appropriate habitat management or control of threats such as wild harvesting or impacts of invasive species. Indeed, as already noted in Chapter 3, if a species is not threatened or endangered, little or no management intervention may be needed; provided the habitat is secure, only monitoring of the area and of the status of the populations will normally be necessary. In such cases, a *species conservation statement* may be made, summarizing the situation (such as the species statements of the UK Biodiversity Action Plan). However, given the continuing pressure on habitats caused by human demographic growth and the consequential need to expand agriculture to feed the growing population, by industrial and building development, the growing threats from invasive species and the impacts of accelerated climate change, it is highly likely that many species that are today regarded as safe will become threatened.

The Species Survival Commission of the International Union for Conservation of Nature (IUCN) has published a handbook on strategic planning for species conservation, primarily intended to provide guidance to IUCN/SSC specialist groups on when and how to prepare and promote species conservation strategies (SCSs). A *species conservation strategy* is defined as a blueprint for saving a species or group of species, across all or part of the species' range. A SCS should contain a status review, a vision and goals for saving the species, objectives that need to be met to achieve the goals, and actions that will accomplish those objectives (IUCN/SSC, 2008). Although largely animal-oriented, this handbook contains much of relevance to CWR conservation. In particular, it adopts a multi-stakeholder participatory approach as recommended in this manual.

As already discussed in Chapter 3, many conservationists and policy-makers would argue against a species-based approach to conservation, largely on the grounds that there are so many species requiring attention that such an approach would not be cost-effective. On the other hand, in many circumstances – and CWR are a case in point – a focus at the species/population level is both deliberate and unavoidable (Kell et al, 2008; see also Box 10.1) whether the species is threatened or not. For CWR, as discussed in Chapter 7, priority may often be given to species that are threatened; in such cases, management interventions will logically address the threats and threat management will be a major component of any management or conservation plan. Given that there is such a high number of CWR and a high probability that many of them are threatened to some degree, it might appear that there is no place for taking conservation action for species that

Box 10.1 The future of species conservation

Paradigms of ecosystem services, pro-poor conservation and rights-based approaches to conservation are taking centre stage, but these approaches all call for continued attention to the fundamental role that species play in underpinning those paradigms. In the brave new world of conservation, species approaches remain core business. We must continue to use all the tools in the species conservation toolbox, from development and implementation of species action plans to reintroduction, *ex situ* management and more.

In the coming decade, no species should knowingly be allowed to become extinct. The conservation community should continue to contribute to monitoring and assessment of status and threat trends in species, including support for indicator development and reporting. Working towards a better understanding of the parameters defining 'sustainable use' of species and encouraging managers of those species to make use of that knowledge will be vital. Similarly, the conservation world should promote all possible efforts to manage and control invasive species.

Source: McNeely and Mainka, 2009

are not threatened. On the other hand, a case can be made for ensuring the future survival of CWR, judged to be of high priority (see Chapter 7), even if the species is not currently threatened, by establishing a genetic reserve.

Species conservation or recovery plans?

The difference between species conservation/action/management plans and recovery plans is a matter of scale and degree, and reflects the extent of management intervention needed (Lleras, 1991).

For species that are not currently threatened or are estimated to have a low probability of extinction, little conservation action is likely to be needed other than to monitor their habitat and populations so that further action can be taken should the situation deteriorate. A species conservation or action plan will not normally be proposed unless the species is regarded for other reasons to be of such high priority that, for example, the setting up of a reserve for it is justified. For CWR that fall into this category that do not occur within protected areas, the setting up of a reserve or a series of reserves while the species still maintains its full range of genetic variability would be appropriate.

For species that are threatened to some extent but are not currently endangered, the removal, mitigation or containment of the factors causing the threat means that some form of intervention is necessary. In such cases a species conservation or action plan will be appropriate, including the setting up of a reserve or some off-site arrangement (see Chapter 11) if the species does not occur in a protected area.

For species that are currently endangered and have already suffered severe population loss or are in rapid decline so that partial or total extinction is likely within decades, a species recovery plan is the appropriate action.

Is the *in situ* conservation of CWR different from that of other wild species?

Another critical issue is whether the nature of CWR changes the focus and methods of *in situ* conservation, i.e. is the aim of genetic conservation of CWR different from that of other species? What is specific about it? As discussed in Chapter 2, the terms genetic conservation or genetic reserve conservation are often used in the case of CWR because of the focus on the maintenance of the genetic diversity in the target species that may be of actual or potential use in plant breeding and improvement and making it available (Maxted et al, 1997; Iriondo and De Hond, 2008). To achieve this, the following actions have been suggested by Iriondo and De Hond (2008):

- minimize the risk of genetic erosion from demographic fluctuations, environmental variation and catastrophes;
- minimize human threats to genetic diversity;
- support actions that promote genetic diversity in target populations;
- ensure access to populations for research and plant breeding;

- ensure availability of material of target populations that are exploited and/or cultivated by local people.

The concept of genetic reserve conservation (see Chapter 3; Maxted et al, 2008) is considered to be one of the major differences between species management plans for CWR and other wild plants. In practice, however, the distinction breaks down and the difference is largely one of objectives or motivation rather than practice. In all cases of *in situ* species conservation or recovery of a wild species, the aim must be to ensure the species' survival and this requires that as much genetic variation as possible be maintained; in this respect there is nothing intrinsically different about CWR conservation or about a genetic reserve. It is primarily the use that may be made of the genetic diversity of the CWR that distinguishes a genetic reserve and, in deciding on the location of areas to be set aside as genetic reserves, the set of populations that maximizes the representation of genetic diversity, both within-population and between-population, should be selected (Maxted et al, 2008). The same considerations also apply to reserves for other target species such as medicinal plants. The management plans for CWR are essentially the same as those for other wild species although actions may be included that are directed at maintaining or enhancing particular sectors of genetic variation within populations, again as would apply to conserving medicinal plant species.

Experience derived from species recovery programmes

Until recently, our experience of targeted *in situ* species conservation has, in fact, mainly been gained from the extensive programmes of recovery plans for threatened or endangered wild species undertaken by a number of European countries (including EU LIFE-Nature projects), Australia, New Zealand, and the US (Boxes 10.3 and 10.4). This has been underpinned by extensive research on conservation biology and conservation genetics (e.g. Simmons et al, 1976; Synge, 1981; Falk and Holsinger, 1991; Bowles and Whelan, 1994; Frankel et al, 1995; Falk et al, 1996; Reynolds et al, 2001).

The United States Fish and Wildlife Service Endangered Species Recovery Program[1] is the largest of these and works in partnership with federal, state and local agencies, tribal governments, conservation organizations, the business community, landowners and other concerned citizens. It has also established a national partnership with the Center for Plant Conservation, which is primarily devoted to *ex situ* conservation, although several of its member gardens are engaged in restoration and recovery actions (see Box 10.2). This programme, along with 27 other federal agencies and most state agencies, reported their expenditures for federally protected species in the 2007 fiscal year: the total expenditure reported was US$1.66 billion, of which US$1.57 billion was reported by federal agencies and US$95.3 million was reported by state agencies.

In the majority of cases, these recovery plans do not refer to species of agrobiodiversity interest, and the focus is not so much on genetic conservation as on survival and recovery of viable populations. The genetic resources sector has focused its attention mainly on *ex situ* conservation until recently, and its involve-

Box 10.2 Center for Plant Conservation (CPC)

Founded in 1984, the Center is dedicated solely to preventing the extinction of native plants in the US. It is supported by a nationwide consortium of 36 leading US botanic institutions, gardens and arboreta. With about one in every ten plant species in the US facing potential extinction, the Center is the only national organization dedicated exclusively to conserving *ex situ* material. Live plant material is collected from nature under controlled conditions and then carefully maintained as seed, rooted cuttings or mature plants. The collection contains more than 600 of America's most endangered native plants and ensures that material is available for restoration and recovery efforts for these species. Network institutions conduct horticultural research and carefully monitor these materials so that endangered plants can be grown and returned to natural habitats. Several CPC institutions are also involved in restoration projects in the field (*in situ*). Scientists are stabilizing current populations of threatened plants and reintroducing new populations in appropriate habitats.

Source: http://www.centerforplantconservation.org

ment in *in situ* conservation has been largely in the area of 'on-farm' conservation of landraces. Its limited involvement in genetic conservation of CWR has not, until very recently, taken this experience of recovery planning into account.

Likewise, the extensive experience of the forestry sector in *in situ* conservation has not been fully acknowledged. The challenge for those involved in CWR conservation is to draw on this accumulated experience and adapt it to the special requirements of genetic conservation.

A detailed global survey of *in situ* conservation of wild species (Heywood and Dulloo, 2005) revealed, not surprisingly, that very few species recovery or management plans have been prepared or implemented for tropical species, highlighting the enormous gulf that exists been actions to conserve tropical and temperate species. Some of the management plans that have been implemented in the tropics are aimed at making sustainable resource extraction economically viable, and improving the economic conditions of the local families involved, rather than at conservation as such, as in the case of a recent project in Peru's Pacaya Samiria National Reserve, involving community-based resource management of palms and aquatic resources. Management plans were created for enmoriche palms (*Mauritia flexuosa*), yarina palms (*Phytelephas macrocarpa*) and huasaí palms (*Euterpe precatoria*) and addressed deleterious harvest practices. Not only did the implementation of the management plans lead to improvements in the availability of resources, but there is strong evidence to suggest that they have helped the recovery of the species concerned (Gockel and Gray, 2009). With an increasing focus on community-based conservation and sustainable use, such examples are likely to become more common, but they do not alter the imbalance between targeted *in situ* species conservation in the tropics and temperate regions. This needs to be addressed as an urgent priority, although there are few indications that there is any political will to do so. In the particular case of CWR, many,

Box 10.3 Species recovery in New Zealand

The New Zealand Biodiversity Strategy (NZBS) 2000 funding package committed NZ$16.5 million (US$11.5 million) between 2000 and 2005 for the Department of Conservation's work on species recovery programmes and mainland islands. This work is focused on enhancing the recovery of threatened indigenous plant and animal species in coastal, land and freshwater ecosystems and will be achieved through intensive management of both threatened species and predators. This work addresses two of the main themes of the NZBS: (i) to ensure that a net gain has been made in the extent and condition of natural habitats and ecosystems important for indigenous biodiversity; and (ii) to ensure populations of all indigenous species and subspecies are sustained in natural or semi-natural habitats, and their genetic diversity is maintained.

Specific objectives

The specific objectives of the programme are to:

- expand freshwater fish, plant, invertebrate and reptile and amphibian recovery work;
- improve planning for priority species;
- provide technical support through the development of new management techniques and databases.

Source: http://www.biodiversity.govt.nz/land/nzbs/habitat/species/index.html

if not most, of them are not charismatic or flagship species and are unlikely to attract public interest or concern.

Species recovery plans

Given the extensive experience available in preparing and implementing recovery plans and because they are essentially a form of management plan, they are considered here in some detail.

Recovery is the process by which the decline of an endangered or threatened species is arrested or reversed and threats removed or reduced so that the species' long-term survival in the wild can be ensured. In terms of the conservation of CWR, Iriondo et al (2008) consider recovery as broadly referring to 'the act of assisting populations of plant species or habitats in the process of returning from a non-self-sustaining (or unstable) state to a self-sustaining (or stable) one'. The restoration or rehabilitation of habitats (also known as revegetation or reclamation) is a major and highly complex topic that is not addressed in detail in this manual as it is unlikely to be undertaken on any substantial scale as part of an *in situ* management project for CWR species.

Recovery plans may involve both habitat recovery actions and population recovery actions. For example, habitat restoration can assist in the recovery of endangered species, some of which may require restoration of degraded habitat for their eventual recovery (Bonnie, 1999). However, these recovery actions are often challenging, costly and difficult operations that involve management actions that may need to be carried out over a number of years. They require teamwork, involving specialists from a number of disciplines as well as concerned stakeholders and the general public.

In the case of CWR genetic conservation, as Kell et al (2008) point out, the focus is on the target species with a view to conserving its variability, not on the habitat. Of course, as discussed in detail in Chapter 2, the species and habitat are intimately linked and mutually dependent. In practice, the effective conservation of any species *in situ* depends critically on identifying the habitats in which they occur and then protecting both the habitat and the species' populations through various kinds of management and/or monitoring. Thus, although *in situ* species conservation is essentially a species-driven process, it also necessarily involves habitat protection.

Consequently, the management plan of a CWR may call for some actions at the habitat level, such as ensuring its effective management (although that is essentially the responsibility of the reserve or protected area manager), weeding to remove competitors, control or removal of invasive species, control of disturbance or fencing to exclude herbivores. However, full-scale habitat or ecological restoration is not normally part of the business of CWR conservation; although, when this is carried out for other reasons, and one or more CWR are known to occur in the restored habitat, then advantage can be taken to develop an appropriate CWR species management plan, provided the conditions are appropriate and the genetic variability of the species is represented. Kell et al (2008) cite examples of habitat restoration where regeneration of the vegetation is combined with a targeted species approach. For example, in Spain on the 8-hectare island of Columbrete Grande (L'illa Grossa), the largest of the Islas Columbretes (Province of Castellón), a mixed recovery programme for habitats and rare and endangered species was started in 1994; since 1997, efforts have focused on recovery of the local endemic leguminous shrub *Medicago citrina*.

A *species recovery plan* is a document stating the research and management actions necessary to stop the decline, support the recovery and enhance the chance of long-term survival in the wild, of a stated species or community of protected wildlife. The goal is the recovery of target species to levels where protection is no longer necessary.

Species recovery plans are mainly used to:

- stabilize and halt the decline in existing populations of threatened species;
- increase, reinforce or rejuvenate existing populations through adding individuals to them (*reinforcement* or *enhancement*);
- transfer material from one part of the existing species' range to another (*translocation*);
- Reintroduce plants of endangered species to locations outside its current range, but within its historic range similar to ones where they previously existed (*reintroduction inter situs*).[2]

Reintroduction is often a controversial process because of fears that it will lead to undesired ecological or genetic consequences; it requires detailed knowledge of an ecosystem functioning on the one hand and of the biology and ecological tolerances of the species on the other. It may also face legal challenges. Reintroduction has been employed in Hawaii by the National Tropical Botanical Garden in collaboration with local landowners for the conservation of rare plant species (Burney and Burney, 2007). For a discussion of the issues, see Akeroyd and Wyse Jackson (1995) and Burney and Burney (2009). A recently proposed method of *human-assisted translocation* or *migration* as a means of responding to the problem that some species may not be able to track changing climatic conditions quickly enough is discussed in Chapter 16.

The overall objectives of a recovery plan are to prevent further loss of individuals, populations, pollinator species and habitat critical for the survival of the species; and to recover existing populations to normal reproductive capacity to ensure viability in the long term, prevent extinction, maintain genetic viability and improve conservation status. The general aim in threatened species' recovery is to establish sufficient self-sustaining healthy populations for the species to be no longer considered as threatened.

The *contents of a species recovery plan* will vary according to the circumstances but should include:

- an evaluation and description of the species' current situation, including any relevant scientific data;
- a recovery objective (for example, a target population number) and a list of criteria for indicating when the objective has been achieved;
- the detailed specific actions that will be required to secure the species;
- implementation procedures using scientific techniques;
- the organizations that will play a part in the recovery process (e.g. botanic gardens, national/regional/local conservation institutions, community bodies, etc);
- an implementation schedule, including priorities of tasks and cost estimates; arrangements for external reviews.

Of these, the first three points are essential for any species recovery plan. The assessment of the status of the CWR will have already been undertaken as part of the selection process already described in Chapter 7 and, once selected, during the ecogeographic survey (Chapter 8).

Species recovery plans vary widely in their scope and extent. Unfortunately, there are not yet any clearly established protocols for species recovery for plants and anyone planning to develop a species recovery plan for a CWR is advised to consult a range of published plans to find those most relevant to their particular species. For examples, see Box 10.4. The model used by the Australian government for recovery plans is given in Box 10.5.

A range of examples of recovery planning in Australia, where recovery plans have been used as a basis for managing a growing number of the country's threatened species since 1989, is given in Box 10.6.

Box 10.4 Examples of recovery plans

The United States Fish and Wildlife Service Threatened and Endangered Species System website lists the species for which recovery plans have been prepared: http://www.fws.gov/endangered/species/recovery-plans.html

For UK species action plans, see the UK Biodiversity Action Plan site which lists numerous examples: http://www.ukbap.org.uk/SpeciesGroup.aspx?ID=31

For the Swiss flora, summary species action/data sheets for over 140 priority species have been prepared (Fiches pratiques pour la conservation Plantes à fleurs et fougères). See: http://www.cps-skew.ch/english/data_sheets.htm; http://www.crsf.ch/index.php?page=fichespratiquesconservation

An example of a Spanish species recovery plan for *Cheirolophus duranii* (as published in the Official State Bulletin) is available at: http://www.uam.es/otros/consveg/documentos/Cheirolophus%20duranii%20Plan%20Recup.pdf

Australia: Conservation and recovery profile for *Haloragodendron lucasii*: http://www.environment.nsw.gov.au/resources/nature/tsprofileHaloragodendron Lucasii.pdf

Australia: Recovery plan for the endangered vascular plant *Alectryon ramiflorus* Reynolds: http://www.derm.qld.gov.au/register/p00174aa.pdf

Species conservation management/action plans

> *Genetic conservation plans must be firmly based on the available scientific information if they are to be the basis of effective policies and practices* (Rogers, 2002).

If the species selected as targets are found to be threatened – and about one in four plant species probably is – then the critical factor at the species or population level is to control, mitigate or eliminate the threat(s) to the populations. This must be addressed in the species management plan.

Conservation management/action plans should be prepared for those species that require some form of management intervention to ensure the continued maintenance of viable populations. As already noted, they are essentially similar to species recovery plans, but the degree or intensity of management intervention is lower, reflecting the lower degree of threat to the population(s). The detailed composition of a management plan will vary from species to species, depending on the biological characteristics of the species, its population status, the location, the aim of the plan and so forth. As pointed out by Heywood and Dulloo (2005), there is no single approach for the genetic conservation of target species that is appropriate for all situations or even generally applicable. On the other hand, Maxted et al (1997) have proposed a practical model that they consider suitable

Box 10.5 Summary of content requirements for a recovery plan of the Australian government

Part A: Species/ecological community information and general requirements

Species/community name
Conservation status/taxonomy/description of community
International obligations
Affected interests
Role and interests of indigenous people
Benefits to other species/ecological communities
Social and economic impacts

Part B: Distribution and location

Distribution
Habitat critical to the survival of the species/community
Mapping of habitat critical to the survival of the species/community
Important populations

Part C: Known and potential threats

Biology and ecology relevant to threatening processes
Identification of threats
Areas under threat
Populations under threat

Part D: Objectives, criteria and actions

Recovery objectives and timelines
Performance criteria
Evaluation of success or failure
Recovery actions

Part E: Management practices

Part F: Duration of recovery plan and estimated costs

Duration and costs
Resource allocation

Source: http://www.environment.gov.au/biodiversity/threatened/recovery.html

for widespread application; the model is being tested in several projects. Common features that should be included in a species management plan are given in Box 10.7 (see also Sutherland, 2000: Box 7.1).

As in the case of recovery plans, the three essential components are: an evaluation of the current status of the species; the aims and objectives of the plan; and the actions proposed.

It is critically important to agree on and include in the management plan a statement on what the objectives are; in other words, what it is hoped that the

Box 10.6 Species recovery planning: Some Australian case studies

Community involvement in the species recovery process: Insights into successful partnerships – Stephanie Williams

Involving the general public in the recovery of threatened species and ecological communities provides discrete short-term benefits for conservation programmes as well as long-term gains in developing social responsibility for Australia's natural heritage. Guidelines for successful engagement of the community in the species recovery process, based on personal experience, are outlined. It is suggested that government agencies provide community endeavours with honesty, support, expertise and sensitivity to the community's concerns for conservation. This will help to develop effective partnerships in species recovery initiatives.

Conservation of the endangered plant Grevillea caleyi *(Proteaceae) in urban fire-prone habitats – Tony D. Auld and Judith A. Scott*

The endangered plant *Grevillea caleyi* (Proteaceae) occurs in bush land that is adjacent to urban areas in the Sydney region. Within these areas, repeated and frequent fire threatens not only the endangered flora but life and property as well. These threats were well illustrated by the impact of the fires that occurred in January 1994 in Sydney. Management of urban fire-prone areas needs to identify those fire regimes likely to drive the endangered flora to extinction, as well as identifying if any populations of endangered flora occur in locations that pose a fire hazard for the protection of life and property. Research into the population dynamics of *G. caleyi*, as part of the development of a recovery plan for the species, indicates that a regime of frequent fire will lead to local population decline and extinction. Consequently, burning on a frequent basis for hazard reduction to protect property assets in the vicinity of *G. caleyi* is inappropriate for the conservation of this plant. Instead, a minimum fire-free interval of 8 to 12 years is recommended for the conservation of *G. caleyi*. Additionally, areas not burnt for 20 to 25 years should be monitored for adult plant survival and seedling recruitment. If all or most adults have died and there is no seedling recruitment then consideration should be given to burning such sites.

Rediscovery programme for the endangered plant Haloragodendron lucasii *– Marita Sydes, Mark Williams, Rob Blackall and Tony D. Auld*

The *Haloragodendron lucasii* rediscovery team was established to try and find new locations of this plant in the wild. Prior to the initiation of the team, only three sites were known with a total of four genetically distinct individuals. Each of these individuals is effectively male sterile. Finding more locations of this endangered plant will lead to the protection of more individuals, the possibility of discovering male fertile plants, as well as assisting the planning of conservation measures. The rediscovery team involved joint coordinating efforts by New South Wales National Parks and Wildlife Service, the Australian National University and Ku-ring-gai Council. Community involvement was encouraged through the use of volunteer groups to search for *H. lucasii* in the field.

Box 10.6 continued

Instruction to community groups involved an evening session, where the details of the recovery of *H. lucasii* and associated genetic research were discussed, through to field days where the public were shown what the plant looks like in the wild. The value of the involvement of the community groups for the rediscovery programme is highlighted by the discovery of a new location for *H. lucasii* in late September 1995.

Threatened by discovery: research and management of the Wollemi pine* Wollemia nobilis *Jones, Hill and Allen – John Benson

The discovery of the Wollemi pine *Wollemia nobilis* in 1994 not only brought to light a new genus in the Araucariaceae and a conifer with at least a 91 million-year-old Gondwanan history, it also increased the threat to the two known wild populations of 40 adults and about 130 seedlings. Although growing in an inaccessible, warm temperate rainforest-lined gorge in a large national park, the impacts of visitation, and indeed researchers, could prove costly to the species. The main threats from people are trampling of seedlings, compaction of the ground and introduction of pathogens. Another threat is wildfire, which has the potential to destroy much of the population in one catastrophic event. A range of *in situ* ecological research and *ex situ* botanical and horticultural research is being conducted on the species to aid its conservation. A species recovery plan has also been prepared. In the short term, a key research programme aims to discover the most efficient way to propagate and cultivate the species to meet market demand for garden plants. This would remove the pressure of illegal seed collection from the fragile wild populations. Since the Wollemi pine is a relic species, 'recovery' is not the question. Management should aim to maintain the current population and genetic variation. Translocation may arise as an issue in the long run, but there would need to be sound reasons for it to be undertaken.

Source: Stephens and Maxwell, 1996

management plan will achieve and how it is intended to fulfil these aims. This will reflect the key decisions made on which populations and how many will be included in the management plan and how many individuals are needed to ensure a minimum viable population. This, in turn, will depend on the distribution pattern of the species, its demography and the distribution of genetic variation within its populations. The information on the species and its status and the ecogeographical information will be available from the ecogeographic surveys already undertaken for the target species, and the threats to the species will also have been identified (Chapter 7). The actions prescribed will vary considerably from plan to plan.

In the case of a species with a narrow or restricted distribution, the aim will normally to be to include all the population(s) within the management plan. In the case of species with a wide distribution, and in which the variation is partitioned into races or ecotypes, a choice must be made as to how many populations and

Box 10.7 Common features of a species management plan

- a description of the species, including its scientific name, essential synonyms, common names, its reproductive biology, phenology and its current conservation status (see Chapter 7);
- ecogeographical information – location of the CWR populations, their habitat, ecology, soil preferences, demography size and viability, genetic variation, population viability analysis (see Chapter 8);
- the nature of the threats affecting the conservation status of the species (see Chapter 7);
- a summary of existing conservation actions that are already being undertaken and by whom;
- the objectives of the management plan;
- the detailed actions that will be required to contain, reduce or eliminate the threats and ensure the maintenance of viable populations of the species;
- the actions that may be needed to safeguard and manage the site;
- the management objective(s) and targets (both short term and long term), and a set of criteria for indicating when the objective(s) are achieved;
- a statement on how the plan will be implemented and what scientific techniques will be adopted;
- identification of any policy or legislative actions that need to be undertaken;
- identification of the lead agency or party and a list of the organizations that will play a part in the management actions (e.g. national/regional/local conservation institutions, botanic gardens, community bodies, etc);
- arrangements for negotiation with the site authorities and other interested parties or stakeholders regarding management interventions;
- an implementation schedule, including prioritization of the various actions or tasks;
- a detailed budget with annual cost estimates for the various actions involved;
- monitoring programme and schedule;
- arrangements for external reviews;
- plans for communication and publicity.

how much of the variation is to be selected for conservation and inclusion in the management plan. For example, for the Monterey pine, *Pinus radiata*, field and laboratory studies have revealed strong genetic differentiation among the five populations studied, each having some unique features, and the implications for genetic conservation, according to Rogers (2004), are that specific conservation efforts must be directed at the population (or lower) level as there is 'no "representative subset" of populations that could effectively conserve the genetic and ecological diversity of the species' (Box 10.8). This, of course, has major implications for the amount of effort, time and costs involved.

Another complicating factor that applies in wide-ranging species is that if the total range of species, or those parts of it that are critical for effective *in situ* genetic conservation, occurs in more than one jurisdiction, there will be additional management and planning challenges in dealing with the operative laws, policies

Box 10.8 Problems of genetic conservation in Monterey pine (*Pinus radiata*)

The Monterey pine is a forest tree species that is widely commercialized outside its native range. Native forests are represented by only five fragmented populations: three along the central coast of California and two on Mexican islands off the coast of Baja California.

> *Current Monterey pine protected areas have not been selected with genetic values in mind, and thus do not necessarily contain representative genetic variation, represent sufficient habitat size or effective population size, or reflect conditions that allow ongoing regeneration and adaptation. There is little information available on within-population genetic structure, but given the steep gradient expressed in various soil and microclimatic features of coastal-to-inland environments, and some indication of within-population genetic structure … it is prudent to assume that several* in situ *reserves per population would be needed to adequately conserve genetic diversity unless (yet to be collected) evidence suggests otherwise. … Thus, current protected areas are not necessarily* in situ *genetic reserves, but some may offer the potential for including genetic values in their management. More information is required to ascertain which currently protected areas may also serve as genetic conservation areas.*

Source: Rogers, 2004

and ordinances of the different jurisdictions' planning cycles, even assuming that all parties agree on the need for coordinated conservation action (Rogers, 2004). In the case of the Monterey pine, just for the three Californian populations, the ownership and management was very diverse, 'including federal, state, county, and city governments; land trusts; universities and other nongovernmental organizations; and private owners (including home owners with some Monterey pine habitat, ranchers, forest companies, and recreation-oriented businesses)'.

A management plan may be concise and just a few pages long or extensive and up to 100 pages or more (see Box 10.4 for examples), depending on the range of activities involved. Ideally, plans should contain photographs or other illustrations of the plant and its habitat, maps and other graphic material. In some countries plans must be published officially once approved – for example, the recovery plan (Plan de Recuperación) for *Crambe sventenii*, *Salvia herbanica* and *Onopordon nogalesii* was published in the Boletín Oficial de Canarias, 5 February 2009 (Nbr. 024) by DECRETO 8/2009. They are occasionally published in journals (e.g Bañares et al, 2003) or as free-standing publications (e.g. the Recovery plan for *Silene hifacensis*, published as a booklet by the Environment Agency of the government of Valencia, Spain (Conselleria de Medi Ambient, Aigua, Urbanisme i Habitatge 2008)).

The successful implementation of a management plan may take many years to achieve and it is usual to include short-, medium- and long-term objectives.

Species management versus area management

Although this has been discussed in detail in Chapter 3, it is important to reiterate that effective *in situ* conservation of a target species is, on the one hand, dependent on the secure and effective management of the area(s) in which the species occurs and, on the other, requires management interventions at the population/species level different from those needed to maintain the area(s); these interventions may even be in conflict with the management policy of the area(s). Thus, a distinction must be made between protected area management plans and species management plans. Both are needed to achieve the successful *in situ* conservation of species or their populations. If the protected area in which a species occurs is extensive and several to many populations occur within it, management of the area and management of the species will most likely require quite different actions and management plans. If, on the other hand, the area is small with only one or two populations, the species and area management requirements will probably coincide to a considerable extent, and it should be relatively easy to make any changes to the area management plan as required, provided the area management authority agrees (see Chapter 9).

It also needs to be re-emphasized that if the target species is threatened, its presence in a protected area will not, in itself, ensure its protection unless the factors causing it to be threatened are addressed.

Single-species versus multi-species plans

One of the basic decisions that must be made in genetic conservation is whether to plan for the conservation of single species or multiple species. Genetic reserve conservation (Chapter 3), as practised so far,[3] has tended to focus more on groups of species occurring together in selected areas rather than on a single target species, largely on the grounds of cost-effectiveness, given that the number of target species is likely to exceed available resources for a species-by-species approach. This parallels the multi-species approach recently adopted for recovery programmes by Australia, Canada, the US and some European Union countries (through the EU Habitats Directive), although previously the single-species approach has been the norm.

The scientific rationale behind the use of multi-species plans is based mainly on the assumption that the target species share the same or similar threats. While the effectiveness of multi-species recovery conservation programmes for CWR has yet to be sufficiently assessed, there is evidence from surveys of multi-species plans for wild species undertaken in Australia, Canada and the US, that insufficient attention to detail is given to individual species within multi-species plans; for these plans to be effective, as much effort must be given to each species as in a series of single-species plans. One report found that nearly half of the multi-species plans failed to display threat similarity greater than that for randomly selected groups of species. The report concluded that, as currently practised, multi-species recovery plans are less effective management tools than single-species plans (Clark and Harvey, 2002). Multi-species planning can be a very complex, time-consuming and expensive process (Canadian Wildlife Service, 2002) and the effectiveness of multi-species plans may be limited because less money and effort is spent per species (Boersma et al, 2001) and they are often poorly resourced as compared with single-species plans.

The advantages of multi-species approaches are summarized in Box 10.9. Comparisons of the strengths and weaknesses of multi-species and ecosystem-based approaches to recovery planning have been made by several authors such as Clark and Harvey (2002), Hoekstra et al (2002), Sheppard et al (2005: Table 1) and Moore and Wooller (2004: Table 3.14). As Kooyman and Rossetto (2008) note, some of the key problems in implementing multi-species plans are:

- they are less likely than single-species plans to include species-specific biological and ecological information, and adaptive management criteria;
- the lumping of species does not appear to be based on any biologically logical criteria (i.e. similarity of habitats or threats);
- multi-species plans have fewer recovery tasks implemented during the life of the plan; and
- species included in multi-species plans have been found to be four times less likely to exhibit positive status trends.

There is too little experience in the case of CWR conservation to judge the relative effectiveness of single- versus multi-species approaches but there is no reason to believe that it will differ significantly from what has been found for other examples of threatened wild species.

Box 10.9 Strengths of multi-species approaches

Multi-species approaches can:

- address common threats in a concise and focused manner (Boyes, 2001);
- streamline the public consultation process;
- reduce duplication of effort in describing the habitats of, and threats to, each species;
- provide a good format for environmental impact statements;
- promote thinking on a broader scale;
- reduce conflicts between listed species occurring in the same area;
- benefit other species not at risk;
- provide an approach that can restore, reconstruct or rehabilitate the structure, distribution, connectivity and function upon which a group of species depends.

Source: Canadian Wildlife Service, 2002

Stakeholders

The successful preparation and implementation of a management plan will involve a wide range of stakeholders. Just as in the creation of a protected area, the local population must be fully consulted and involved so that their interests and concerns are taken into account, considering that the formulation of a species management plan will affect the way in which the area is managed[4] and possibly access to populations of the target species and restrictions on their use. As already noted, the increasing focus on community-based conservation initiatives reinforces the emphasis on the requirement of the broad-scale participation of those most affected by conservation and management interventions.

Species management plans prepared by the UNEP/GEF CWR Project countries

The main source of problems faced by the countries in preparing management plans was the almost total lack of previous experience in this area. Not only had no species management plans been prepared before the initiation of the CWR Project, but knowledge of what was involved was lacking and there was a general failure to appreciate the distinction between preparing a protected area management plan and a species management or recovery plan. Such confusion is widespread and there was little available literature until very recently to give any guidance.

A fully detailed management plan for the Erebuni Reserve has, in fact, been prepared and its action plan includes both habitat and species management actions (see Chapter 9, Box 9.8).

A management plan for the selected priority cereals (*Triticum boeoticum*, *T. araraticum*, *T. urartu*, *Aegilops tauschii*) has been developed. The following state agencies participated in the development process: Ministry of Nature Protection (GEF and CBD focal point agency), Ministry of Agriculture, Institute of Botany, Yerevan State University and Armenian Agrarian University. All the main institutions involved in conservation activities in Armenia were contacted to nominate experts who could be engaged in the development process. There were a number of meeting sessions before and during the preparation process of the plan. A draft was sent for comment to the aforementioned institutions and the feedback received was discussed with project partners. The draft plan was also presented through Aarhus Convention Centres[5] in Armenia to local communities. An outline of the management plan is given in Table 10.1.

Sri Lanka has prepared a species management plan for *Cinnamomum capparu-coronde* in the Kanneliya Forest Reserve (see Chapter 9).

Uzbekistan has developed a management plan for *Amygdalus bucharica* within the protected territory of Chatkal State Biosphere Reserve. There were no problems with implementation of this plan in the protected territory. The reserve administration is cooperating as a partner and has agreed to include the

Table 10.1 *Outline content of the Management Plan for* In Situ *Conservation of* Triticum boeoticum, T. araraticum, T. urartu *and* Aegilops tauschii *in Armenia*

1 Introduction
2 Description
  2.1 Morphological characteristics of *Triticum urartu, T. boeoticum, T. araraticum, Aegilops tauschii*
  2.2 Taxonomy of the target species
  2.3 Current distribution (in the country, inside and outside of protected areas; distribution maps and any other relevant information)
  2.4 Habitat and ecology
  2.5 Biological characteristics (life cycle, life form), seed characteristics, phenology, pollination, dispersers, pest and diseases
  2.5 Conservation status
3 Evaluation
  3.1 Importance
    3.1.1 Cultural value of the CWR for local community
    3.1.2 Potential value of the CWR for research, breeding or other functions
  3.2 Threats
    3.2.1 For conserved population in Erebuni Reserve
    3.2.2 Outside protected areas
      3.2.2.1 Land privatization
      3.2.2.2 Uncontrolled grazing and hay harvesting
      3.2.2.3 Road construction
      3.2.2.4 Industrial and agricultural waste pollution
4 Identification of stakeholders
5 Goals/objectives
6 Management of threats
7 Strategic actions
8 Actions to ensure protection in protected area(s)
9 Actions to ensure protection outside protected areas
10 Improvement of *ex situ* collections
11 Research and monitoring
12 Public awareness and education
13 Action Plan (2009 to 2013); the management plan for wild wheats in the Erebuni State Reserve is available on the Crop Wild Relatives portal at: http://www.cropwildrelatives.org/index.php?id=3263

management plan developed in the frame of the CWR Project into the management plan of the reserve.

Management plans for walnut, pistachio and apple tree for insufficiently protected territories of the Ugam-Chatkal National Park are being developed. As they become available in English the management plans developed by each country will be made available through the CWR Global Portal at: http://www.cropwildrelatives.org/index.php?id=3263.

Conclusions

To date, few species management plans have been prepared or implemented for CWR. We have to rely mainly on the extensive experience that has been gained from the recovery plans for endangered wild species that have been prepared in a number of countries, mostly, however, in the temperate world.

Although the aim and focus of conserving CWR *in situ*, sometimes termed genetic conservation, is on maintaining the genetic diversity in the species for use in plant breeding, management or recovery plans for CWR are essentially similar to those for other wild species. Globally, very few such plans have been made for CWR and no specific, generally agreed protocols are yet available.

The level of management intervention required will depend on the status of the CWR in question, ranging from little or no intervention other than monitoring, in the case of species that are not currently at risk, to full-scale recovery, for species that are critically endangered and in rapid decline.

A critical decision that has to be made is whether to prepare single-species or multi-species plans. There is little or no evidence as to the relative effectiveness of these two approaches in the case of CWR.

The detailed composition of a species management or recovery plan will depend on the biology of the species, its conservation status, its location and other local circumstances. The essential elements are: (a) a full evaluation and description of the current status of the species; (b) a clear statement of the goals and objectives; and (c) an indication of the specific actions that are proposed.

The CWR Project countries have, in most cases, prepared a species management plan for one of their priority CWR, but none of these have been fully implemented due to the limited length of the Project.

Further sources of information

Frankel, O.H., Brown, A.H.D. and Burdon, J.J. (1995) *The Conservation of Plant Biodiversity*, Cambridge University Press, Cambridge, 'Chapter 6: The conservation *in situ* of useful or endangered wild species'.

Heywood, V.H. and Dulloo, M.E. (2005) In Situ *Conservation of Wild Plant Species – A Critical Global Review of Good Practices*, IPGRI Technical Bulletin no 11, FAO and IPGRI, IPGRI, Rome, Italy.

Iriondo, J.M. and De Hond, L. (2008) 'Crop wild relative *in situ* management and monitoring: The time has come', in N. Maxted, B.V. Ford-Lloyd, S.P. Kell, J.M. Iriondo, M.E. Dulloo and J. Turok (eds) *Crop Wild Relative Conservation and Use*, pp319–330, CAB International, Wallingford, UK.

Iriondo, J.M., Maxted, N. and Dulloo, M.E. (eds) (2008) *Conserving Plant Diversity in Protected Areas*, CAB International, Wallingford, UK.

Notes

1. http://www.fws.gov/endangered/species/recovery-plans.html
2. Commonly (although incorrectly) referred to as *inter situ* (Burney and Burney, 2009).
3. Most genetic reserve conservation has been undertaken in Turkey and other countries in the Middle East/SW Asia. See, for example, Al-Atawneh et al (2008) and Tan and Tan (2002).
4. The terms conservation and management are used interchangeably given that conservation, in this context, normally involves essentially management interventions to a greater or lesser degree.
5. Aarhus Convention on Access to Information, Public Participation in Decision-Making and Access to Justice in Environmental Matters.

References

Akeroyd, J. and Wyse Jackson, P. (1995) *A Handbook for Botanic Gardens on Reintroduction of Plants to the Wild*, Botanic Gardens Conservation International (BGCI), Richmond, UK

Al-Atawneh, N., Amri, A., Assi, R. and Maxted, N. (2008) 'Management plans for promoting *in situ* conservation of local agrobiodiversity in the West Asia centre of plant diversity', in N. Maxted, B.V. Ford-Lloyd, S.P. Kell, J. Iriondo, E. Dulloo and J. Turok (eds) *Crop Wild Relative Conservation and Use*, pp340–361, CABI Publishing, Wallingford, UK

Bañares, Á., Marrero, M., Carqué, E. and Fernández, Á. (2003) 'Plan de recuperación de la flora amenazada del Parque Nacional de Garajonay. La Gomera (Islas Canarias). Germinación y restituciones de *Pericallis hansenii*, *Gonospermum gomerae* e *Ilex Perado* ssp. *Lopezlilloi*', *Botanica Macaronésica*, vol 24, pp3–16

Boersma, P.D., Kareiva, P., Fagan, W.F., Clark, J.A. and Hoekstra, J.M. (2001) 'How good are endangered species recovery plans?', *BioScience*, vol 51, pp643–649

Bonnie, R. (1999) 'Endangered species mitigation banking: Promoting recovery through habitat conservation planning under the Endangered Species Act', *The Science of the Total Environment*, vol 240, pp11–19

Bowles, M.L. and Whelan, C. (eds) (1994) *Restoration of Endangered Species: Conceptual Issues, Planning and Implementation*, Cambridge University Press, Cambridge

Boyes, B. (2001) 'Multi-species local recovery planning: Benefits and impediments', in B.R. Boyes (ed) *Biodiversity Conservation From Vision to Reality*. Lockyer Watershed Management Association. Lockyer Landcare Group, Forest Hill, Australia. http://bruceboyes.info/wp-content/uploads/2010/01/2000_SE_QLD_Biodiversity_Conference_Proceedings.pdf

Burney, D.A. and Burney, L.P. (2007) 'Paleoecology and 'inter situ' restoration on Kaua'i, Hawai'i', *Frontiers in Ecology and Environment*, Ecological Society of America, vol 5, no 9, pp483–490

Burney, D.A. and Burney, L.P. (2009) '*Inter situ* conservation: Opening a 'third front' in the battle to save rare Hawaiian plants', *BGjournal*, vol 6, pp17–19

Canadian Wildlife Service (2002) 'Special report: Custom-designing recovery', *Recovery: An Endangered Species Newsletter*, Canadian Wildlife Service

Clark, J.A. and Harvey, E. (2002) 'Assessing multi-species recovery plans under the Endangered Species Act', *Ecological Applications*, vol 12, no 3, pp655–662

Falk, D.A. and Holsinger, K.E. (eds) (1991) *Genetics and Conservation of Rare Plants*, Oxford University Press, New York and Oxford

Falk, D.A., Millar, C.I. and Olwell, M. (eds) (1996) *Restoring Diversity*, Island Press, Washington, DC

Frankel, O.H., Brown, A.H.D. and Burdon, J.J. (1995) *The Conservation of Plant Biodiversity*, Cambridge University Press, Cambridge

Gockel, C.K. and Gray, L.C. (2009) 'Integrating conservation and development in the Peruvian Amazon', *Ecology and Society*, vol 14, no 2, p11, available at: http://www.ecologyandsociety.org/vol14/iss2/art11/

Heywood, V.H. (2005) 'Master lesson: Conserving species *in situ* – a review of the issues', *Planta Europa IV Proceedings*, http://www.nerium.net/plantaeuropa/proceedings.htm, accessed 10 May 2010

Heywood, V.H. and Dulloo, M.E. (2005) In Situ *Conservation of Wild Plant Species – A Critical Global Review of Good Practices*, IPGRI Technical Bulletin, no 11, FAO and IPGRI, International Plant Genetic Resources Institute (IPGRI), Rome, Italy

Hoekstra, J.M., Clark, J.A., Fagan, W.F. and Boersma, P.D. (2002) 'A comprehensive review of Endangered Species Act recovery plans', *Ecological Applications*, vol 12, pp630–640

Iriondo, J.M. and De Hond, L. (2008) 'Crop wild relative *in situ* management and monitoring: The time has come,' in N. Maxted, B.V. Ford-Lloyd, S.P. Kell, J.M. Iriondo, M.E. Dulloo and J. Turok (eds) *Crop Wild Relative Conservation and Use*, pp319–330, CAB International, Wallingford, UK

Iriondo, J.M., Maxted, N. and Dulloo, M.E. (eds) (2008) *Conserving Plant Diversity in Protected Areas*, CAB International, Wallingford, UK

IUCN/SSC (2008) *Strategic Planning for Species Conservation: A Handbook*, Version 1.0, International Union for Conservation of Nature (IUCN) Species Survival Commission, Gland, Switzerland

Kell, S.P., Laguna, L., Iriondo, J. and Dulloo, M.E. (2008) 'Population and habitat recovery techniques for the *in situ* conservation of genetic diversity', in J. Iriondo, N. Maxted and M.E. Dulloo (eds) *Conserving Plant Genetic Diversity in Protected Areas*, Chapter 5, pp124–168, CAB International, Wallingford, UK

Kooyman, R. and Rossetto, M. (2008) 'Definition of plant functional groups for informing implementation scenarios in resource-limited multi-species recovery planning', *Biodiversity and Conservation*, vol 17, pp2917–2937

Lleras, E. (1991) 'Conservation of genetic resources *in situ*', *Diversity*, vol 7, pp72–74

Maxted, N., Hawkes, J.G., Ford-Lloyd, B.V. and Williams, J.T. (1997) 'A practical model for *in situ* genetic conservation', in N. Maxted, B.V. Ford-Lloyd and J.G. Hawkes (eds) *Plant Genetic Conservation: The* In Situ *Approach*, pp545–592, Chapman and Hall, London, UK

Maxted, N., Kell, S.P. and Ford-Lloyd, B. (2008) 'Crop wild relative conservation and use: Establishing the context', in N. Maxted, B.V. Ford-Lloyd, S.P. Kell, J.M. Iriondo, M.E. Dulloo and J. Turok (eds) *Crop Wild Relative Conservation and Use*, pp3–30, CAB International, Wallingford, UK

McNeely, J.A. and Mainka, S.A. (2009) *Conservation for a New Era*, International Union for Conservation of Nature (IUCN), Gland, Switzerland

Moore, S.A. and Wooller, S. (2004) *Review of Landscape, Multi- and Single-Species Recovery Planning for Threatened Species*, World Wide Fund for Nature (WWF) – Australia

Reynolds, J.D., Mace, G.M., Redford, K.H. and Robinson, J.G. (eds) (2001) *Conservation of Exploited Species*, Cambridge University Press, Cambridge

Rogers, D.L. (2002) '*In situ* genetic conservation of Monterey pine (*Pinus radiata* D. Don): Information and recommendations', Report No. 26, University of California Division of Agriculture and Natural Resources, Genetic Resources Conservation Program, Davis, CA, USA

Rogers, D.L. (2004) '*In situ* genetic conservation of a naturally restricted and commercially widespread species, *Pinus radiata*', *Forest Ecology and Management*, vol 197, pp311–322

Sheppard, V., Rangeley, R. and Laughren, J. (2005) *Multi-Species Recovery Strategies and Ecosystem-Bases Approaches*, World Wide Fund for Nature (WWF) – Canada, http://assets.wwf.ca/downloads/wwf_northwestatlantic_assessmentrecovery strategies.pdf, accessed 17 May 2010

Simmons, J.B., Beyer, R.I., Brandham, P.E., Lucas G.L. and Parry, V.T.H. (eds) (1976) *Conservation of Threatened Plants: The Function of Living Plant Collections in Conservation and Conservation-Oriented Research and Public Education*, Plenum Press, New York, NY, USA

Stephens, S. and Maxwell, S. (eds) (1996) *Back from the Brink: Refining the Threatened Species Recovery Process*, Australian Nature Conservation Agency, Surrey Beatty & Sons, Chipping Norton NSW, Australia

Sutherland, W.J. (2000) *The Conservation Handbook: Techniques in Research, Management and Policy*, Blackwell Science Ltd, Oxford, UK

Synge, H. (ed) (1981) *The Biological Aspects of Rare Plant Conservation*, Wiley, Chichester, UK

Tan, A. and Tan, A.S. (2002) '*In situ* conservation of wild species related to crop plants: The case of Turkey', in J.M.M. Engels, V. Ramantha Rao, A.H.D. Brown and M.T. Jackson (eds) *Managing Plant Genetic Diversity*, pp195–204, CAB International, Wallingford, UK

Chapter 11

Conservation Strategies for Species/Populations Occurring Outside Protected Areas

More than 90 per cent of the terrestrial surface of the earth is not covered by any form of protected area category. If this situation does not change, there will be severe loss of biological wealth in the next few decades (Halladay and Gilmour, 1995).

Aims and purpose

Given that national parks and other conservation areas cover only 12 to 13 per cent of the earth's surface in total, it is clear that these areas alone will not ensure the survival of species and ecological communities, even without the impacts of accelerated global change. It is crucial, therefore, that lands outside national reserve networks be managed in ways that allow as much biodiversity as possible to be maintained. The *in situ* conservation of species outside protected areas, where the majority of them occur, is a seriously neglected aspect of biodiversity conservation and in the face of global change it must demand much further attention from governments and conservation agencies. This approach is also known as *off-reserve management* (Hale and Lamb, 1997).

This approach should also be seen within the context of integrating protected areas within wider landscapes, seascapes and natural resource policies (Ervin et al, 2010), one of the benefits being to achieve additional conservation benefits outside of protected areas (Box 11.1).

Other reasons for paying more attention to the conservation of resources in land outside protected areas are given by Torquebiau and Taylor (2009):

- Farming and land management practices strongly influence available natural resources and biodiversity.

Box 11.1 Achieving additional conservation benefits outside of protected areas

A significant proportion of biodiversity is located outside of protected areas – working with other interest groups and sectors across the wider land/seascape matrix can significantly improve biodiversity conservation, even without protected status being achieved. For example, ecologically friendly practices can be pursued in agriculture and extractive industries, while actors involved in agroforestry and sustainable tourism can adjust their practices so they are more compatible with biodiversity conservation. Regeneration and reforestation schemes can also help, potentially with funding from initiatives such as the Clean Development Mechanism of the Kyoto Protocol.

Source: Ervin et al, 2010

- Agricultural (or useful) biodiversity – the plants and animals domesticated or used by man, together with associated ecosystems, land-use systems, wild species and indigenous practices – is the foundation of sound farming practices and is under threat from large-scale 'industrial' agriculture. This also applies to natural forest biodiversity, including the extraction of non-timber forest products, and exotic plantation (or industrial forestry).
- There is strong evidence that biodiversity can contribute to improved development, although there is continuing debate about the relations between conservation, food security and poverty reduction.

It follows logically that many CWR will be numbered among the species that grow outside protected areas, and for these, off-reserve management can be an important strategy. We need to address what actions may be proposed so that many areas that are currently not protected, but house target species, will be maintained in a manner that ensures their conservation at the ecosystem or landscape level by positive management policies or the prevention of certain forms of activity. In addition, it may be possible to take actions through various forms of agreement with landowners to ensure such areas outside formal protection, whether on public or private land, can provide a sufficient degree of protection to target species and ensure the maintenance of viable populations.

Several authors have noted that many CWR occur in disturbed, pre-climax plant communities such as roadsides, field margins and orchards, which tend not to be included in protected areas (Jain, 1975; Maxted et al, 1997; Maxted and Kell, 2009). For example, Al-Atawneh et al (2008) observed that in the Wadi Sair Reserve in Palestine, the wild pear species, *Pyrus syriaca* Boiss., is only found as scattered trees, never as continuous populations, and the largest populations are found near the borders of fields and in areas not grazed as they receive some protection by being surrounded by fruit tree orchards. Conservation of this species must take place primarily outside of the existing protected areas, supplemented by *ex situ* measures. CWR may also occur as weeds in agricultural, horticultural and silvicultural agroecosystems, and as Maxted and Kell (2009) note, they are often associated with traditional cultural practices or with marginal environments. The abandonment of such traditional agricultural systems will place many weedy CWR at risk.

In view of the scale of the problem and the large numbers of CWR for which formal protection is unlikely to be achieved, we need to invest heavily in a range of actions outside of, and complementary to, the formal protected area system in order to afford some degree of protection to CWR species and their habitats. Many of these actions depend on engaging private landowners in the conservation process. A wide range of indirect means exist through agreements, such as conservation easements, to reduce the level of exploitation of areas or to contain threats. These agreements include:

- conservation easements, including covenants, trusts, partnerships, with or without financial or tax incentives;
- incentive-based schemes, including agro-environmental schemes;
- local conservation strategies;
- public and private collaboration for conservation;
- special cases such as conservation in vegetation fragments and micro-reserves;
- habitat conservation planning (HCP) and mitigation banking.

Conservation easements

Conservation easements are legal agreements that allow landowners to voluntarily restrict or limit the kinds of development that may occur on their land (TNC 2003, 2008; Merenlender et al, 2004). Generally, conservation easements are voluntary agreements between landowners and another party, usually a private local or national conservation organization, for the preservation and protection of land in its natural, scenic, historic, agricultural, forested or open space condition. They may be negotiated in conjunction with an international conservation organization such as the United States Nature Conservancy (see below) and may be acquired through purchasing from the landowner, given as a gift or inherited. Title to the land remains with the owner who may receive tax benefits, depending on the country and national or regional legislation.

Easements can serve as a means of helping protect biodiversity in cases where purchase of the land is not possible or even as an interim measure while purchase is being negotiated. The agreements are legally binding and can afford long-term protection. The restrictions of the easement, once agreed, are perpetual and apply to all future owners of the land. They are detailed in a legal document recorded in the local land records; the easement becomes a part of the chain of title on the property.

Easements can be used to conserve land that is of biologically significant value while, at the same time, the landowner can continue to own and use the property. An example is the Grassland Reserve Program administered by the United States Department of Agriculture (USDA) Natural Resources Conservation Service (NRCS) and USDA Farm Service Agency (FSA) in cooperation with the USDA Forest Service. It is a voluntary programme that helps landowners and operators restore and protect grassland, including rangeland and pastureland, and certain other lands, while maintaining the areas as grazing lands. The effectiveness of buying easements as a conservation strategy is reviewed by Armsworth and Sanchirico (2008).

In the US, The Nature Conservancy (TNC), one of the world's leading conservation charities, has been a major player in conservation easements, which it regards as one of the most powerful, effective tools available for the permanent conservation of private lands in the US. TNC has negotiated easements in 20 states[1] in the US and has been granted easements on roughly 30,000 acres in Latin America (see Box 11.2), the Caribbean and Canada.

TNC has adopted a broad approach to easements, to protect land and water, directly or indirectly, as habitats for plant and animal biodiversity. It notes that easements can be designed to:

- protect natural habitat from destruction by conversion to other uses such as subdivision and development;
- protect open space of varying kinds from development or other disturbance;
- protect natural habitat from destruction by intensive agriculture;
- conserve forests through limitations on forest management and development;
- preserve agriculture and grazing lands from subdivision and development;
- protect water resources by limiting disturbance of lands in the watershed;
- provide for public use and access, such as through trail easements.[2]

Box 11.2 TNC role in conservation easement at Cuatro Ciénegas, Mexico

In 2000, The Nature Conservancy (TNC) and its Mexican partner organization, Pronatura Noreste, A.C., purchased the 7000-acre Rancho Pozas Azules (Ranch of the Blue Pools), situated in a 200,000-acre valley in the northern state of Coahuila. The area contains 77 endemic species found nowhere else in the world. The purchase was one of the largest private land purchases for conservation purposes in Mexico. Pronatura holds the title to the property and is responsible for its management as a nature preserve. As part of the transaction, Pronatura accepted a conservation easement over the 200-acre parcel that the seller retained. The easement was the first in north-eastern Mexico. TNC is helping Pronatura expand the reserve by purchasing Rancho Pasta de Garza, a 2964-acre private ranch located to the north of the reserve. More than 300 of the valley's 883 plant species are also found here.

Source: See http://www.nature.org/wherewework/northamerica/states/texas/files/chihuahuan_desert_1008_lowres.pdf

Off-reserve management

Various types of off-reserve management are practised, such as in production forests, agricultural landscapes and urban landscapes, roadsides and transport corridors.

Conservation easements and forestry

Conservation easements can be an effective tool for maintaining working forests, preserving environmental values and protecting communities from excessive development pressure according to the Society of American Foresters (2007), which supports easements as one tool for ensuring sustainable forest management. But, as they observe, easements are not appropriate for all forest lands and should only be entered into with full understanding of their consequences. 'Selling or donating conservation easements may allow landowners who are committed to sustainable management to resist pressure to sell their property to developers. Similarly, in the face of pressure to withdraw working forests from active management, conservation easements offer a way to provide adequate environmental and open-space benefits while allowing continued timber harvesting.' In the US, conservation easements are negotiated and run by federal agencies, state natural resources agencies, and nearly 1700 local, regional and national land trusts. An overview of current efforts and summaries of the various programmes involved are given in a recent report (US Endowment for Forestry and Communities, 2008).

Forest genetic conservation outside protected areas

The maintenance of genetic resources outside protected areas has been carried out traditionally in forestry, albeit neither consistently, nor in all cases consciously, as an act of conservation (Palmberg-Lerche, 1993, pers. comm. to V. Heywood). Kanowski (2001) points out that the conservation of many rare and threatened species continues to depend on the management of production forests or on private land outside the protected area system, highlighting the need to adopt forest conservation strategies that extend beyond protected areas if biodiversity conservation goals are to be achieved.

> *The broader vision for* in situ *forest conservation recognizes that achieving and sustaining forest conservation also requires the integration of social and economic goals into conservation planning processes. It therefore recognizes the development of more collaborative participatory modes of conservation planning and management as essential to achieving and sustaining forest conservation goals. New forms of partnership between many of the actors with interests in forests, which recognize the diversity of their roles and contributions, are especially important in delivering conservation outcomes* (Kanowski, 2001).

Box 11.3 Conservation fields for forest genetic resources in Indonesia

In Indonesia, to promote *in situ* conservation of forest tree genetic resources in areas where concessions have been granted, the National Committee on Genetic Resources works together with the Association of Forest Concessionaires to design conservation fields within concession areas. It was agreed that around 200ha of forests should be left uncut in each concession area. In this way, there is a remnant of original forest in each locality, which will serve as a reference for future studies, as well as a place where seeds of native trees can be collected.

Source: Sastrapradja, 2001

It is estimated that approximately 90 per cent of the global forest area lies outside of public protected areas and a World Bank study notes that while existing parks and protected areas are the cornerstones of biodiversity conservation, they are insufficient on their own to ensure the continued existence of a vast proportion of tropical forest biodiversity. Promoting more biodiversity-sensitive management of ecosystems outside protected areas, especially of those known to contain target species, needs to be given high priority. This is especially applicable to forests that are already subject to some form of management such as for timber production.

As Kanowksi (2001) indicates, off-reserve management can make a significant contribution to regional biodiversity conservation, provided appropriate management systems and processes are in place, and may contribute to the conservation of those values that cannot be fully protected in conservation reserves and existing protected areas, largely because of land-tenure and land-use patterns.

The setting aside of areas within forestry concessions as a means of conserving original forest and providing a seed source is another approach that has been adopted, for example in Indonesia (Box 11.3).

Conservation of CWR in traditional agroecosystems

CWR are frequently found in disturbed, pre-climax plant communities such as roadsides, field margins or orchards and often occur in traditionally managed agroecosystems and agroforestry systems or in marginal environments. Their conservation in such areas is incidental and not a result of deliberate policy. As such, their conservation is far from secure, especially when traditional cultivation systems are abandoned in favour of more modern agricultural practices. But as Maxted and Kell (2009) note, these areas often contain large thriving populations of CWR and can act as important corridors for CWR gene flow and dispersal and as reservoirs to bolster genetic reserve populations. We need to consider whether any effective steps can be taken to enhance or reinforce such incidental conservation of CWR, such as the creation of micro-reserves as described below.

Set-aside schemes

The majority of wild species have, of course, managed to survive, at least up to now, outside protected areas, but the chances of their long-term survival in the face of global change and worldwide habitat loss and fragmentation will be enhanced if the areas in which they occur are managed or *set aside* for some non-conservation purpose that does not cause harm to their ecosystems.

Examples include land that is set aside for military use, airport protection zones and grounds of public and private institutions such as hospitals, universities and commercial companies. Some of the side effects of war may also be beneficial for conservation, including demilitarized zones or 'no-man's lands', some of which can be very rich in biodiversity. Such survival is subject to the prevailing dynamics of the system and may not result in a sufficiently broad or representative sample of the species being maintained. Nonetheless, in a broad biodiversity conservation context it is valuable and, although it cannot be regarded as fully effective *in situ* species conservation, it is probably as much as can be expected for the majority of CWR, given the large numbers involved and the lack of massive investment in this area.

In Europe, set-aside is a term that was applied to land that farmers were not permitted to use for any agricultural purpose. Although introduced by the European Economic Community in 1988 as part of a set of measures to prevent overproduction, it was soon realized that this practice often had beneficial effects on the biodiversity of the land concerned. Some farmers chose to set aside those areas that would provide the most benefit to wildlife. In some cases, for example, farmers converted the land taken out of production to woodland. The scheme was abolished in 2008.

Agricultural conservation easements are designed to keep land available for farming and prevent its use for building or other urban influences but are of little value for CWR conservation.

Public and private collaboration for conservation

As González-Montagut (2003) observes, 'limited funds, and the requirement for counterpart funds, leave no room for competition between institutions interested in financing protected areas'. Synergies between the public and private sectors need to be developed. Various models of private–public cooperation for conservation of biodiversity have been adopted by different countries. An action plan for private protected areas is described in Langholz and Krug (2004) (see also Box 11.4).

In Costa Rica, the Legislative Assembly approved a law in 1992 that allows the legal designation of private wildlife reserves. Under this legislation, private wildlife refuges consist of informally protected private nature reserves that qualify for designation as government-approved and officially recognized wildlife refuges. Under this programme, landowners must develop and adhere to a government-approved management plan specifying restrictions on land and resource use. In return, refuge owners receive three incentives:

> **Box 11.4 Private protected areas: An emerging issue**
>
> Privately owned protected areas continue their quiet proliferation throughout much of the world. Despite this expansion, little is known about them. Preliminary evidence suggests that private parks number in the thousands and protect several million hectares of biologically important habitat. They serve as increasingly important components of national conservation strategies. In a time when many governments are slowing the rate at which they establish new protected areas, the private conservation sector continues its rapid growth. Conservationists desperately need to examine this trend closely, assessing its overall scope and direction, and determining ways to maximize its strengths while minimizing its weaknesses.
>
> *Source:* Langholz and Krug, 2003

1 an exemption from property taxes for land declared as a refuge;
2 access to technical assistance for managing the protected area; and
3 assistance in the event of a squatter invasion.

Voluntary and legal, covenants, trusts and partnerships, with or without financial or tax incentives or payment for management and associated costs

Incentive-based schemes

Incentive-based schemes whereby landowners or tenants are offered payments in return for helping conserve or protect areas such as native forests and other vegetation, watersheds or wetlands or ecosystem services have been introduced by a number of countries. Examples are the CapeNature Stewardship Programme in the Western Cape province of South Africa (Box 11.5), the Conservation Partners Programme in New South Wales, Australia, the BushTender scheme in Victoria, Australia (see Box 11.6), the Grain-for-Green Programme in China (SFAB, 2000; Gee, 2006; Liu and Wu, 2010) for converting steep cultivated land to grassland and forest, and the informally protected wildlife reserves in Costa Rica approved by Costa Rica's Legislative Assembly in 1992 (Langholz et al, 2000). In Catalonia, Spain, the Xarxa de Custòdia del Territori, a network for land stewardship, was established in 2003. It is a not-for-profit organization working to foster land stewardship as a conservation strategy for the natural, cultural and landscape resources and values of the region and its environment. The network comprises over 150 associations, foundations, city councils, enterprises and persons working in land stewardship. It works with networks within Europe, such as the Réseau de Coopération Eurorégionale pour la Gestion Conservatoire, and with Latin America.

In recent years, the concept of payment schemes for environmental services (PES) has received considerable attention in various Latin American countries as

Box 11.5 The CapeNature Stewardship Programme, South Africa

The vision of the stewardship programme is threefold:

- to ensure that privately owned areas with high biodiversity value receive secure conservation status and are linked to a network of other conservation areas in the landscape;
- to ensure that landowners who commit their property to a stewardship option will enjoy tangible benefits for their conservation actions;
- to expand biodiversity conservation by encouraging commitment to, and implementation of, good biodiversity management practices on privately owned land, in such a way that the private landowner becomes an empowered decision-maker.

The three stewardship options that the CapeNature Conservation Stewardship Programme are promoting include:

1 Contract nature reserves – legally recognized contracts or servitudes on private land to protect biodiversity in the long term.
2 Biodiversity agreements – negotiated legal agreements between the conservation agency and a landowner for conserving biodiversity in the medium term.
3 Conservation areas – flexible options with no defined period of commitment (includes conservancies).

Source: Langholz et al, 2000

an innovative tool for the financing of sustainable management of land and water resources (FAO/FLD/IPGRI, 2004).

Some of these schemes have been viewed with suspicion, largely on the grounds that they allow foreigners to buy up huge tracts of land as in the case of the Conservation Land Trust (CLT) of Douglas Tompkins or the Conservación Patagónica (CP) of Kris Tompkins, through which large areas of forest land were acquired for conservation purposes. Clearly, governments need to maintain strict vigilance of such schemes, but it is widely agreed that they have so far proved beneficial. A review of biodiversity offsets is given by Bayon (2008).

Habitat conservation plans and endangered species mitigation

In an attempt to resolve conflicts that had arisen regarding the conservation of endangered species on private lands, the United States Fish and Wildlife Service has been promoting the use of 'habitat conservation plans', whereby the 'take' of some individuals of endangered species or adverse modification of part of their habitat is allowed in exchange for an undertaking to minimize and mitigate the loss of such habitat to the 'maximum extent practicable' (Bonnie, 1999). The

Box 11.6 Conservation outside protected areas in Australia

Roadside Conservation Committee, Western Australia

Established by the Western Australia government in 1985, its terms of reference are to coordinate and promote the conservation and effective management of rail and roadside vegetation for the benefit of the environment and the people of Western Australia. Roadsides often contain remnant native vegetation that has an important role in the conservation of native flora, particularly the case with rare flora, as in some cases it is their only remaining habitat. It publishes a series of guidelines on topics such as assessing the conservation values of roadsides, designating and managing flora roads and managing and harvesting native flowers, seeds and timber from roadsides. For further information see:
http://www.dec.wa.gov.au/management-and-protection/off-reserve-conservation/roadside-conservation-committee.html.

The BushTender scheme

The BushTender scheme aims to conserve areas of remnant vegetation on private land by using an auction-based process to allocate biodiversity contracts. Officials receive the bids from potential suppliers and the assessed biodiversity importance of each site, so they can calculate which of those bids offer best value for money in terms of the greatest biodiversity value for least cost per hectare. It pays private landowners to enter into contracts to undertake management to improve the quality or area of native vegetation on their land. Landowners identify what management activities they will undertake, prepare a management plan and submit a bid indicating what payment they would seek from the government (of Victoria State). The trials have been oversubscribed and they seem to afford appreciable conservation benefits. For a critical evaluation see:
http://een.anu.edu.au/wsprgpap/papers/stoneha1.pdf.

Western Australia Remnant Vegetation Protection Scheme

This scheme provides assistance to landholders to fence remnant vegetation. Landholders apply for a subsidy, which is assessed on the basis of nature conservation value. Funding is tied to entry to a 30-year contract deed for the protection and management of the native vegetation. Funding assistance was originally set at AU$600 (US$497) per kilometre of fencing materials, that is about 50 per cent of the cost of materials. Assistance has now been raised to AU$900 (US$746) per kilometre with another increase to AU$1200 (US$995) being considered. This is equivalent to 100 per cent of material costs. Under the scheme, over 1094 projects have been funded with in excess of 38,000ha of remnant vegetation being fenced at a cost of approximately AU$2.25 (US$1.87) million.
See http://www.myoung.net.au/water/publications/motivating_people.pdf.

Land for Wildlife, State of Victoria
Land for Wildlife is a voluntary, non-binding scheme that allows landholders to register their properties if areas within the property are actively managed for nature conservation. Participation in the scheme is voluntary and a landholder can remove their property from the register at any time. The programme provides recognition of conservation effort, a network of other interested landholders and extension support and management advice. Over 3500 properties are registered with Land for Wildlife, making it the most successful programme, in terms of participation, in Australia.

Off-reserve conservation of natural grasslands
A range of mechanisms is available to help protect natural temperate grassland remnants located outside of conservation reserves. These include memoranda of understanding (MOU), regional plans, joint management agreements, voluntary conservation agreements, local environment plans and other planning mechanisms such as designation as public land categories where permitted activities are compatible with conservation of the grassland values. For further information, see: Natural Temperate Grassland of the Southern Tablelands of NSW and the Australian Capital Territory, http://www.environment.gov.au/cgi-bin/sprat/public/publicshowcommunity.pl?id=14.

underlying principle is that some individuals of an endangered species or parts of their habitat may be expendable over the short term so long as enough protection is provided to ensure the long-term recovery of the species. This is known as endangered species mitigation and had proved highly controversial (Wilhere, 2009). Bonnie (1999) has suggested the adoption of 'mitigation banking' for wetlands whereby landowners would be allowed to seek 'a permit to destroy endangered species habitat and mitigate the loss by buying mitigation credits from other private landowners who restore and/or protect important habitats'.

Community/participatory conservation areas

In a review of protected areas and people, Kothari (2008) observes that two changes have been revolutionizing protected area policy and management in an increasing number of countries: first, the increased participation of local communities and others in what were once solely government-managed protected areas, transforming them into collaboratively managed protected areas (CMPAs); and second, the increasing recognition of indigenous and community conserved areas (ICCAs), many different kinds of which occur across the world but have so far remained outside the scope of formal conservation policies and programmes. According to a recent report on the role of indigenous people in biodiversity conservation, traditional indigenous territories encompass up to 22 per cent of the world's land surface and coincide with areas that hold 80 per cent of the planet's biodiversity (Sobrevila, 2008).

Collaboratively managed protected areas (CMPAs)

There is already extensive literature on collaborative management and its benefits (Kothari, 2006a). A good example of this is the Venezuela–Expanding Partnerships for the National Parks System Project, the objective of which is to implement a co-management model that guarantees the sustainable management of the Canaima National Park through an alliance between indigenous peoples, private sector institutions and government agencies. Another is the Kaa-Iya del Gran Chaco National Park, Bolivia's largest protected area with an area of 3,440,000 ha, is managed collaboratively by the Capitania de Alto y Bajo Isoso indigenous people's organization, the Wildlife Conservation Society (WCS) and the Bolivian National Park Service (SERNAP). The park is the only national protected area in the Americas created as the result of an initiative by an indigenous organization. Further examples can be found in a range of both developed and developing countries such as Canada, Indonesia, France, the Philippines and South Africa.

Indigenous and community conserved areas (ICCAs)

A considerable part of the world's biological diversity is located in territories whose ownership, control and use is in the hands of indigenous and local communities, including nomadic peoples. Despite this, conservation policies have often largely ignored the fact that these people and communities conserve many of these sites, actively or passively, through traditional and modern ways. This is partly due to lack of knowledge, and partly to the suspicion that such methods of conservation are not sufficiently effective. Some conservationists would argue that effective conservation needs a new approach whereby on-the-ground agencies, both government and local, set the broad agenda for research and decide how to implement the results (Smith et al, 2009) – in other words, 'let the locals lead' (see Chapter 5).

The term *indigenous and community conserved* areas (ICCAs) is applied to such areas (Kothari, 2006a) defined as 'natural and modified ecosystems, containing significant biodiversity values, ecological services, and cultural values, voluntarily conserved by indigenous and local communities, through customary laws or other effective means' (Pathak et al, 2004). They are extremely diverse in terms of their governance institutions, their management objectives, and ecological and cultural impacts. They can range from a tiny forest patch of less than a hectare, as in the case of sacred sites or forests, to several million hectares, as in the case of indigenous protected areas in some South American countries.

There is also an increase in the number of indigenous protected areas and reserves that are incorporated into the official protected area system. According to Kothari (2008), indigenous reserves account for one-fifth of the Amazon forests and have been shown to be effective against illegal logging, mining and other threats impacting forests outside these reserves. These include reserves that have been integrated into national protected area systems, such as the 68,000ha Alto Fragua–Indiwasi National Park of Colombia. The government of Madagascar has

Box 11.7 The key benefits of ICCAs

ICCAs are critical from an ecological and social perspective in many ways. They often (though not always):

- help conserve critical ecosystems and threatened species;
- maintain essential ecosystem functions, including water security and gene pools;
- sustain the cultural and economic survival of tens of millions of people, not only in countries of the tropics but also industrialized nations;
- provide corridors and linkages for animal and gene movement, including often between two or more officially protected areas (as illustrated by examples from Southern Africa, North America and South America);
- synergize links between agricultural biodiversity and wildlife, providing larger land/waterscape-level integration;
- offer crucial lessons for participatory governance, useful even in government-managed protected areas;
- offer lessons in integrating customary and statutory laws, and formal and non-formal institutions, for more effective conservation;
- build on and validate sophisticated ecological knowledge systems, elements of which have wider positive use;
- aid in community resistance to destructive development, saving territories and habitats from mining, dams, logging, tourism, overfishing and so on;
- help communities in empowering themselves, especially to reclaim or secure territories, tenure and rights to or control over resources;
- aid communities to better define their territories, e.g. through mapping, such as in Central America (see Solis et al, 2006);
- help create a greater sense of community identity and cohesiveness, and also a renewed vitality and sense of pride in local cultures, including among the youth who are otherwise alienated from these by modern influences;
- create conditions for other developmental inputs to flow into the community;
- lead to greater equity within a community and between the community and outside agencies;
- conserve biodiversity at relatively low financial cost (though often high labour inputs), with costs of management often covered as part of normal livelihood or cultural activities, through existing systems and structures; and
- provide examples of relatively simple administration and decision-making structures, avoiding complex bureaucracies.

Source: Kothari, 2006b

also diversified its types of protected area governance as part of its commitment to triple the area under protection.[3]

Areas conserved by communities are characterized by being voluntarily established and their management in the hands of the communities; in turn, the local

Box 11.8 An example of local co-management and its impact on CWR in Madagascar

The tapia forest is a type of forest that is only found on the western slopes of the Madagascar high plateaux (at around 1000m in height). It is home to the tapia, *Uapaca bojeri* (Euphorbiaceae), and several species of the endemic family Sarcolaenaceae. Economic activities in the region are based on agriculture. In addition, local populations collect a certain number of resources from the forest, such as tapia fruits for local use and marketing, dead tapia trees for firewood, wild mushrooms and tubers of two species of yam (*D. hexagona* and *D. heteropoda*) for food supplementation. The tapia forest also hosts the wild silkworm species *Boroceras madagascariensis* which is used in the weaving of much appreciated wild silk. Thus, the tapia forest is of essential role in the local communities' economy.

Gestion Locale Sécurisée (GELOSE) contracts for the transfer of management of the tapia forest were signed by several communities in the rural municipality of Arivonimamo (about 50–90km west of the capital Antananarivo). Among the clauses of the contracts, local communities obtained exclusive rights to the exploitation of the transferred forests and the legal right to protect their forests and resources from predators, mainly people who were not members of the community. They also were required to set up *Uapaca* nurseries and proceed with reforestation. Fire protection was also built around the transferred ecosystem because the region undergoes annual bushfires that contribute to the reduction of forest lands.

The communities benefited from several training sessions from the technical departments on topics such as identifying donors and asking for small project funding, silkworm raising and silk weaving. The communities also expressed interest in the cultivation of *D. alata* (cultivated species), received training and have started to set up yam fields.

Management transfer has been shown to contribute significantly to an increase in the income of the local community. As one of the consequences observed, the pressure on wild yams was reduced.

This approach has, however, some shortcomings; one of the most important being the failure, in some cases, to respect the agreements. Also, the sanctions for non-compliance to be applied by the community itself based on what is called 'fihavanan' (roughly translated as based mainly on friendly and family relationships), are not always effectively implemented, with the result that the management transfers sometimes fail.

communities have the obligation to conserve and sustainably use the resources of the areas based on their traditional knowledge, practices and customary laws. The main benefits of ICCAs are listed in Box 11.7.

An example is the Parque de la Papa (Potato Park), Peru, an Indigenous Biocultural Heritage Area (Área de Patrimonio Biocultural Indígena: APBCI).[4] In 2002, the six Quechuan agrarian communities, known as Chawaytiré, Sacaca, Kuyo Grande, Pampallaqta, Paru Paru and Amaru, declared some 10,000ha of their lands the *Parque de la Papa*, which was soon followed by an agreement with the International Potato Centre (CIP) in Lima, Peru that allowed the repatriation of some 420 varieties of potatoes previously collected by CIP for the purposes of plant breeding (see also Box 5.6). The Potato Park focuses on protecting and preserving the critical role and interdependence of the indigenous biocultural heritage (IBCH) for the maintenance of local rights and livelihoods and the conservation and sustainable use of agrobiodiversity.

In Madagascar, a system of secured local management of natural resources, known as GELOSE (Gestion Locale Sécurisée) was introduced in 1996. It is a legal framework for introducing the sharing of responsibility over natural resource management among users and the transfer of rights from central government to the local community. GELOSE allows communities to define their own goals and develop regulations for resource use and management in the form of by-laws, provided they are consistent with national policy (Antona et al, 2004). An example of GELOSE relating to CWR is given in Box 11.8.

Off-site agreements and species recovery

Off-site agreements can be negotiated as part of a recovery strategy for endangered species – see Box 11.9 for an Australian example.

Special cases

Conservation in vegetation fragments

Fragmentation of vegetation is a widespread phenomenon (Saunders et al, 1987) and, in the temperate world, most habitats are small fragments or remnants of previously much larger and more continuous ecosystems. This is now becoming more common in tropical areas, largely as a result of deforestation, which poses problems for the design of protected areas for CWR, especially in increasingly non-steady-state environments as a result of global change. Vegetation fragments also include a wide variety of specialized habitats that may be important for conservation. These include field boundaries such as hedgerows, hedge banks, lines of trees, stone walls, ditches and stream banks, which may play a role in maintaining habitat mosaics and providing connectivity as well as housing rare or scarce species (Marshall and Moonen, 1998). Road verges and unmowed powerline strips (Russell et al, 2005) may play a similar role. The questions needing to be addressed are: How far can species and populations survive in vegetation remnants? Is conservation of vegetation fragments worthwhile? What action is possible? One approach is to accept the facts of the situation and try and establish small-scale reserves, as in the case of the micro-reserves created in Spain and other parts of Europe discussed below. Small reserves are inherently unstable and difficult to maintain and manage but may be judged worthwhile, at least in the short term, especially for CWR of high importance. For a discussion of these issues see Heywood (1999).

Box 11.9 Example of off-site negotiations for recovery of endangered species in Australia

The National Multi-species Recovery Plan for the Cycads negotiates conservation agreements to secure significant known populations of cycads on freehold and leasehold property. It is desirable that the populations of cycads are secured with perpetual arrangements that ensure continued appropriate management in the long term. For cycads, a conservation agreement between the landholders and the Queensland Parks and Wildlife Service (QPWS) is an appropriate model for significant populations not currently existing in national park, state forest or conservation reserves. These voluntary agreements are negotiated with landholders to create a nature refuge over part or all of a property and are attached to the land title. They allow for production and land management activities compatible with conservation of the values of the land such as sustainable grazing but generally prohibit further destruction or removal of individuals. QPWS extension officers undertake property assessments, negotiate the conservation agreement and provide follow-up advice and assistance with management of the nature refuge.

Nature refuge landholders may be eligible for Queensland government incentives. In addition, lessees of state land may be entitled to benefits under proposed changes under the Land Act (1994) and may be advantaged in seeking grants for conservation works such as fencing through natural resource management funding bodies. A conservation agreement will provide access to volunteer groups to assist with conservation work, for example fencing on grazing properties where cycads are a threat to stock.

Where significant populations occur on private land, some controlled harvesting of cycad seeds and foliage for commercial sale by the landowner may provide a significant incentive for entering into a conservation agreement and providing on-ground management of populations.

Source: Queensland Herbarium, 2007 – National Multi-species Recovery Plan for the Cycads

Conservation fields

A German project called '100 Fields for Biodiversity' aims at establishing a nationwide network of conservation fields for wild arable plant species. The project is financially supported by the Deutsche Bundestiftung für Umwelt (DBU)[5] and seeks to counter the ongoing loss of species by implementing a network of conservation fields. In these fields, the areas are managed without using herbicides and in tune with the growth preferences of the wild arable plants. It is hoped that the conservation fields will act as future centres for potential recolonization of rare species.[6]

Micro-reserves

Small-scale reserves, frequently referred to as *micro-reserves*, have been established in various parts of the world to afford protection to threatened species,

Box 11.10 Spanish plant micro-reserves

A network of plant micro-reserves (PMR) was pioneered in Spain by Emilio Laguna of the environment agency (Conselleria de Medio Ambiente) of the regional government of Valencia, Spain and the first one was established in 1997. By the end of 2008, the Valencian community held 273 officially protected plant micro-reserves that house populations of more than 1625 species of vascular plants. Of these, 1288 populations of 527 species are targeted for long-term monitoring. The sites are protected by orders of the environment agency. The management plan designates a few priority plants in each PMR, which are targeted for conservation actions (census, management projects, population reinforcement if required, etc). Only two actions are designated for all the PMRs: census of priority species and the collection of their seeds to be transferred to the germplasm bank of the botanic garden at the University of Valencia. More than 1050 populations, belonging to 450 taxa, have been targeted for census and seed collection; however, both actions are still at the starting point for most PMR, so their implementation represents an important challenge for the coming years.

Source: Laguna, 2004 and http://microreserve.blogspot.com/

usually in fragmented vegetation (Saunders et al, 1991; Turner and Corlett, 1996; Heywood, 1999). In the last 10 to 15 years, a great deal of interest has been generated by the network of plant micro-reserves established in the Valencia region in Spain (see Box 11.10). Micro-reserves in Spain are small-scale protected areas, usually less than one or two hectares as in the Valencian examples, but up to 200ha in other regions. They often maintain a high concentration of endemic, rare or threatened species. Micro-reserves may be considered as an option in areas where the vegetation has been subjected to fragmentation and the species populations within these areas are similarly reduced or fragmented. Because of the small area occupied by micro-reserves and their frequent simplicity in legal and management terms, it may be possible for them to be established in great number and to complement the larger, more conventional protected areas. On the other hand, their long-term viability remains in question, especially in the light of global change.

Micro-reserves have also been established in others parts of Spain such as Castilla y León, Castilla-La Mancha, Murcia and Menorca. The model is being introduced with modifications in some other European countries. A pilot network of micro-reserves in Western Crete was set up under the European Union LIFE Nature 2004 Programme. One of the species targeted was *Phoenix theophrasti*, a wild relative of the date palm, at Preveli beach.[7]

An innovative use of micro-reserves is being developed for Lima beans (*Phaseolus lunatus*) in the Central Valley of Costa Rica. Because of their patchy and fragmented distribution, the usually small population size and other factors, two types of micro-reserve were designed (Meurrens et al, 2001; Baudoin et al, 2008), either in original sites of the existing natural populations (provided these sites are sufficiently protected from any human disturbance) or in artificially established micro-conservation reserves for synthetic populations created from seeds of four nearby populations collected in their sites of origin.

Need for monitoring

As with CWR populations within protected areas, routine monitoring of various elements or activities at the sites of various forms of off-site conservation is necessary to see how far the site management is actually maintaining the target CWR populations. This may cover:

- evaluation of compliance with the management plan and implementation mechanisms;
- evaluation of the biological performance of the management plan;
- determining whether the management objectives remain appropriate;
- resource monitoring;
- monitoring plant and animal population counts;
- undertaking phenology studies;
- monitoring human activities such as wild-harvesting; and
- monitoring the spread of invasive species and the effectiveness of the actions to counter-control them.

Off-site conservation in the GEF/UNEP CWR Project countries

Armenia: Conservation of CWR outside protected areas[8]

According to current legislation in Armenia, plants growing in forests, pastures, hay meadows and other lands of special importance are afforded some degree of *in situ* conservation in that their use is subject to regulation. Exploitation of the plant resources on these lands must be conducted in a way that allows natural regeneration to take place.

The rare and endangered plants listed in the Red Data Book of Armenia are a special case. According to a recent study, about 70 per cent of plants in the Red Data Book are CWR. As stipulated by the Law on Flora, landowners must make provisions to ensure conservation of the rare and endangered (Red-listed) species growing on their lands. Any activity that can lead to the decline in the number of these species or deteriorate the habitats is prohibited.

The policy framework regulating conservation and use of wild plants (including CWR) outside protected areas is far from ideal in Armenia. Neither is it adequately enforced. Certain reforms took place during the last decade to improve the regulatory framework: in particular, the Law on Flora (1999), Land Codex (2002), Forest Codex (2005) and other legal acts arising from these have been adopted. These norms are, however, mainly limited to the wild plants growing on the state-owned lands. It is up to the landowners to decide the fate of the plants growing on private lands. One possible solution to ensure conservation of plants on private lands would be adoption of incentive schemes, but this is not possible during the present stage of economic development in the country. It can be inferred, therefore, that the populations of CWR occurring on private lands are more threatened. At present, however, the conservation status of plants on these lands is relatively satisfactory in that private lands are abandoned in many rural areas of Armenia since their exploitation would require significant investment such as expensive fertilizers and equipment. The same is true for highland rural areas and villages located close to the state border. Agricultural activities are limited on these lands, as the younger generation leaves the villages for the cities. Wild plants, especially CWR (among them many weedy species), thrive on the abandoned lands.

Further sources of information

Hale, P. and Lamb, D. (eds) (1997) *Conservation Outside Nature Reserves*, Centre for Conservation Biology, University of Queensland, Brisbane, AU.

Merenlender, A.M., Huntsinger, L., Guthey, G. and Fairfax, S.K. (2004) 'Land trusts and conservation easements: Who is conserving what for whom?', *Conservation Biology,* vol 18, pp67–75.

The Nature Conservancy (TNC) (2003) *Conservation Easements – Conserving Land, Water and a Way of Life,* available at: http://www.nature.org/aboutus/howwework/conservation-methods/privatelands/conservationeasements/files/consrvtn_easemnt_sngle72.pdf.

The Nature Conservancy (TNC) (2008) *Conservation Easements: All About Conservation Easements,* http://www.nature.org/aboutus/howwework/conservationmethods/private-lands/conservationeasements/about/allabout.html.

Sobrevila, C. (2008) *The Role of Indigenous Peoples in Biodiversity Conservation: The Natural but Often Forgotten Partners,* The World Bank, Washington, DC

Notes

1. Conservation easements across the US: http://www.nature.org/aboutus/howwework/conservationmethods/privatelands/conservationeasements/about/art15087.html
2. Conservation easements at The Nature Conservancy: http://www.nature.org/aboutus/howwework/conservationmethods/privatelands/conservationeasements/about/tncandeasements.html
3. http://news.mongabay.com/2006/0117-madagascar.html
4. http://www.parquedelapapa.org/
5. www.dbu.de
6. www.schutzaecker.de
7. CRETAPLANT: A Pilot Network of Plant Micro-Reserves in Western Crete: http://cretaplant.biol.uoa.gr/docs/A5_Interim_Report.pdf (accessed 24 September 2009).
8. Contributed by Siranush Muradyan.

References

Al-Atawneh, N., Amri, A., Assi, R. and Maxted, N. (2008) 'Management plans for promoting *in situ* conservation of local agrobiodiversity in the West Asia centre of plant diversity', in N. Maxted, B.V. Ford-Lloyd, S.P. Kell, J. Iriondo, E. Dulloo and J. Turok (eds), *Crop Wild Relative Conservation and Use,* pp340–361, CABI Publishing, Wallingford, UK

Antona, M., Bienabe, E.M., Salles, J.M., Péchard, G., Aubert, S. and Ratsimbarison, R. (2004) 'Rights transfers in Madagascar biodiversity policies: achievements and significance', *Environment and Development Economics,* vol 9, pp825–847

Armsworth, P.R. and Sanchirico, J.N. (2008) 'The effectiveness of buying easements as a conservation strategy', *Conservation Letters,* vol 1, pp182–189

Baudoin, J.P., Rocha, O.J., Degreef, J., Zoro, Ni, I., Ouédraogo, M., Guarino, L. and Toussaint, A. (2008) '*In situ* conservation strategy for wild Lima bean (*Phaseolus lunatus* L.) populations in the Central Valley of Costa Rica: A case study of short-lived perennial plants with a mixed mating system', in N. Maxted, B.V. Ford-Lloyd, S.P. Kell, J.M. Iriondo, M.E. Dulloo and J. Turok (eds) *Crop Wild Relative Conservation and Use,* pp364–379, CAB International, Wallingford, UK

Bayon, R. (2008) 'Chapter 9: Banking on biodiversity', in L. Starke (ed) *2008 State of the World: Innovations for a Sustainable Economy,* The Worldwatch Institute, W.W. Norton and Co., New York and London

Bonnie, R. (1999) 'Endangered species mitigation banking: Promoting recovery through habitat conservation planning under the Endangered Species Act', *The Science of the Total Environment,* vol 240, pp11–19

Ervin, J., Mulongoy, K. J., Lawrence, K., Game, E., Sheppard, D., Bridgewater, P., Bennett, G., Gidda, S.B. and Bos, P. (2010) *Making Protected Areas Relevant: A Guide to Integrating Protected Areas into Wider Landscapes, Seascapes and Sectoral Plans and Strategies,* CBD Technical Series No. 44, Convention on Biological Diversity, Montreal, Canada

FAO/FLD/IPGRI (2004) *Forest Genetic Resources Conservation and Management, Vol 1: Overview, Concepts and Some Systematic Approaches,* International Plant Genetic Resources Institute (IPGRI), Rome, Italy

Gee, C. (2006) 'Grain for green', *Ecosystem Marketplace,* 24 February 2006

González-Montagut, R. (2003) 'Private-public collaboration in funding protected areas in Mexico', Paper presented at the Fifth World Parks Congress, September 2003, Durban, South Africa

Hale, P. and Lamb, D. (eds) (1997) *Conservation Outside Nature Reserves,* Centre for Conservation Biology, University of Queensland, Brisbane, AU

Halladay, P. and Gilmour, D. A. (eds) (1995) Conserving biodiversity outside protected areas: The role of traditional agro-ecosystems, IUCN, Gland and Cambridge

Heywood, V.H. (1999) 'Is the conservation of vegetation fragments and their biodiversity worth the effort?' in E. Maltby, M. Holdgate, M. Acreman and A.G. Weir (eds) *Ecosystem Management: Questions for Science and Society,* pp65–76, Royal Holloway Institute for Environmental Research, Royal Holloway, University of London

Jain, S.K. (1975) 'Genetic reserves', in O.H. Frankel and J.G. Hawkes (eds) *Crop Genetic Resources for Today and Tomorrow,* pp379–396, Cambridge University Press, Cambridge

Kanowski, P. (2001) '*In situ* forest conservation: A broader vision for the 21st Century', in B.A. Thielges, S.D. Sastrapradja and A. Rimbawanto (eds) In Situ *and* Ex Situ *Conservation of Commercial Tropical Trees,* pp11–36, Faculty of Forestry, Gadjah Mada University and International Tropical Timber Organization, Yogyakarta, Indonesia

Kothari, A. (2006a) 'Community conserved areas', in M. Lockwood, G. Worboys and A. Kothari (eds) *Managing Protected Areas: A Global Guide,* Earthscan, London, UK

Kothari, A. (2006b) 'Community conserved areas: Towards ecological and livelihood security', *Parks,* vol 16, no 1, pp3–13

Kothari, A. (2008) 'Protected areas and people: The future of the past', *Parks,* vol 17, no 2 DURBAN+5, pp23–34

Laguna, E. (2004) 'The plant micro-reserve initiative in the Valencian Community (Spain) and its use to conserve populations of crop wild relatives', *Crop Wild Relatives,* vol 2, pp10–13

Langholz, J. and Krug, W. (2003) 'Emerging issue: "Private Protected areas"', WPC Governance Stream, Parallel Session 2.5. Protected Areas Managed by Private landowners, 13 September 2003, http://www.earthlore.ca/clients/WPC/English/grfx/sessions/PDFs/session_2/PPA_action_plan.pdf

Langholz, J. and Krug, W. (2004) 'New forms of biodiversity governance: Non-state actors and the private protected area action plan', *Journal of International Wildlife Law and Policy,* vol 7, pp9–29

Langholz, J., Lassole, J. and Schelhas, J. (2000) 'Incentives for biological conservation: Costa Rica's private wildlife refuge program', *Conservation Biology,* vol 14, pp1735–1745

Liu, C. and Wu, B. (2010) *'Grain for Green Programme' in China: Policy Making and Implementation*? China Policy Institute, University of Nottingham, Briefing Series – Issue 60, April 2010

Marshall, E.J.P. and Moonen, C. (1998) *A Review of Field Margin Conservation Strips in Europe,* A report for the UK Ministry of Agriculture, Fisheries and Food, IACR – Long Ashton Research Station, Department of Agricultural Sciences, University of Bristol, UK

Maxted, N. and Kell, S.P. (2009) *Establishment of a Global Network for the* In Situ *Conservation of Crop Wild Relatives: Status and Needs,* FAO Commission on Genetic Resources for Food and Agriculture, Rome, Italy

Maxted, N., Hawkes, J.G., Ford-Lloyd, B.V. and Williams, J.T. (1997) 'A practical model for *in situ* genetic conservation', in N. Maxted, B.V. Ford-Lloyd and J.G. Hawkes (eds) *Plant Genetic Conservation: The* In Situ *Approach,* pp545–592, Chapman and Hall, London, UK

Merenlender, A.M., Huntsinger, L., Guthey, G. and Fairfax, S.K. (2004) 'Land trusts and conservation easements: Who is conserving what for whom?', *Conservation Biology,* vol 18, pp67–75

Meurrens, F., Degreef, J., Rocha, O.J. and Baudoin, J.P. (2001) 'Demographic study in micro-conservation sites with a view to maintain *in situ* wild Lima beans (*Phaseolus lunatus* L.) in the Central Valley of Costa Rica', *Plant Genetic Resources Newsletter,* no 128, pp45–50

Pathak, N., Bhatt, S., Balasinorwala, T., Kothari, A. and Borrini-Feyerabend, G. (2004) 'Community conserved areas: A bold frontier for conservation', Briefing Note 5, TILCEPA/IUCN, CENESTA, CMWG and WAMIP, Tehran, Iran, http://cmsdata.iucn.org/downloads/cca_briefing_note.pdf

Russell, K.N., Ikerd, H. and Droege, S. (2005) 'The potential conservation value of unmowed powerline strips for native bees', *Biological Conservation,* vol 24, pp133–148

Sastrapradja, S.D. (2001) 'The role of *in situ* conservation in sustainable utilization of timber species', in B.A. Thielges, S.D. Sastrapradja and A. Rimbawanto (eds) In Situ *and* Ex Situ *Conservation of Commercial Tropical Trees,* pp37–51, Faculty of Forestry,

Gadjah Mada University and International Tropical Timber Organization, Yogyakarta, Indonesia

Saunders, D.A., Arnold, G.W., Burbidge, A.A. and Hopkins, A.J.M. (1987) 'The role of remnants of native vegetation in nature conservation: future directions', in *Nature Conservation: The role of remnants of native vegetation*, pp387–392, Surrey Beatty in association with CSIRO and CALM, Chipping Norton, NSW, AU

Saunders, D.A., Hobbs, R.J. and Margules, C.R. (1991) 'Biological consequences of ecosystem fragmentation: A review', *Conservation Biology*, vol 5, pp18–32

SFAB (2000) 'Guojia jiwei he linyeju di 111 hao wenjian-Guanyu jinyibu zuohao tuigeng huanlin huancao shidian gongzuo de jianyi' (The 111th document issued Department of Planning, Forestry Administration Bureau: Appendix: Implementation proposals for Grain-for-Green policy in the upper reaches of the Yangtze River and the upper and middle reaches of the Yellow River), State Forestry Administration Bureau (SFAB), China

Smith, R.J., Verissimo, D., Leader-Williams, N., Cowling, R.M. and Knight, A.T. (2009) 'Let the locals lead', *Nature*, vol 462, pp280–281

Sobrevila, C. (2008) *The Role of Indigenous Peoples in Biodiversity Conservation: The Natural but Often Forgotten Partners*, The World Bank, Washington, DC

Society of American Foresters (2007) *Conservation Easements – A Position Statement of the American Foresters*, initially adopted on 9 December 2001 and revised and renewed on 10 June 2007, Society of American Foresters, Bethesda, MD, USA

Solís, V., Cordero, P.M., Borrás, M.F., Govan, H. and Varel, V. (2006) 'Community conservation areas in Central America: Recognising them for equity and good governance', *Parks*, Special issue on *Community Conserved Areas*, vol 16, no 1, pp21–27

TNC (2003) *Conservation Easements – Conserving Land, Water and a Way of Life*, The Nature Conservancy (TNC), http://www.nature.org/aboutus/howwework/conservationmethods/privatelands/conservationeasements/files/consrvtn_easemnt_sngle72.pdf

TNC (2008) *Conservation Easements: All About Conservation Easements*, The Nature Conservancy (TNC), http://www.nature.org/aboutus/howwework/conservationmethods/privatelands/conservationeasements/about/allabout.html, accessed 20 May 2010

Torquebiau, E. and Taylor, R.D. (2009) 'Natural resource management by rural citizens in developing countries: Innovations still required', *Biodiversity and Conservation*, vol 18, no 10, pp2537–2550

Turner, J.M. and Corlett, R.T. (1996) 'The conservation value of small, isolated fragments of lowland tropical rainforest', *Trends in Ecology and Evolution*, vol 11, pp330–333

US Endowment for Forestry and Communities (2008) *Forest Conservation Easements: Who's Keeping Track?*', US Endowment for Forestry and Communities, Greenville SC, USA

Wilhere, G.F. (2009) 'Three paradoxes of habitat conservation plans', *Environmental Management*, vol 44, pp1089–1098, doi:10.1007/s00267-009-9399-0

Chapter 12

Complementary Conservation Actions

Adopting a complementary conservation strategy means that a range of methods are employed, each appropriate to a specific component part of the overall conservation programme and taken together, these methods complement each other in order to achieve the most efficient and safest conservation in the long term (Sharrock and Engels, 1996).

Aim of this chapter

As this book often makes clear, *in situ* conservation is the favoured approach to CWR conservation, as it has the distinct advantage that target species are continuously exposed to a changing natural environment that allows new diversity to be generated. However, such exposure can often dramatically threaten the very existence of these species. For this reason, *in situ* conservation approaches will often need to be supported by complementary conservation approaches for the sake of security. Such complementary conservation approaches also have the advantage that they help facilitate access by plant breeders to important genetic materials for crop improvement. Some level of complementary conservation will need to be practised for the optimal conservation of CWR. It is beyond the scope of this manual to provide an in-depth examination of the various complementary conservation approaches available for CWR. Moreover, it is the aim of this chapter to provide the reader with a general overview of the types of approaches and techniques that are available and to highlight how these might be used to complement *in situ* conservation, such as the provision of a safety net for genetic diversity, which is difficult to conserve *in situ* or threatened in the wild. Further, the potential role of *ex situ* collections in facilitating the recovery and reintroduction of CWR populations *in situ* is highlighted.

Introduction

Conserving CWR *in situ* is not sufficient in itself. While *in situ* conservation is essential to maintain the evolution of the species and allow new diversity to be created through natural selection processes, it presents many disadvantages for conservation and has severe limitations in facilitating the use of CWR for crop improvement (see Box 12.1) (for reviews see Maxted et al, 1997 and Engels et al, 2008). Although *in situ* conservation is an efficient tool for conservation of CWR, in order to make CWR more accessible for crop improvement and other human uses and to ensure that the maximum genetic diversity of target species is safely conserved, other approaches will need to be applied. It is important to back up any *in situ* interventions with complementary *ex situ* conservation in genebanks as seed, pollen, living plants (in field genebanks or in botanic gardens), tissue culture, or cryopreservation, depending upon the biology of species to be conserved.

As discussed in Chapter 1, CWR have increasingly provided new genes for crop improvement (Hajjar and Hodgkin, 2007) when they are readily accessible to plant breeders and, in such cases, have been extensively used as a source of useful genetic traits for disease resistance as well as abiotic (temperature and drought) stress tolerance, crop yield and improved quality. With the anticipated impacts of climate change on agricultural production, climate-adapted traits most likely to be found within CWR will become even more in demand by plant breeders. Consequently, back-up samples of CWR in *ex situ* collections to facilitate access and use in plant breeding programmes is becoming a high priority. However, wild relatives of some crops are still poorly represented in *ex situ* collections, despite the fact that over the last decade there has been an increase of 3 per cent in their collection, as shown in the recent update of the *State of the World Report for Plant Genetic Resources for Food and Agriculture* (FAO, 2010).

It should be noted that the *ex situ* conservation of some CWR presents major challenges for genebank managers both from technical and management aspects. Often, the storage conditions which have been established mainly for major crops are not well adapted for some of their wild relatives, on which limited research has been undertaken to refine their conservation *ex situ*. Some may have dormancy problems or be simply difficult to germinate, while other species may have recalcitrant seeds. In fact, the seed storage behaviour can vary among, and even within, species, and different provenances may not adapt to the field condition in the case of field collection. For example, among *Coffea* species, a range of storage behaviours from orthodox to recalcitrant may be found. Protocols may not exist for specific CWR to be considered for *in vitro* or cryopreservation conservation. Some CWR may be easier to store *ex situ* than the crop species, e.g. seed-bearing *Musa* species. There can also be restraints in accessing germplasm from genebanks due to government policies relating to the exchange of germplasm, property rights, access and benefit-sharing regulations, or phytosanitary regulations. Further, the cost of maintaining a genebank should be taken into account as it can be prohibitive in many countries and lack of sustained funding in such instances can threaten collections.

Box 12.1 Advantages and disadvantages of *in situ* and *ex situ* conservation of CWR

Advantages	*Disadvantages*
***In situ* conservation**	
• Avoids storage problems associated with field genebanks and recalcitrant seeds • Allows evolution to continue through exposure to pests and diseases and other environmental factors • Indirect benefits, including ecosystem services • Sustainable use by local people	• Requires extensive areas for effective conservation • Exposes natural populations to a wide range of natural catastrophic events (storms, hurricanes, cyclones) and other threats • Materials cannot be readily used and are very difficult to access • Subject to conflict with management by landowners (CWR may not have high priority) • Expensive to maintain
***Ex situ* conservation**	
• Rescue of threatened germplasm • Requires limited space to conserve large numbers of accessions • Conserves an adequate representative sample of CWR populations • Ease of accessibility and exchange of germplasm; use can be promoted • Evaluation facilitated • Ease of documentation • No exposure to pests, diseases and other hazards (except for field collections and botanic gardens) • Indefinite maintenance of germplasm • More cost-effective compared to *in situ* conservation	• Freezes the evolutionary process • Difficult to ensure adequate sampling (intra-specific variability) • Total genetic integrity cannot be ensured due to human error, selection pressure during regeneration • Only limited accessions can be conserved in field genebanks • Natural catastrophes in field genebanks • *In vitro*-somaclonal variation

What do we mean by a CWR complementary conservation strategy?

The concept of a complementary conservation strategy for CWR involves the combination of different conservation actions, which together lead to an optimum sustainable use of genetic diversity existing in a target gene pool, now and in the future (Dulloo et al, 2005). Complementary conservation strategies are also known as integrated or holistic and the principle is that the full range of conservation options available should be considered and the appropriate combination applied in particular situations (Falk and Holsinger, 1991; Given, 1994). The two main conservation approaches (*ex situ* and *in situ*) are both important in the conservation and use of genetic diversity. In addition, it may be appropriate to attempt other techniques such as *inter situs* (see below) and assisted migration or colonization (see Chapter 14).

The ultimate purpose of germplasm conservation is use and, consequently, any conservation strategy should include mechanisms that will also ensure access to the germplasm by relevant stakeholders. Other important issues that must be addressed in a conservation strategy include issues related to policy and legal frameworks, documentation, socio-economic aspects, infrastructure and networks. Since the needs of users and the conservation technologies may change over time, a complementary conservation strategy should be flexible enough to allow such changes to be taken into consideration. Dulloo et al (2005) proposed a framework for developing a complementary conservation strategy using coconuts as an example. The process involves first defining the options for conservation of the target species, taking into account the feasibility of conserving it *in situ*, its seed storage behaviour, whether or not the species can be conserved as seeds, whether or not protocols for *in vitro* or cryopreservation are developed or whether they can only be conserved as live plants in field genebanks or botanic gardens and, finally, if options such as translocation or *inter situs* approaches are necessary.

The choice of the complementary conservation actions should also take into account the intended use of the conserved germplasm, available infrastructure and human resources, space availability, accessibility and so on. Nevertheless, in the case of CWR, one must keep in mind that their conservation is not always based on their availability for immediate use. Based on these elements, state of knowledge and the options available to date, a framework for a complementary conservation strategy can be developed. Thus a complementary conservation strategy can be seen as a logical process and not just a selection of appropriate conservation methods. The framework can be seen as a series of steps (see Figure 12.1); at each step information is gathered, specific actions taken and/or decisions made. It is important that proper consultation be held with all stakeholders in developing the complementary conservation strategy (see Chapters 4 and 5 for in-depth discussion on how to involve stakeholders). This could be done by establishing a network of stakeholders, facilitated by a lead agency. It would be the role of this network or committee to then define the complementary conservation strategy objectives and sub-objectives. These could be, for example, the necessity for creating a back-up of the *in situ* population, for implementing a reintroduction/recovery programme, carrying out research, use in evaluation/breeding programmes or increasing the awareness of the public on the importance of CWR (see Chapter 16) or for training and education (see Chapter 15). For each specific objective, the complementary conservation strategy options available should then be analysed in terms of their feasibility and requirements in infrastructure, human resources, land, costs, accessibility and the risks involved. The pros and cons of

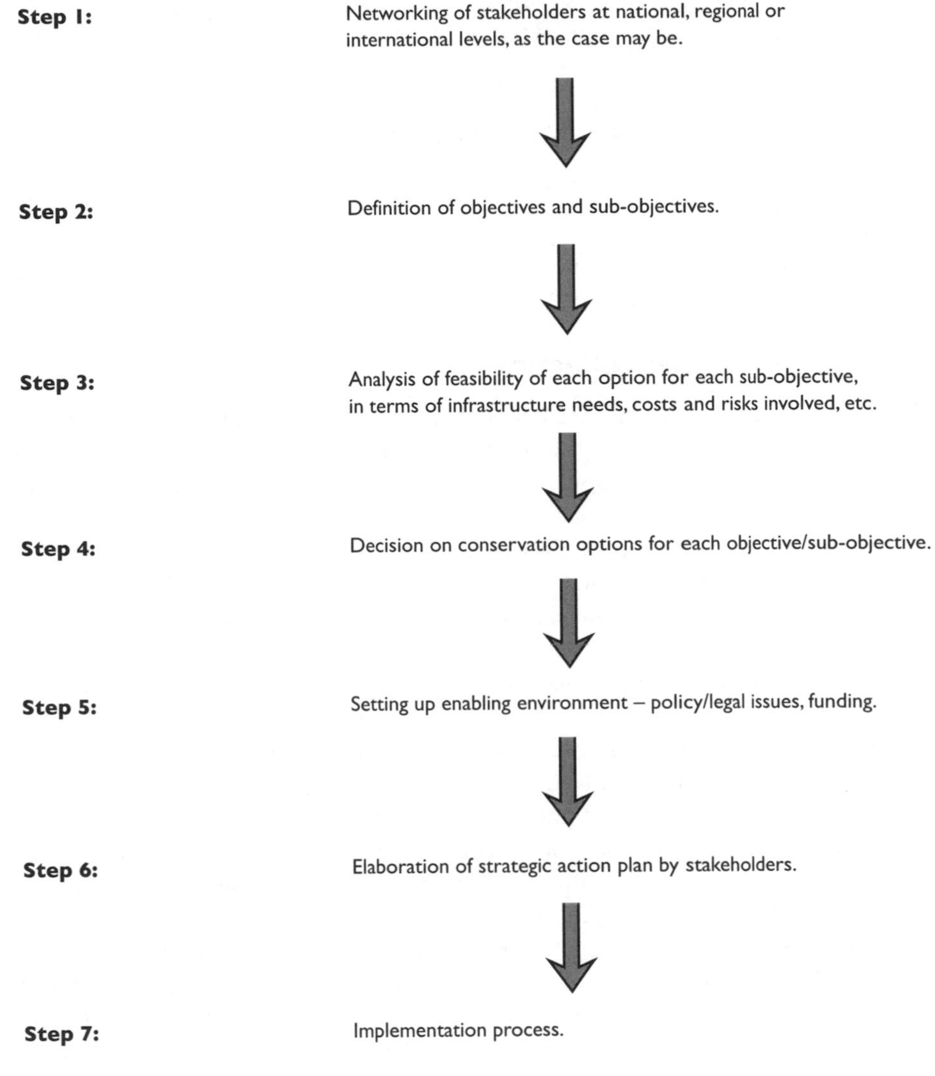

Figure 12.1 *Framework for developing a complementary conservation strategy*

each of the possible options must be weighed against each other and decisions made on the complementary conservation strategy options to be followed for specific objectives.

The next important step in the process would be to ensure there is an adequate policy and regulatory framework in place, which will allow implementation of the complementary conservation strategy options. This would involve an analysis, and possible revisions, of policy issues in terms of legislation, germplasm exchange and benefit-sharing. Consideration must also be given to sources of funding. Once these issues are addressed and put into place, a strategic action plan can be developed and implemented (steps 6 and 7 in Figure 12.1). For each step, the stakeholder network should be consulted before relevant decisions are made and responsibilities assigned to the various relevant players.

Ex situ conservation options

This section provides some technical guidelines for establishing an *ex situ* collection and a brief description of the different *ex situ* conservation options. For a general review of complementary *ex situ* methods see Guerrant et al (2004), Thormann et al (2006) and Engels et al (2008) .

Guidelines for seed collecting

Collecting seeds or other propagules is obviously the first activity for establishing an *ex situ* collection. This needs to be well planned and well prepared to maximize the genetic diversity of the population. It is not intended here to provide a detailed account of collecting. There are a number of excellent technical guidelines written about how to plan and prepare for collecting for *ex situ* conservation (Guarino et al, 1995; Schmidt, 2000; Smith et al, 2003; Guerrant et al, 2004; ENSCONET, 2009). Given that seeds are the easiest and most amenable material to collect and conserve, most of these guidelines focus on seeds. However, Guarino et al (1995) also provide guidelines on how to collect vegetatively propagated germplasm (see their Chapters 21 and 22). They also provide guidelines for collecting *in vitro* and pollen material (see their Chapters 24 and 25). In addition, there is much information available for download on the internet. For example, there are good seed collecting summaries and field manuals for download from the Millennium Seed Bank site.[1] A documentary on wild chickpeas seed collecting by Ken Street, available to view at http://www.seedhunter.com/, provides an excellent insight into the practical realities of collecting for diversity.

It may be desirable to target collecting sites that contain the highest species and genetic diversity. The use of predictive tools (such as FloraMap, DIVA-GIS (Hijmans et al, 2001)) based on geographic information systems (GIS) can help identify such potential collecting sites (see Chapter 8). Guarino et al (2001) provides a general discussion on the application of species distribution models in the conservation and use of plant genetic resources. Many of the GIS methods use the climatic variables as the principal drivers of geographic distribution and can be used to predict sites of high species diversity. For example, Hijmans and Spooner (2001) used DIVA-GIS to describe the geographical distribution of wild relatives of potatoes and identified Peru as a location for high numbers of wild potato species, including rare wild species. Their study also allowed the identification of areas of high species richness which facilitated the design of *in situ* conservation reserves to protect them. Another good example is the study by Jarvis et al (2005), where GIS was used to optimize a collecting mission for a rare wild pepper species (*Capsicum flexuosum* Sendtn.) in Paraguay. The species was found at five out of the seven points predicted to harbour the species and was not found at four of the five points predicted not to harbour the species. Such approaches allow the collecting of germplasm to be carried out in a much more systematic and efficient manner.

The GapAnalysis website, http://gisweb.ciat.cgiar.org/GapAnalysis/, developed by Bioversity International, the International Rice Research Institute (IRRI) and the International Centre for Tropical Agriculture (CIAT), is a useful tool to enable plant collectors to target areas that contain traits and taxa that are under-represented in *ex situ* collections. A detailed methodology on how to carry out a crop wild relative gap analysis is provided at http://gisweb.ciat.cgiar.org/GapAnalysis/?p=139.

In addition, Maxted et al (2008) use a gap analysis methodology to identify areas for conservation, taking into consideration ecogeographic characteristics of the target taxon, as well as elements of the diversity effectively represented by existing *in situ* and *ex situ* conservation actions. The methodology is illustrated by its application to the African *Vigna* species.

In the context of this manual, some key activities that must be implemented when collecting samples include the following:

- Gather the information about the species to be collected so as to develop the *ex situ* conservation strategy. This should include information about seed storage behaviour, plant phenology (flowering, fruiting times) and reproductive biology, as well as ecogeographic information including botanical nomenclature, synonyms, historical location data and as full a data set as possible from local and regional herbaria. This is discussed in more detail in Chapter 8.
- Liaise with the relevant stakeholders and organize a stakeholder network.
- Undertake a gap analysis to identify the populations most in need of collecting due to threats, but also to identify areas rich in diversity.
- Obtain the necessary authorization for collection. Collection should be undertaken in line with national and international laws and regulations.
- Devise a sampling strategy for collecting that optimizes the highest level of genetic diversity, including the number of plants to be sampled. According to ENSCONET (2009), it is recommended to collect from five populations across the range of the species and to try and collect from at least 50 plants (preferably 200 plants) per population, but this should be used solely as a guide. The actual number to collect may depend on local circumstance and the plant collector needs to use his or her judgement so that the maximum genetic diversity can be captured while not endangering the population. Another consideration is the proposed use of the material, e.g. long-term safety back-up or reintroduction. See also the Guidelines on the Conservation of Medicinal Plants (1986), jointly published by the WHO, IUCN and WWF: http://apps.who.int/medicinedocs/documents/s7150e/s7150e.pdf (accessed 23 November 2010).
- Collect seeds and other materials in the field, including a herbarium specimen to verify the taxonomic identity. This is important since, very often, seeds of unknown species are collected and remain in collections as such for a long

time. Such collections then have very limited value and use (see Miller and Nyberg, 1995).

Guidelines for the proper handling of the seeds in the field include:

- Seeds should be extracted from fruits where possible and pre-cleaned.
- Seeds should be prepared for safe transportation in a paper bag, envelope or cloth bag.
- If transportation to the genebank is expected to take a long time, it is best to dry seeds over silica gel or other appropriate desiccant in plastic containers.
- Avoid exposing seeds to direct sunlight and high humidity (especially at night).

For more details see Smith, 1995; Schmidt, 2000 (see Sections 3 to 5); Smith et al, 2003 (see Section 1); and ENSCONET, 2009.

Ex situ conservation methods

Box 12.2 summarizes the different methods available for *ex situ* conservation.

Seed genebanks

Very few seed genebanks are dedicated to wild species, such as CWR (Heywood, 2009). The first to be established, in 1966, was the seed bank (ETSIA-UPM, now BGV-UPM) for native Spanish species at the Polytechnic University of Madrid, founded by the late Professor C. Gómez Campo; it now contains samples of 350 threatened Spanish species and subspecies, representing almost a quarter of the threatened flora of Spain. An even more notable exception is the Millennium Seed Bank (MSB) at Wakehurst Place, Royal Botanic Gardens, Kew, UK, which aims to house up to 10 per cent of the world's seed-bearing flora, principally from arid zones, by 2010. Recently, the MSB celebrated this target by the collection of a crop wild relative of banana from China, *Musa itinerans*, which may provide valuable genetic material for breeding new varieties of banana with resistance to diseases. The National Center for Genetic Resources Preservation (NCGRP) of the United States Department of Agriculture, based in Fort Collins, Colorado, also aims to systematically preserve a national collection of genetic resources including many CWR.

The representation of CWR in agricultural seed banks is very patchy and often the accessions are very small and collected incidentally rather than as part of a deliberate policy. The second *State of the World Report on Plant Genetic Resources for Food and Agriculture* (FAO, 2010) reports 10 per cent of the global germplasm holdings are wild species. Of these, forages and industrial crops account for a relatively high proportion of CWR. However, Maxted and Kell (2009), highlight that only between 2 and 6 per cent of global genebank *ex situ* collections are CWR, and of the total number of CWR species, only about 6 per cent have any accessions conserved *ex situ*. The discrepancy between these figures may also be a consequence of how CWR are defined.

Box 12.2 *Ex situ* conservation methods

Seed genebanks: This involves the drying of seeds to low moisture content (generally between 3 and 7 per cent) and storing in moisture-proof containers at low temperature (4°C for short-term conservation and –20°C for long-term conservation; FAO and IPGRI, 1994). Only taxa with orthodox seeds that can support drying to low moisture content and are cold-tolerant can be conserved in seed genebanks.

Field genebanks: This consists of growing living plants in a field, or very often in pots, in a screen or greenhouse. Field genebanks offer easy access to the plant material for characterization, evaluation and subsequent utilization, but are often difficult and expensive to maintain, time-consuming, labour-intensive, vulnerable to bad weather conditions, may mingle with adjacent plants, hybridize and blend with each other and can only conserve limited genetic material because of space issues.

Botanic Gardens: This covers the maintenance of (usually) small numbers of living plants in the garden collections and landscapes; extensive samples grown in field plots or under glass as conservation collections or as temporary collections for use in reintroduction experiments. Many botanic gardens have a strong focus on growing wild origin material, including CWR. They also play an important public awareness and education function.

Tissue culture: This involves the maintenance of explants in a sterile, pathogen-free environment with a synthetic nutrient medium. Different *in vitro* conservation methods are available: (1) slow growth conservation by limiting the environmental conditions and/or the culture medium; (2) synthetic seed technique, which aims to use somatic embryos as true seeds by encapsulating embryos in alginate gel, which can then be stored after partial dehydration and sown directly.

Cryopreservation: This involves the storage of a range of living tissues, including cell suspension, calluses, shoot tips, embryo and even whole seeds, at extremely low temperatures, usually at –196°C in liquid nitrogen, at which cell metabolism is effectively suspended. The material has to survive the freezing procedure before storage and the thawing procedure after storage. A range of cryopreservation techniques, including controlled rate cooling, vitrification, encapsulation-dehydration, encapsulation-vitrification, dormant bud preservation, pre-growth desiccation and droplet freezing.

Pollen storage: Pollen can be stored in the same way as described above for seeds and used as a conservation method for genetic resources, especially for perennial species of fruit and forest trees. It has a relatively short viability when conserved under classical storage conditions (partial desiccation followed by storage at sub-zero temperatures) and has therefore been used only to a limited extent in germplasm conservation.

Over 200 botanic gardens around the world also have seed banks (Laliberté, 1997; BGCI, 1998), ranging from small numbers of accessions stored in a domestic or commercial deep freezer to large-scale custom-built facilities, such as the Germplasm Bank of the Environmental Agency of Andalucía (Banco de

Germoplasma Vegetal Andaluz de la Consejería Andaluza de Medio Ambiente) at the Jardín Botánico de Córdoba, Spain, which stores more than 7000 accessions or propagules, mainly seeds, of more than 1500 different species of Andalusian plants and about 500 other Iberian endemic species. The Fletcher Jones Education Centre for the Preservation of Biodiversity complex at Rancho Santa Ana, California, USA, includes cold storage for seeds, climate-controlled growth chambers that facilitate germination studies and graduate programme research, seed-processing equipment and ample laboratory space.

There is an extensive amount of literature on the state of the art of *ex situ* seed conservation. Among them, the key information sources include Engels and Wood (1999); Hawkes et al (2000); Engels and Visser (2003); Smith et al (2003); Rao et al (2006); Thormann et al (2006); Engels et al (2008). An interactive self-learning module on seed handling in genebanks has been prepared by Bioversity International to help genebank technicians to process and prepare seeds for conservation (http://cropgenebank.sgrp.cgiar.org/images/flash/seed_handling_elearning_module/index.htm). In addition, the MSB technical information sheets (http://www.kew.org/msbp/scitech/publications/info_sheets.htm) contain information of relevance to *ex situ* conservation of CWR.

Field genebanks

For many species that produce no seeds (clonally propagated) or have seeds that are desiccation- and cold-sensitive, such as cacao, rubber, oil palm, coffee, banana and coconut, field genebanks are the best means for their conservation. For example, in Madagascar, the wild relatives of coffee are conserved in an important field genebank located at Kianjavato, first established in the early 1960s; to date, the collection holds 171 accessions (Dulloo et al, 2009). One of the advantages of field genebanks is that materials can be easily characterized and evaluated. For example, a project on tropical fruit trees in the Philippines led to the establishment of field collections of wild relatives of four tropical fruits trees including durian (*Durio zibethinus* Murray), mangosteen (*Garcinia mangostana* L.), jackfruit (*Artocarpus heterophyllus* L.) and pili nuts (*Canarium ovatum* Engl.). The germplasm collected were characterized and evaluated and, as a result, two new commercial varieties, a jackfruit variety, officially named 'Baybay Sweet', and a mangosteen accession, named 'UPLB Sweet', have been approved and registered with the National Seed Industry Council (NSIC) of the Philippines and are being marketed.

Key references for the management of field collection include those of Engelmann, 1999; Hawkes et al, 2000; Reed et al, 2004; Thormann et al, 2006.

Living conservation collections in botanic gardens

Historically, botanic gardens have played a key role in the collection and exchange of seed and other propagules with other gardens (Heywood, 2009). The roles played by botanic gardens such as Bogor, Howrah (Calcutta), Pamplemousses (Mauritius) and Singapore in introducing and developing plantation crops such as tea, oil palm, rubber, coffee and various spices are fully recognized (Heywood, 1991). Botanic gardens are now much more involved in the conservation of plant genetic resources, particularly for non-crop, medicinal and wild species, focusing on rare and endangered species (Du Puy and Wyse Jackson, 1995; Maunder et al, 2004). For example, the Royal Botanic Garden in Edinburgh, UK, developed an International Conifer Conservation Programme in 1991, and has been engaged in activities on assessing the conservation status of endangered conifers and has developed a Conifer Action Plan for the IUCN. It has also been active in carrying out applied research activities on conifers and on establishing a network of *in situ* and *ex situ* sites to protect the threatened species.

The conservation role of botanic gardens has often been the subject of much debate. Given that botanic gardens have limited amounts of space, the number of accessions of individual species is limited and, thus, their value for genetic diversity conservation is often questioned. However, it has been demonstrated that for rare species, botanic garden collections may help to conserve a higher genetic diversity than wild populations and can be used to augment the genetic diversity of wild populations. An example is the wild populations of *Brighamia insignis* A. Gray, an endemic species of Hawaii, which is represented by only 20 individuals in the wild but is widely cultivated in botanic gardens. Using isozymes, Gemmill et al (1998), were able to show that the collections held at the National Tropical Botanic Garden (NTBG) in Hawaii, were a good representation of the diversity found in the wild and would therefore serve as suitable stock population to augment natural populations. Botanic gardens also have considerable horticultural expertise that can help with the propagation of rare species and subsequently their reintroduction back into the wild. An example of the use of botanic gardens for *ex situ* conservation in Sri Lanka is given in Box 12.3.

Tissue culture

The problems associated with field genebanks as described above have prompted much research into the development of alternative techniques, notably *in vitro* culture or tissue culture techniques for recalcitrant seed and vegetatively propagated species (see Box 12.2). The major prerequisites for conservation by tissue culture are the availability of skilled personnel and reasonably equipped laboratory facilities (see Reed et al, 2004, for the physical requirements for a plant tissue culture laboratory). Examples of crop wild species that are conserved *in vitro* include the Global *Musa* Germplasm Collection under the management of the International Network for the Improvement of Banana and Plantain (INIBAP)/Bioversity hosted by the Katholieke Universiteit Leuven (KULeuven). The collection contains nearly 1200 accessions and represents the single centralized holding of a large proportion of the known gene pool. About 15 per cent of the collection includes wild *Musa* species (INIBAP, 2006). Other examples of tissue culture collection containing wild relatives include the cassava collection at CIAT, potatoes at CIP, Peru, and wild apples (NCPGR-USDA).

Box 12.3 *Ex situ* conservation initiatives in botanic gardens of Sri Lanka

Under the National Botanic Gardens (NBG), are the Royal Botanic Gardens (RBG) at Peradeniya, Hakgala, Gampaha (Henerathgoda), Sitawake (Awissawella) and Mirijjawila (Hambantota District), which provide coverage of all major climatic zones. The medicinal plant gardens at Ganewatte (on 23ha) in the North-Western Province and a Biodiversity Complex at Gampola, also function under the NBG. The RBG at Peradeniya, located on 59ha has over 4000 species under cultivation. It is mandated for *ex situ* conservation and has pioneered floriculture in Sri Lanka. However, only a fraction of the species in the botanic gardens at present are endemic to Sri Lanka, and the role of these institutions as reservoirs of indigenous biodiversity is not well established, due to historical reasons. This trend has been reversed somewhat in recent times, and the RBG now has 1471 specimens from local species, while the more recently developed herbarium at the Hakgala Botanic Gardens has about 2000 specimens from local species. One of the main objectives of the NBG is for development of technologies related to exploitation of lesser-known and under-utilized plants and development of ornamental and amenity horticulture. There are several medicinal plant gardens located in the wet zone of Sri Lanka (i.e. in Navinna and Meegoda). The Ayurvedic Garden in Navinna harbours around 200 species of medicinal plants, with more than 1500 individual plants.

Source: Fourth Country Report from Sri Lanka to the United Nations Convention on Biological Diversity, 2009

Cryopreservation

Cryopreservation is one of the most promising conservation methods for long-term conservation. One of its most important advantages is that it requires very little space as compared to *in vitro* and field genebanks. It is also the more cost-effective method for long-term conservation requiring very little maintenance (Dulloo et al, 2009). The maintenance of the collection is reduced to mainly topping up the liquid nitrogen, as there is no need for re-culturing, as is the case for *in vitro* conservation. However, cryopreservation protocols, like *in vitro* culture techniques, need to be developed for each and every species, a factor that limits application to a wide diversity of CWR. To date, very few, if any, cryo-collections exist for CWR. Kew has developed protocols for cryopreservation of wild plants, especially ferns, mosses, orchids, shrubs and herbs (see: http://www.kew.org/ksheets/pdfs/K31_cryopreservation.pdf).

Research on cryopreservation has made much progress and protocols for conserving over 200 plant species are now available (Engelmann and Takagi, 2000; Engelmann, 2004). Research work involving CWR undertaken in Australia has led to the development of cryopreservation protocols for *Carica papaya* and a wild relative *Vasconcellea pubescens* (Ashmore et al, 2007) and for certain *Citrus* species (Hamilton et al, 2005, 2008). For a review of cryopreservation techniques, see also Engelmann (2000), Thormann et al (2006) and Reed (2008). Of these,

Reed (2008) provides a practical guide on plant cryopreservation and gives step-by-step instructions for the transfer of cryopreservation technology in conservation of important plants materials.

Pollen storage

Pollen is another plant material that can be stored and used as a conservation method for genetic resources, especially for perennial species of fruit and forest trees, and can be of much interest for CWR. It is commonly used by plant breeders, particularly for production of haploids in breeding programmes, to bridge the gap between male and female flowering time and to improve fruit setting in orchards (Towill, 1985; Alexander and Ganeshan, 1993). For example, the main use of coffee pollen is for breeding, since crosses may have to be made between trees that do not flower simultaneously or that grow far apart (Walyaro and van der Vossen, 1977). Collection and storage of pollen could be a way to obtain a more representative sample of genetic diversity in wild populations (Panella et al, 2009). For this reason alone, pollen can be an effective way of conserving, as well as using, CWR in breeding activities.

Pollen is also used for distributing and exchanging germplasm among locations, since transfer of pests and diseases through pollen is rare (except for some virus diseases) and is subjected to less stringent quarantine restrictions. Other uses are preserving nuclear genes of germplasm, studies in basic physiology, biochemistry and fertility, and studies for biotechnology involving gene expression, transformation and *in vitro* fertilization (Towill and Walters, 2000).

Pollen storage also has several disadvantages. Many species produce small amounts of pollen, which are not sufficient for effective pollen collection and processing. Because of its low viability, pollen needs to be replenished periodically. In this context, it is obvious why pollen preservation is supplemental: the seed or clone must also be conserved to yield the pollen. Multiple generations introduce the risk of population genetic problems, such as loss of alleles through random drift or splitting of adaptive complexes. Only paternal material is conserved and regenerated, and in order to utilize the germplasm, a recipient female plant must always be available for fertilization.

Use of *ex situ* collection in the recovery and reintroduction of CWR populations

Wild populations of CWR are often depauperate genetically to the point of near extinction as a result of habitat degradation and other threats. *In situ* conservation of these populations would require the development of a recovery plan and active interventions to reconstitute these populations. It is important to ensure a broad genetic base of the wild populations to guarantee its survival in the long term, especially with rapidly changing environmental conditions including climate change.

Ex situ collections can be used in recovery programmes in two main ways:

1 To reintroduce a species that has disappeared from its natural site. While the species may have become extinct from one of its sites, if accessions from the same site have been collected in the past and conserved in genebanks or in botanic gardens, these can provide valuable materials for restoration. However, the reintroduction of *ex situ* materials to the wild can be a complex activity and needs to be undertaken with great caution. One must ensure that the stock or accessions introduced are really native to the site, that the plants are free of diseases and that they have adequate genetic diversity to ensure their survival, etc. To assist conservationists in thinking through and taking all factors into account, the IUCN/SSC Re-introduction Specialist Group have developed policy guidelines for reintroduction (IUCN/SSC, 1995). These guidelines are applicable to both animals and plants and are therefore rather general. The IUCN technical guidelines on the management of *ex situ* populations for conservation (IUCN, 2002), also discusses the increasing value of *ex situ* conservation in *in situ* ecosystem and habitat conservation. The *Handbook for Botanic Gardens on the Reintroduction of Plants to the Wild* (Akeroyd and Wyse Jackson, 1995) published by BGCI contains plant-specific guidelines and provides botanic garden managers with guidance on the reintroduction of plants materials from botanic gardens to the wild and explores the issues of reintroduction and challenges of the reintroduction process.
2 *Ex situ* collections can be used in enrichment planting or reinforcement or supplementation if the population is threatened and is not regenerating in the wild. New plant material may be obtained from *ex situ* collections and planted to reinforce the population at the site. Again, it is important to observe precautions in such practices so as not to disrupt and threaten the genetic integrity of the natural population. In recovery programmes, it is important to consider the provenance of the material, the use of genetically variable reintroduction stock, as well as the potential of loss of genetic diversity (IUCN/SSC, 1995; IUCN, 2002; Guerrant et al, 2004; Kell et al, 2008).

In both these cases, it is important to ensure that the provenance of the materials introduced comes from the same site or as close to it as possible, in order to ensure the genetic integrity of the population. It is also most likely that material from the site would be locally adapted to it and this would ensure higher probability of success of the reintroduction. Often, however, such materials may not be available. In such cases, it is recommended that plant materials come from environments that have matching ecogeographic characteristics.

In practical terms, when there is a need to use *ex situ* collections for *in situ* intervention, the following steps should be followed and be included in the recovery plan:

1 **Site assessment** – a thorough examination of the site should be carried out, documenting not only the status of the target population, including popula-

tion size of the target species, patterns of distribution at the site, competitive plants, associated plants, pollinators, dispersers and predators, but also any threats affecting the population. The latter would need to be resolved prior to any reintroduction of the species. The site assessment would determine the strategy to adopt for replanting, in terms of planting density, pattern of planting, revegetation methods required (see below), etc.

2 **Revegetation method** – there are a number of revegetation methods that can be used to reintroduce the species back into the wild. These could be through direct seeding, planting using naked-rooted seedlings, potted seedlings or planting under nurse crops.
3 **Identification of source material** – the source of the material from *ex situ* collection must be chosen with a great deal of attention. Accessions coming from the same site, or as close to it as possible, should be selected.
4 **Sampling to ensure genetic diversity** – samples from the genebank accession(s) should be taken so as to represent the maximum genetic diversity present in the accessions. It is recommended to sample seeds from as many accessions as possible.
5 **Propagating of materials** – the planting materials (seeds or cuttings) should be multiplied in a nursery, taking into account dormancy and germination difficulties and an equal number of plants from each accession raised to the required number of plants needed for the replanting. It is important to clearly label all the plants with scientific names and accession number for long-term monitoring.
6 **Site preparation and replanting** – the success of the reintroduction will depend on good site preparation. As mentioned above, if there are any competing factors (competing alien plants, predators) that would affect the regeneration of the plants, they would need to be controlled prior to planting. Methods used could be as simple as weeding out competing plants to more elaborate treatments using chemical or biological control agents, depending upon the nature of the problem.
7 **Post-planting treatment** – once planted, the seedlings should be monitored and measures taken to ensure their survival. This may include mulching and weed control, either by hand or using herbicides. If they die, they need to be replaced from the nursery stock. It is important that a nursery stock of the *ex situ* accessions continue to be maintained to provide for some gap filling after planting in the wild.

Inter situs and other conservation approaches

In addition to *in situ* and *ex situ* conservation strategies, a number of other approaches have been developed recently, some of which blur the distinction between *ex situ* and *in situ*. For tree species, for example, the concept of 'forest genebanks' has been introduced (Shaanker et al, 2002): these are *in situ* sites that act as repositories of genes from as many diverse populations as possible, so as to maximize the representation of genes captured. Other strategies involve maintaining *ex situ* populations in artificially created simulations of the ecosystems in which they occur naturally.

The term *inter situs*[2] conservation has been applied to the reintroduction of species to locations outside their current range but within the known recent past range of the species[3] (Burney and Burney, 2009). It contrasts with the 'assisted migration' discussed in Chapter 14 and has been practised with apparent success to save rare Hawaiian plants. It is a procedure which involves considerable risks and should not be practised except in very urgent cases.

Sources of further information

Akeroyd, J. and Wyse Jackson, P. (1995) *A Handbook for Botanic Gardens on Reintroduction of Plants to the Wild*, Botanic Gardens Conservation International (BGCI), p31.

ENSCONET (2009) *ENSCONET Seed Collecting Manual for Wild Species*. ISBN: 978-84-692-3926-1, available at: http://www.maich.gr:9000/PDF/Collecting_protocol_English.pdf.

Engels, J.M.M., Maggioni L., Maxted N. and Dulloo, M.E. (2008) 'Complementing *in situ* conservation with *ex situ* measures', in J. Iriondo, N. Maxted and M.E. Dulloo (eds) *Conserving Plant Genetic Diversity in Protected Areas,* Chapter 6, pp169–181, CAB International, Wallingford, UK.

Guarino, L., Ramanatha Rao, V. and Reid, R. (1995) 'Collecting plant genetic diversity technical guidelines, CAB International, Wallingford, UK.

Notes

1. http://www.kew.org/msbp/scitech/publications/03-Collecting%20techniques.pdf; http://www.kew.org/msbp/scitech/publications/fieldmanual.pdf
2. Usually referred to, incorrectly and ungrammatically, as *inter situ*.
3. This usage differs from that of Blixt (1994) who applies it to the maintenance of domesticates in farmers' fields, more commonly referred to as on-farm conservation.

References

Akeroyd, J. and Wyse Jackson, P. (1995) *A Handbook for Botanic Gardens on Reintroduction of Plants to the Wild,* Botanic Gardens Conservation International (BGCI), Richmond, UK

Alexander, M.P. and Ganeshan, S. (1993) 'Pollen storage', in K.L. Chadha and J.E. Adams (eds) *Advances in Horticulture, vol 1, Fruit Crops: Part I,* Malhotra Publishing House, New Delhi, India

Ashmore S.E., Drew, R.A. and Azimi-Tabrizi, M. (2007) 'Vitrification-based shoot tip cryopreservation of *Carica papaya* and a wild relative *Vasconcellea pubescens', Australian Journal of Botany,* vol 55, pp541–547

BGCI (1998) *Seed Banks,* Botanic Gardens Conservation International (BGCI), http://www.bgci.org/resources/Seedbanks/ (accessed 30 May 2010)

Blixt, S. (1994) 'Conservation methods and potential utilization of plant genetic resources in nature conservation', in F. Begemann and K. Hammer (eds) *Integration of Conservation Strategies of Plant Genetic Resources in Europe*, IPK and ADI, Gatersleben

Burney, D.A. and Burney, L.P. (2009) '*Inter situ* conservation: Opening a 'third front' in the battle to save rare Hawaiian plants', *BGjournal*, vol 6, pp17–19

Dulloo, M.E., Ramanatha Rao V., Engelmann F. and Engels J. (2005) 'Complementary conservation of coconuts', in P. Batugal, V.R. Rao and J. Oliver (eds) *Coconut Genetic Resources*, pp75–90, IPGRI-APO, Serdang, Malaysia

Dulloo, M.E., Ebert, A.W., Dussert, S., Gotor, E., Astorg, C., Vasquez, N., Rakotomalala, J.J., Rabemiafar, A., Eira, M., Bellachew, B., Omondi, C., Engelmann, F., Anthony, F., Watts, J., Qamar, Z. and Snook, L. (2009) 'Cost efficiency of cryopreservation as a long-term conservation method for coffee genetic resources', *Crop Science*, vol 49, pp2123–2138, doi:10.2135/cropsci2008.12.0736

Du Puy B. and Wyse Jackson P. (1995) 'Botanic gardens offer key component to biodiversity conservation in the Mediterranean', *Diversity*, vol 11, no 1 and 2, pp47–50

Engelmann, F. (ed) (1999) *Management of Field and* In Vitro *Germplasm Collection*, Proceedings of a consultation meeting, 15–20 January 1996, CIAT, Cali, Colombia, International Plant Genetic Resources Institute (IPGRI), Rome, Italy

Engelmann, F. (2000) 'Importance of cryopreservation for the conservation of plant genetic resources', in F. Engelmann and H. Tagaki (eds) *Cryopreservation of Tropical Plant Germplasm: Current Research Progress and Application*, Japan International Research Center for Agricultural Sciences, Tsukuba, Japan/International Plant Genetic Resources Institute, Rome, Italy

Engelmann, F. (2004) 'Plant cryopreservation: Progress and prospects', *In Vitro Cellular and Developmental Biology – Plant*, vol 40, pp427–433

Engelmann, F. and Takagi, H. (eds) (2000) *Cryopreservation of Tropical Plant Germplasm: Current Research Progress and Applications*, Japan International Research Centre for Agricultural Sciences, Tsukuba, Japan/IPGRI, Rome, Italy

Engels, J.M.M. and Visser, L. (eds) (2003) *A Guide to Effective Management of Germplasm Collections*, International Plant Genetic Resources Institute Handbooks for Genebanks 6, International Plant Genetic Resources Institute (IPGRI), Rome, Italy

Engels, J.M.M. and Wood, D. (1999) 'Conservation of agrobiodiversity', in D. Wood and J.M. Lenné (eds) *Agrobiodiversity: Characterization, Utilization and Management*, pp355–385, CAB International, Wallingford, UK

Engels, J.M.M., Maggioni, L., Maxted, N. and Dulloo, M.E. (2008) 'Complementing *in situ* conservation with *ex situ* measures', in J. Iriondo, N. Maxted and M.E. Dulloo (eds) *Conserving Plant Genetic Diversity in Protected Areas*, Chapter 6, pp169–181, CAB International, Wallingford, UK

ENSCONET (2009) *ENSCONET Seed Collecting Manual for Wild Species*, European Native Seed Conservation Network (ENSCONET), ISBN: 978-84-692-3926-1, http://www.ensconet.eu/Download.htm, accessed 31 May 2010

Falk, D.A. and Holsinger, K.E. (eds) (1991) *Genetics and Conservation of Rare Plants*, Oxford University Press, New York and Oxford

FAO (2010) *Second Report on the State of the World's Plant Genetic Resources for Food and Agriculture*, Food and Agriculture Organization of the United Nations (FAO), Rome, Italy

FAO and IPGRI (1994) *Genebank Standards*, Food and Agriculture Organization of the United Nations (FAO)/International Plant Genetic Resources Institute (IPGRI), Rome, Italy

Gemmill, C.E.C., Ranker, T.A., Ragone, D., Pearlman, S.P. and Wood, K.R. (1998) 'Conservation genetics of the endangered endemic Hawai'ian genus *Brighamia* (Campanulaceae)', *American Journal of Botany,* vol 85, no 4, pp528–539

Given, D.R. (1994) *Principles and Practice of Plant Conservation,* Timber Press, Portland, OR, USA

Guarino, L., Ramanatha Rao, V. and Reid, R. (1995) *Collecting Plant Genetic Diversity Technical Guidelines,* CAB International, Wallingford, UK

Guarino L., Jarvis A., Hijmans R.J. and Maxted N. (2001) 'Geographic information systems (GIS) and the conservation and use of plant genetic resources', in J. Engels, V. Ramanatha Rao, A.H.D. Brown and M.T. Jackson (eds) *Managing Plant Genetic Diversity,* pp387–404, CAB International, Wallingford, UK

Guerrant Jr., E.O., Havens, K. and Maunder, M. (eds) (2004) *Ex Situ Plant Conservation. Supporting Species Survival in the Wild,* Island Press, Washington, DC

Hajjar, R. and Hodgkin, T. (2007) 'The use of wild relatives in crop improvement: A survey of developments over the last 20 years', *Euphytica,* vol 156, pp1–13

Hamilton, K.N., Ashmore, S.E. and Drew, R.A. (2005) 'Investigations on desiccation and freezing tolerance of *Citrus australasica* seed for *ex situ* conservation', in S.W. Adkins, P.J. Ainsley, S.M. Bellairs, D.J. Coates and L.C. Bell (eds) *Proceedings of the Fifth Australian Workshop on Native Seed Biology,* pp157–161, Australian Centre for Minerals Extension and Research (ACMER), Brisbane, Queensland, AUS

Hamilton, K.N. (2008) 'Protocol 19.7.2 – Cryopreservation of wild Australian citrus seed', in H.W. Pritchard and J. Nadarajan 'Cryopreservation of orthodox (desiccation tolerant) seeds', in B.M. Reed (ed) *Plant Cryopreservation: A Practical Guide,* Springer, Berlin, Germany

Hawkes J.G., Maxted N. and Ford-Lloyd, B.V. (2000) *The* Ex Situ *Conservation of Plant Genetic Resources,* Kluwer Academic Publishers, Dordrecht, The Netherlands

Heywood V.H. (1991) 'Developing a strategy for germplasm conservation in botanic gardens', in V.H. Heywood and P.S. Wyse Jackson (eds) *Tropical Botanic Gardens – Their Role In Conservation and Development,* pp11–23, Academic Press, London, UK

Heywood, V.H. (2009) 'Botanic gardens and genetic conservation', *Sibbaldia* guest essay, *Sibbaldia, The Journal of Botanic Garden Horticulture,* no 7, pp5–17

Hijmans, R.J. and Spooner, D.M. (2001) 'Geographic distribution of wild potatoes species', *American Journal of Botany,* vol 88, no 11, pp2101–2112

Hijmans, R.J., Guarino, L., Cruz , M. and Rojas, E. (2001) 'Computer tools for spatial analysis of plant genetic resources data:1 DIVA-GIS', *Plant Genetic Resources Newsletter,* vol 127, pp15–19

INIBAP (2006) *Global Conservation Strategy for* Musa *(Banana and Plantain),* International Network for the Improvement of Banana and Plantain (INIBAP), www.croptrust.org/documents/web/Musa-Strategy-FINAL-30Jan07.pdf, accessed 23 March 2010

IUCN (2002) *IUCN Technical Guidelines on the Management of* Ex-Situ *Populations for Conservation,* approved at the 14th Meeting of the Programme Committee of IUCN Council, Gland, Switzerland, 10 December 2002, International Union for Conservation of Nature (IUCN), http://intranet.iucn.org/webfiles/doc/SSC/SSCwebsite/Policy_statements/IUCN_Technical_Guidelines_on_the_Management_of_Ex_situ_populations_for_Conservation.pdf

IUCN/SSC (1995) *IUCN/SSC Guidelines for Re-introductions,* SSC Re-introduction Specialist Group, approved by the 41st Meeting of the IUCN Council, Gland, Switzerland, May 1995. International Union for Conservation of Nature (IUCN),

Gland, Switzerland, http://iucnsscrsg.org/policy_guidelines.php
Jarvis A, Williams, K., Williams,D., Guarino, L., Caballero, P.J. and Mottram, G. (2005) 'Use of GIS in optimizing a collecting mission for a rare wild pepper (*Capsicum flexuosum* Sendtn.) in Paraguay', *Genetic Resources and Crop Evolution*, vol 52, no 6, pp671–682
Kell, S.P., Laguna, L., Iriondo, J. and Dulloo, M.E. (2008) 'Population and habitat recovery techniques for the *in situ* conservation of genetic diversity', in J. Iriondo, N. Maxted and M.E. Dulloo (eds), *Conserving Plant Genetic Diversity in Protected Areas,* Chapter 5, pp124–168, CABI Publishing, Wallingford, UK
Laliberté, B. (1997) 'Botanic garden seed banks/genebanks worldwide, their facilities, collections and network', *Botanic Gardens Conservation News,* vol 2, pp18–23
Maunder M., Guerrant Jr, E.O., Havens, K. and Dixon, K.W. (2004) 'Realizing the full potential of *ex situ* contributions to global plant conservation', in E.O. Guerrant Jr., K. Havens and M. Maunder (eds) Ex Situ *Plant Conservation. Supporting Species Survival in the Wild,* Island Press, Washington, DC
Maxted, N. and Kell, S.P. (2009) *Establishment of a Global Network for the* In Situ *Conservation of Crop Wild Relatives: Status and Needs*, FAO Commission on Genetic Resources for Food and Agriculture, Rome, Italy
Maxted, N., Ford-Lloyd, B.V. and Hawkes, J.G. (1997) 'Complementary conservation strategies', in N. Maxted, B.V. Ford-Lloyd and J.G. Hawkes (eds) *Plant Genetic Conservation: The* In Situ *Approach,* Chapman and Hall, London, UK
Maxted, N., Dulloo, M.E., Ford-Lloyd, B.V., Iriondo, J. and Jarvis, A. (2008) 'Gap analysis: A tool for complementary genetic conservation assessment', *Diversity and Distributions*, vol 14, no 6, pp1018–1030
Miller, A.G. and Nyberg, J.A. (1995) 'Collecting herbarium vouchers', in L. Guarino, V. Ramanatha Rao and R. Reid (eds) *Collecting Plant Genetic Diversity Technical Guidelines,* Chapter 27, pp561–573, CAB International, Wallingford, UK
Panella L., Wheeler, L. and McClintock, M.E. (2009) 'Long-term survival of cryopreserved sugarbeet pollen', *Journal of Sugar Beet Research,* vol 46, pp1–9
Rao N.K., Hanson, J., Dulloo, M.E., Ghosh, K., Nowell, D. and Larinde, M. (2006) *Manual of Seed Handling in Genebanks,* Handbooks for Genebanks No 8, Bioversity International, Rome, Italy
Reed, B. (ed) (2008) *Plant Cryopreservation: A Practical Guide,* Springer, New York, USA
Reed, B., Engelmann F., Dulloo M.E. and Engels J.M.M. (2004) *Technical Guidelines on Management of Field and* In Vitro *Germplasm Collections*, Handbook for Genebanks No 7, IPGRI, Rome, Italy
Schmidt, L. (2000) *Guide to Handling Tropical and Subtropical Forest Seed,* Danida Forest Seed Centre, http://en.sl.life.ku.dk/Publikationer/Udgivelser/DFSC/DFSCBook1.asp.
Shaanker, Uma R., Ganeshaiah, K.N., Nageswara Rao, M. and Ravikanth, G. (2002) 'Forest gene banks – a new integrated approach for the conservation of forest tree genetic resources', in J.M.M. Engels, A.H.D. Brown and M.T. Jackson, (eds) *Managing Plant Genetic Diversity*, pp229–235, CAB International, Wallingford, UK
Sharrock, S. and Engels, J. (1996) 'Complementary Conservation', INIBAP Annual Report 1996, pp 8–9, INIBAP, Montpellier.
Smith, R.D. (1995) 'Collecting and handling seeds in the field', in L. Guarino, V. Ramanatha Rao And R. Reid (eds) *Collecting Plant Genetic Diversity Technical Guidelines,* Chapter 20, pp419–456, CAB International, Wallingford, UK
Smith R.D., Dickie, J.B., Linington, S.H., Pritchard, H.W and Probert, R.J. (2003) *Seed Conservation: Turning Science into Practice,* Royal Botanic Gardens Kew, Richmond, UK

Thormann, I., Dulloo, M.E. and Engels, J. (2006) 'Techniques for *ex situ* plant conservation', in R.J. Henry(ed), *Plant Conservation Genetics,* pp7–36, Haworth Press, AUS

Towill, L.E. (1985) 'Low temperature and freeze-/vacuum-drying preservation of pollen', in K.K. Harthaa(ed) *Cryopreservation of Plant Cells and Organs,* pp171–198, CRC Press, Boca Raton, FL, USA

Towill, L.E. and Walters, C. (2000) 'Cryopreservation of pollen', in F. Engelmann and H. Takagi(eds) *Cryopreservation of Tropical Plant Germplasm – Current Research Progress and Applications,* pp115–129, Japan International Centre for Agricultural Sciences, Tsukuba/International Plant Genetic Resources Institute, Rome, Italy

Walyaro, D.J. and van der Vossen, H.A.M. (1977) 'Pollen longevity and artificial cross-pollination in *Coffea arabica* L', *Euphytica,* vol 26, pp225–231

Chapter 13

Monitoring of Areas and Species/ Populations to Assess Effectiveness of Conservation/Management Actions

Introduction: Surveillance and monitoring

> *The primary purpose of monitoring, if not launched purely for scientific interest, is to collect information that can be used for development of conservation policy, to examine the outcomes of management actions and to guide management decisions* (Kull et al, 2008).

Monitoring is a core activity of biodiversity conservation and of conservation biology (Marsh and Trenham, 2008) and has been described as a centrepiece of nature conservation across the globe (Schmeller, 2008). And yet, as often noted, many monitoring programmes do not have a sound ecological basis, are poorly designed, do not lead to management interventions or responses and are disconnected from decision-making. Monitoring is often given low priority because it can be difficult and expensive to implement (Danielsen et al, 2009) and monitoring programmes are often inadequately funded and inadequately implemented.

Essentially, monitoring consists of making reliable observations from nature to detect, measure, assess and draw conclusions as to how species and ecosystems are changing through time and space, either naturally or as a consequence of deliberate or inadvertent human intervention. It is applied in many different ways – to track the status of endangered species, the spread of invasive species, the health of ecosystems, the effectiveness of protected areas and other conservation actions, and more generally to assess the state and main trends of biodiversity through indicators and monitoring at national, regional and global levels. A useful review of the conceptual issues involved in ecological monitoring is given by Noon (2003).

Monitoring is undertaken at various scales: it can be applied from the population and individual level to the whole biosphere. Monitoring is undertaken at a global level by the Convention on Biological Diversity (CBD), United Nations agencies (e.g. the Food and Agriculture Organization of the United Nations (FAO) and the United Nations Environment Programme (UNEP)), by international non-governmental organizations (INGOs) (e.g. the Consultative Group on International Agricultural Research (CGIAR) and the Organisation for Economic Co-operation and Development (OECD)) and by non-governmental organizations (NGOs) (e.g. the International Union for Conservation of Nature (IUCN), the World Wide Fund for Nature (WWF) and the World Resources Institute (WRI)). It is also undertaken at the regional level (e.g. by the European Community) and at national and local levels.

The CBD proposes the following actions under Article 7: Identification and Monitoring:

- identify ecosystems, species and genomes important for conservation and sustainable use;
- monitor the components identified to determine priorities;
- identify and monitor activities that may be harmful to biodiversity;
- maintain and organize data obtained from the above.

The International Treaty on Plant Genetic Resources (ITPGRFA), on the other hand, does not mention the monitoring of biodiversity or agricultural biodiversity, even though it is clearly an important component of actions needed to maintain such biodiversity and use it sustainably (see Box 13.1). A summary of agrobiodiversity monitoring information at a European level is given by Schröder et al (2007) who note that one important precondition of agrobiodiversity indicators is the documentation of genetic resources in national and international inventories.

Biodiversity monitoring is a highly technical and complex area and it is beyond the scope of this manual to go into additional details on this topic. Several major texts and handbooks have been published to which the reader is referred

Box 13.1 Agricultural biodiversity monitoring

The monitoring of agrobiodiversity has two main tasks: to document loss of agrobiodiversity as early as possible and to work as a management tool concerning the objectives, the programmes and the necessary measures for the conservation and sustainable use of agrobiodiversity. In addition, it visualizes the outcomes of a policy that is dedicated to sustainability. Therefore the instruments of monitoring, like regular surveys, indicators and inventories, need to be further developed.

Source: Information System Genetic Resources (GENRES)
http://www.genres.de/genres_eng/agrobiodiv/agrobiodiv_mon.htm, accessed 1 October 2009

(see the section on further information at the end of the chapter). A critical review of biological monitoring and of recent developments is given by Yoccoz et al (2001).

Monitoring is defined by Elzinga et al (1998) as 'the collection and analysis of repeated observations or measurements to evaluate changes in condition and progress toward meeting a management objective'. The term 'surveillance' which originates from the French word meaning to 'watch over' is often used interchangeably with 'monitoring'. Both imply repeated recording of information over time. 'Sampling', 'recording' and 'observation' may be one-off events, or form part of a surveillance or monitoring scheme. A more rigorous definition is given by Hellawell (1991): 'intermittent (regular or irregular) surveillance undertaken to determine the extent of compliance with a predetermined standard or the degree of deviation from an expected norm'. The standard, in this context, according to Tucker et al (2005), can be a baseline position such as the maintenance of a particular area or population or a position set as an objective such as 200ha of a particular habitat or 200 individuals of a population.

Briefly, monitoring can (Tucker et al, 2005):

- establish whether standards are being met;
- detect changes and trigger responses if any of the changes are undesirable;
- contribute to the diagnosis of the causes of change;
- assess the success of actions taken to maintain standards or to reverse undesirable changes and, where necessary, contribute to their improvement.

A distinction may be made between two types of monitoring: *status* and *strategy effectiveness* monitoring (Ervin et al, 2010). As they note, *status monitoring* asks the question, 'What is the status and trend of biodiversity independent of our actions?' while *strategy effectiveness monitoring* asks the question, 'Are our conservation actions achieving the desired results?' (see Ervin et al, 2010, Box 24). Both types are important in monitoring programmes for CWR.

Establishing a baseline

A critical issue in monitoring and the use of indicators is the need to establish a baseline from which to start and compare the data to be collected. This will involve compiling and reviewing existing information on the population, species, habitat or other element, process or action that is the target of monitoring. In practice, this is much more difficult than might appear at first sight. We have already seen how incomplete or inadequate is our knowledge of many aspects of biodiversity that affect the conservation of CWR, for example, the lack of inventory of protected areas, uncertainties about the detailed geographical distribution of species, the existence and pattern of genetic variation within populations, the extent of genetic erosion, the extent to which ecosystems are affected by invasive alien species, etc. Ecogeographic surveys, discussed in detail in Chapter 8, will provide such a baseline for many of the features that one might wish to monitor in a CWR conservation programme.

It is also important that agreed definitions of key terms are used so that measurements are comparable. Alternative values for definition of parameters can have significant impacts. For example, when FAO redefined the term 'forest' between the 1990 and 2000 Forest Resource Assessments, reducing minimum height from 7 to 5m, minimum area from 1.0 to 0.5ha and crown cover from 20 per cent in developed and 10 per cent in developing countries to a uniform 10 per cent, global forest increased by 300 million ha or approximately 10 per cent. Likewise, forests are defined in different ways by different countries – by principal land use (in Bolivia), by forest cover (in Chile) – and the threshold for the definition of forest cover differs from less than 10 per cent in Iran to 75 per cent in South Africa.

Faced with such a situation, the important thing in any monitoring programme is to ensure that terminology is used consistently by all participants, especially when many different actors are involved. As already discussed in Chapter 8, widely agreed standards should be followed, such as those of TDWG (Biodiversity Information Standards). Similarly, accurate taxonomic information is essential, as has been stressed already in Chapters 6 and 8 of this manual.

For guidance on sampling and measuring vegetation characteristics, such as stratification, cover, phytomass and leaf area index, see van der Maarel (2005) and Bonham (1989) and for structural-physiognomic features such as growth form, see the following texts: *Aims and Methods of Vegetation Ecology* by Mueller-Dombois and Ellenberg (1974); *Vegetation Description and Analysis: A Practical Approach* by Kent and Coker (1995) and Dierschke's (1994) classic, *Pflanzensoziologie – Grundlagen und Methoden*. For sampling of species characteristics see van der Maarel (2005). Much useful information on many aspects of sampling and census methods applicable to monitoring can be found in *Ecological Census Techniques – A Handbook* by Sutherland (2006) and information on monitoring can be obtained from Sutherland (2000), *The Conservation Handbook: Techniques in Research, Management and Policy*.

CWR and monitoring: Identification and selection of variables to be measured

For the effective conservation of CWR, a range of monitoring activities may need to be undertaken. These include monitoring the key characteristics of a species and its habitat to ensure that management interventions and actions are in fact meeting their objectives.

For example, one may wish to monitor:

- changes to population/species abundance, trends in population size and structure, so as to assess the health and viability of the population, both before and after any management intervention;

- changes in genetic diversity;
- predator numbers, to assess the effectiveness of control programmes;
- the spread or control of invasive species to assess their impact on the species populations and the habitat or area as a whole;
- changes in vegetation cover or soil condition, to assess the state of the CWR's habitat;
- the effects of management interventions undertaken as part of a species management or recovery plan.

Most schemes monitor both the distribution (range, area) and the species composition of the target habitats or ecosystems.

Species and population monitoring

What is it?

Species and population monitoring is the regular observation and recording of changes in status and trend of species or their populations in a certain territory. The primary purpose of such monitoring is to collect information that can be used to examine the outcomes of management actions and to guide management decisions. This is frequently carried out for species that have been assessed as threatened so as to determine when conservation actions are necessary or when existing ones need to be intensified.

In the case of CWR, it may be necessary to monitor population numbers, size, density, structure and demographic variables as part of the assessment of their conservation status, as well as the subsequent impacts on the CWR populations of any management interventions that are prescribed in the species management plan so as to judge their effectiveness.

Species and population monitoring programmes, like biological monitoring in general, are remarkably variable and diverse in scale, coverage and aims. Marsh and Trenham (2008) attempted to detect trends in plant and animal population monitoring, and the goals and strategies showed signs of diversifying with some approaches becoming more frequent, such as area occupied and presence/absence approaches, while others have yet to be widely applied, such as risk-based monitoring and linking the results of monitoring directly to management decisions. It is important, therefore, that the objectives of any proposed population monitoring are clearly defined in advance (see Yoccoz et al, 2001).

Many sampling and analytical techniques are available for species and population monitoring (see Chapter 8) and are reviewed by Stork and Samways (1995) and specifically for CWR by Iriondo et al (2008).

Which attributes to monitor?

The attributes of species for which monitoring goal targets may be set include range, abundance, demography, population dynamics and habitat requirements (Tucker et al, 2005):

Quantity

- presence/absence;
- range;
- population size;
- frequency;
- number/density;
- cover.

Population dynamics

- recruitment;
- mortality;
- emigration;
- immigration.

Population structure

- age;
- sex ratio;
- fragmentation or isolation;
- genetic diversity.

Habitat requirements

Demographic monitoring

Demographic monitoring is the most common form of population monitoring, especially for rare or endangered species, and will often be found to be an appropriate approach for CWR where the focus is often on the maintenance of viable populations and their genetic variability.

Demographic monitoring is the assessment of population changes and their causes throughout the life cycle, and measures attributes such as germination and mortality rates, growth, size, density and distribution. It can also be used to help establish the factors determining the distribution and abundance of species and predict the future structure of populations. Demographic monitoring may involve frequent measurements or mapping if the necessary level of resolution is to be achieved (Given, 1994). The main demographic approaches are: (1) population and availability analysis; (2) single age/stage class investigations; and (3) demographic structure (Elzinga et al, 1998). Demographic approaches to monitoring are often time-consuming and expensive procedures and therefore not always feasible. Moreover, a demographic approach may not be appropriate for particular situations: Elzinga et al (1998), for example, caution against the inappropriate use of demographic monitoring for certain types of species, notably those with long-lived seed banks, dense vegetative reproduction, very short or very long lifespan, episodic reproduction, multiple stems and mat-like morphology, high densities and large populations in heterogeneous habitats (see Elzinga et al, 1998, Figure 12.13).

Genetic monitoring: What is it and when to use it?

The conservation of CWR focuses, as we have seen, on the genetic diversity found within the target species as a possible source of traits that may be used in breeding. But, at the same time, the long-term aim of *in situ* conservation of CWR is to ensure that sufficient genetic variation is maintained so as to ensure the survival of the species and allow the evolutionary processes to continue, thereby generating new variation that may allow the species to adapt to changing conditions. This is best done by protecting the environment and habitats in which the target species occur and controlling or limiting the threats that affect both the habitats and the species.

As we have seen, population monitoring can be a laborious and expensive exercise. The monitoring of genetic diversity can be an even more costly approach, especially if molecular methods are employed, so that its widespread use is not possible or even to be recommended. There may, however, be circumstances in which it is important to undertake genetic monitoring of high priority CWR. As already indicated when considering field ecogeographic surveying, information on the distribution of genetic variation in populations of CWR will normally have to be obtained through surrogate measures such as morphological (also 'visible') markers, which themselves are phenotypic traits or characters such as leaf shape, flower colour, growth habit, or through the use of biochemical markers (including allelic variants of isozymes detected through electrophoresis).

The circumstances in which genetic monitoring is likely to be used are discussed by Iriondo et al (2008, pp118–120), who give a series of examples. In particular it may be used:

- To assess the genetic diversity within the target populations in terms of total numbers of genotypes or alleles (richness) or the frequency of different genotypes or alleles (evenness). Such information on genetic diversity can be used to help to compare populations and determine which should be selected for *in situ* conservation and to decide which populations should be monitored so as to follow changes in genetic diversity over time.
- To estimate geneflow between populations, trends in the extent of inbreeding within populations and differentiation between populations or subpopulations.

Genetic analysis software

A wide range of software packages for undertaking genetic analyses is available. For a comprehensive alphabetical listing containing nearly 500 programmes see: *An Alphabetic List of Genetic Analysis Software*, currently maintained at North Shore Long Island Jewish Research Institute, New York, USA (2002 to date).[1]

Molecular markers for genetic analysis

An extensive range of molecular markers for use in genetic monitoring is available, but developments in this area are rapid and frequent so the reader should search the internet for the latest technologies. Molecular or DNA markers are loci (sites) in the genome of an organism at which the DNA base sequence varies among the different individuals of a population. They have the advantage over morphological or biochemical markers that they are not affected by environmental factors or the state of development of the plant.

Most of the reviews of molecular markers assume that the reader has a good level of knowledge of plant genetics and molecular biology; however, a very useful introduction to markers (and their use in marker assisted selection) for those with only basic knowledge has been written by Collard et al (2005).

Desirable features of DNA markers that have been suggested by various authors (e.g. Joshi et al, 1999; Iriondo et al, 2008) include:

- highly polymorphic in nature;
- co-dominant inheritance (determination of homozygous and heterozygous states of diploid organisms);
- frequent occurrence and scattered throughout the genome;
- selectively neutral behaviour (can be focused on expressed genes);
- readily available;
- easy to use, rapid assay and cheap;
- high reproducibility; and
- data can be easily and reliably exchanged between laboratories.

Unfortunately, no single marker type matches all these criteria, although some come close such as SSRs (simple sequence repeats).

A comparison of four molecular markers: inter-retrotransposon amplified polymorphism (IRAP); retrotransposon-microsatellite amplified polymorphism (REMAP); sequence-specific amplified polymorphism (SSAP); and amplified fragment length polymorphism (AFLP) for genetic analysis in *Diospyros* L. (Ebenaceae) in terms of information value and effectiveness is given by Du et al (2009). A comparison of molecular markers for genetic analysis of *Macadamia* in terms of the type, amount and cost-efficiency of the information generated, using data from published studies, is given by Peace et al (2004).

A summary of 'dos' and 'don'ts' regarding the user of molecular markers is given by Iriondo et al (2008):

- *Do not* plan to do molecular population genetic monitoring first in any *in situ* conservation assessment.
- *Do not* undertake molecular population genetic assessment/monitoring without very good reason, or without specific questions to answer, and until other proxy genetic assessments have been fully examined.
- *Do not* necessarily plan for routine sequential population genetic monitoring.
- *Do* use molecular population genetic assessment as a last resort and for fine-tuning to:
  - select the most suitable and fittest populations for *in situ* conservation;

- measure inbreeding/outbreeding in a species as a pilot survey;
- monitor populations or critical situations;
- select for conservation among candidate populations of inbreeding species;
- select the 'best' small isolated populations for protection;
- determine the effects of a severe drop in actual population size on genetic diversity;
- establish whether gene flow is occurring between fragmented populations.

Habitat monitoring

Habitat/protected area monitoring can be defined as 'the collection and analysis of repeated observations or measurements to evaluate changes in condition and progress toward meeting a management objective' (Elzinga et al, 2001).

Habitat monitoring (sometimes known as ecosystem monitoring) involves making repeated recordings of the condition of the target habitats or ecosystems so as to detect or measure changes from a predetermined standard, target state or previous status (Hellawell, 1991). It may cover the range and distribution of habitat types and the area occupied and often their species composition and, in some cases, abundance. It may also provide information on the status of some of the components of the habitat such as species or populations and has been suggested as a cost-effective substitute for the simultaneous monitoring of several species (Gottschalk et al, 2005).

The features of a habitat that may be monitored, including aspects of quantity, structure, function or dynamics (Tucker et al, 2005) are:

Quantity

- area;
- quality: physical attributes;
- geological (e.g. presence of bare rock or deep peat);
- water (e.g. presence of open water or depth of water table).

Quality: composition

- communities;
- richness or diversity;
- typical, keystone or indicator species;
- presence–absence;
- frequency;
- number or density;
- cover;
- biomass.

Quality: structure

- inter-habitat (landscape) scale (e.g. fragmentation, habitat mosaics);

- intra-habitat scale;
- macro-scale;
- horizontal (e.g. plant community mosaics);
- vertical (e.g. ground-, shrub- and tree-layer topography);
- micro-scale;
- horizontal (e.g. patches of short and tall vegetation);
- vertical (e.g. within-layer topography).

Quality: dynamics

- succession;
- reproduction or regeneration;
- cyclic change and patch dynamics.

Quality: function

- physical and biochemical (e.g. soil stabilization, carbon sinks;)
- ecosystem processes.[2]

Habitat monitoring covers a wide variety of approaches. The traditional way of acquiring information on habitats is through field recording and mapping of vegetation, plants communities or habitat types. More recently, remote sensing which uses computer-aided interpretation and visualization of satellite imagery has been applied (Turner et al, 2003). Aerial photography may be used in either approach.

Developing a monitoring programme

Whatever the object of monitoring – species, habitat or policy – a monitoring programme or strategy should be prepared that sets out: the objectives; the methodology to be employed for each feature to be monitored; a sampling strategy, if appropriate; a review of the resources and equipment needed; a review of any legal aspects such as licences that may be necessary; a system and methodology for recording and storing data; a process for analysing and interpreting the data; and an implementation schedule.

It is important to ensure that monitoring programmes are properly designed, the baseline established and the sampling is adequate; otherwise, it will be difficult to detect trends accurately. There is evidence that much of current practice is far from satisfactory (Yoccoz et al, 2001; Noon, 2003; Kull et al, 2008).

The development of monitoring programmes is often considered a step-wise process (e.g. Elzinga et al, 1998; Noon, 2003). As summarized by Noon (2003):

- specify goals and objectives;
- characterize system stressors;
- develop conceptual models of the system;
- select monitoring indicators;

Box 13.2 The main steps involved in a monitoring programme

1. Complete background tasks.
2. Develop objectives.
3. Design and implement management.
4. Design monitoring methodology.
5. Implement monitoring as a pilot study.
6. Implement and complete monitoring.
7. Report and use results.

Source: Elzinga et al, 1998

- establish sampling design;
- define response criteria; and
- link monitoring results to decision-making.

Elzinga et al (1998) give an overview of the steps involved in setting up a monitoring programme for plant populations (see Box 13.2). Each of the steps can, in turn, be broken down into a series of sub-steps, thus the background tasks comprise:

- completion and review of existing information (see Chapters 6 and 8 of this manual);
- review of planning documents of the relevant land management/conservation agencies to ensure the monitoring is in harmony with their established goals;
- identification of priority species and/or populations (see Chapter 7 of this manual);
- assessment of the resources needed and available for monitoring – management support, people with appropriate skills, suitable equipment both low-tech, such as vehicles and measuring instruments, and high-tech, such as GIS, GPS and satellite imagery;
- determination of the scale of the monitoring actions – what part of the range of the species or populations;
- determination of the intensity and frequency of the monitoring;
- review what is being proposed with the management agency(ies) and seek external review if appropriate.

Details of the methodologies involved in each step for population monitoring are given by Elzinga et al (1998) and specifically for CWR by Iriondo et al (2008: Chapter 4).

Selection of monitoring sites

One of the key decisions that must be made is how many and which populations of CWR will be targeted, thus influencing the sites to be selected for monitoring. The selection of sites will depend on the nature, pattern and extent of the habitat, as well as the number, size and distribution of the populations of the target CWR and on the availability of resources for the monitoring programme.

Selection of indicators for populations and threats

In the species management plan for a CWR, the key components are the actions proposed to combat, mitigate or eliminate the threatening processes. These will have been identified during the ecogeographic survey stage (see Chapter 8). When undertaking monitoring of the effectiveness of these management actions, appropriate indicators need to be devised.

Sampling

A census may be undertaken of the populations to be monitored, although this may not be feasible or practical in a species with very large numbers of individuals. In cases where information is needed on the overall habitat or population but where it is not practical to conduct all the individual measurements this would imply, sampling can be employed. Basically, sampling is a means whereby a part of the habitat, population or other unit is selected so as to provide an overall assessment of its status, nature or quality. Questions regarding the sampling design, sampling objectives, size of the sampling unit, population parameters, such as the number of individuals (population size), density and cover, number of features of the plant, such as leaves and flowers and confidence limits are discussed by Elzinga et al (1998) and by Iriondo et al (2008, Chapter 8).

Timing and frequency of monitoring

Accurate monitoring of results will depend, to a large degree, on the timing and frequency of the monitoring. This will depend partly on the life history of the plant, its phenology, its growth form and the season when it is most easily measured. Life form also affects the frequency of monitoring needed, as will the rate at which population and habitat change is occurring. The more threatened a population, the more frequently it may need to be monitored. If the timing of monitoring is not appropriate to the circumstances, valuable information may be missed.

Reporting

A monitoring report may take many forms but is likely to include:

- an executive summary;
- background information on the project;
- maps, illustrations, photographs or drawings showing locations of the baseline monitoring locations;
- monitoring methodology employed and any standards used;
- equipment used and calibration details;
- parameters monitored;
- monitoring locations;
- frequency and intensity of monitoring;
- date, time, frequency and duration;
- results of monitoring;
- analysis; and
- conclusions and recommendations.

Costs of monitoring programmes: Involving the local population

Monitoring programmes, as we have seen, range from simple field surveys to complex procedures that can involve very considerable costs to cover staff salaries of professionals and significant material and/or equipment such as permanent sampling sites, satellite imagery, remote sensing, advanced computing facilities, data analysis and interpretation. However, the budgets normally allocated by countries to biodiversity conservation are limited; the use of professionals alone is seldom possible and use must be made of volunteers coordinated by the experts. In addition, it is important to involve local stakeholders.

Every effort should be made to involve local people and organizations in monitoring, as they have a vested interest in the areas and the species concerned. The CBD guidelines for creating a management plan (2008) note that local groups will be more likely to collect information which they can analyse and use themselves in managing the ecosystem. This information can be complemented by other monitoring activities. However, in practice, as Danielsen et al (2009) comment, 'most of the literature on methods of natural resource monitoring covers an externally driven approach in which professional researchers from outside the study area set up, run, and analyse the results from a monitoring programme funded by a remote agency'.

Causes of monitoring failures

In practice, monitoring often fails to meet expectations. For example, an assessment was undertaken by Kull et al (2008) of 63 plant monitoring schemes from

Europe (collected into a database, DaEuMon), and 33 schemes found through a literature search, covering 354 vascular plant species in total, of which 69 are listed in Annex II of the European Union Habitats Directive. They found that current schemes collect insufficient data, particularly on the dynamics of the extent and distribution pattern of species, and concluded that the quality and general effectiveness of monitoring programmes would be improved if the publication of monitoring data was planned when designing a scheme. Another aspect that needed to be given strong emphasis in developing monitoring schemes was taxonomic diversity and the integration of different scales, as well as the context of different types of sustainable management.

A summary of the most common causes of failure of monitoring are given by Elzinga et al (2001), including technical reasons such as poor project design, use of multiple observers or unreliable data collectors, poor analysis of results and institutional problems such as lack of support to monitoring programmes or analysis of data, and failure to implement results.

Monitoring and climate change

As Lepetz et al (2009) note:

> *it is generally difficult to predict long-term biological responses, as we have little knowledge concerning lag-times between a given effect and its related responses. To show and understand climate change impacts on biodiversity, it is essential to monitor individuals/populations/species over a long time period, usually spanning several decades, as effects are detectable only after many years.*

A critical issue that will arise as climate change takes hold is the alteration in the dynamics of the habitats in the protected areas, the migration patterns of some of its component species and possibly of target CWR themselves, as discussed in Chapter 14. Monitoring requirements might therefore include habitat change and population movements. This is discussed in more detail in Chapter 14.

Experience from Armenia, Madagascar and Uzbekistan

Armenia

A monitoring system was developed in 2007 and then jointly tested and fine-tuned in 2008 with the protected area authorities. The system was applied to monitoring the state of the populations of four target species – wild relatives of wheat, namely *Triticum boeoticum* Boiss., *Triticum urartu* Thum. ex Gandil., *Triticum araraticum* Jakubz., *Aegilops tauschii* Cosson. – within the Erebuni State Reserve. The following factors were selected for regular observations and recording:

- climate;
- soils (contamination);
- natural and human-induced disturbances;
- phenological observations;
- population size and area occupied;
- pests and diseases;
- invasive species.

Protocols and field forms were developed for each of the above factors. In addition, a stand-alone software tool was developed to record and store the monitoring data, developed in Visual Basic 6.0, using MS Access as a database. Either viewing or editing modes can be chosen. Although developed for the Erebuni State Reserve, it can be easily customized for any other protected area. The modules on *phenological observations*, *population size and area occupied*, and *pests and diseases* are currently linked to the target species; however, the number of species can be increased.

The procedures adopted for monitoring of wild cereal species in the Erebuni State Reserve are given in Annex II.

Technical difficulties

Certain difficulties were encountered in mapping the distribution of target species within the protected areas and, subsequently, calculating the area occupied. The distributions of *T. urartu* and *A. tauschii* are not uniform. They occur in small patches that are not spatially static but vary from year to year. However, their identification is possible only after close on-site inspection by an expert; sometimes further examination in the lab is required. As for *T. araraticum* and *T. boeoticum*, they are more abundant and more uniformly distributed within the protected area; however, there are certain areas where none of the species of interest can be found. These areas are rather small and can only be identified after intensive field-work by qualified experts during the spike-bearing stage. To solve these problems, a sampling methodology was developed using GIS software functions: it was successfully tested.

Madagascar

Dioscorea spp. population monitoring protocols have been established and tested by national partners. Monitoring is being carried out jointly with park personnel, forest commission from local community and CWR national partners.

Uzbekistan

The monitoring methodology was developed in the framework of the CWR Project. Pilot plots measuring 37m by 83m were established in areas with high CWR distribution for four priority target species: wild almond in the Chatkal Biosphere Reserve, wild pistachio in Pistalisay, and wild apple and walnut in Aksarsay. Three pilot plots were established in each location and monitoring was

Figure 13.1 *General review of pilot plot 2 (walnut) – strictly protected territory. Ugam Chatkal National Park, Uzbekistan*

Figure 13.2 *Walnut population on pilot plot 2 – strictly protected territory. Ugam Chatkal National Park, Uzbekistan*

carried out. Results were then provided to the management authorities of the Ugam Chatkal National Park, where these CWR species occur. Monitoring will be carried out every five years in both spring and summer. The results of the monitoring exercise are available online (www.cwr.uz) in Russian and are being translated into English.

Sources of further information

Elzinga, A.L., Salzer, D.W. and Willoughby, J.W. (1998) *Measuring and Monitoring Plant Populations*, Bureau of Land Management, Denver, CO, USA.

Elzinga, C.L., Salzer, D.W., Willoughby, J.W. and Gibbs, D.P. (2001) *Monitoring Plant and Animal Populations*, Blackwell Scientific Publications, Abingdon, UK.

Hill, D., Fasham, M., Tucker, G., Shewry, M. and Shaw, P. (eds) (2005) *Handbook of Biodiversity Methods: Survey, Evaluation and Monitoring*, Cambridge University Press, Cambridge.

Iriondo, J.M., Maxted, N. and Dulloo, M.E. (eds) (2008) *Conserving Plant Diversity in Protected Areas*, CAB International, Wallingford UK, Chapter 3.

Stork, N.E. and Samways, M.J. (1995) 'Section 7: Inventorying and monitoring', in V.H. Heywood(ed) *Global Biodiversity Assessment*, Cambridge University Press, Cambridge.

Sutherland, W.J. (2000) *The Conservation Handbook: Techniques in Research, Management and Policy*, Blackwell Science Ltd, Oxford, UK.

Tucker, G., Bubb, P., de Heer, M., Miles, L., Lawrence, A., Bajracharya, S.B., Nepal, R.C., Sherchan, R. and Chapagain, N.R. (2005) *Guidelines for Biodiversity Assessment and Monitoring for Protected Areas*, KMTNC, Kathmandu, Nepal.

Yoccoz, N.G., Hichols, J.D. and Boulinier, T. (2001) 'Monitoring of biological diversity in space and time', *Trends in Ecology and Evolution*, vol 16, pp446–453.

Notes

1. http://linkage.rockefeller.edu/soft/.
2. Tucker et al (2005) warn that such processes are difficult to define and even more difficult to assess and monitor so that it may not be practical to use them to monitor habitat conditions.

References

Bonham, C.D. (1989) *Measurements for Terrestrial Vegetation*, John Wiley and Sons, New York, NY, USA

Collard, B.C.Y., Jahufer, M.Z.Z., Brouwer, J.B. and Pang, E.C.K. (2005) 'An introduction to markers, quantitative trait loci (QTL) mapping and marker-assisted selection for crop improvement: The basic concepts', *Euphytica*, vol 142, pp169–196, doi:10.1007/s10681-005-1681-5

Danielsen, F., Burgess, N.D., Balmford, A., Donald, P.F., Funder, M., Jones, J.P., Alviola, P., Balete, D.S., Blomley, T., Brashares, J., Child, B., Enghoff, M., Fjeldså, J., Holt, S.,

Hübertz, H., Jensen, A.E, Jensen, P.M., Massao, J., Mendoza, M.M., Ngaga, Y., Poulsen, M.K., Rueda, R., Sam, M., Skielboe, T., Stuart-Hill, G., Topp-Jørgensen, E. and Yonten, D. (2009) 'Local participation in natural resource monitoring: A characterization of approaches', *Conservation Biology,* vol 23, pp31–42

Dierschke, H. (1994) *Pflanzensoziologie – Grundlagen und Methoden,* Verlag Eugen Ulmer, Stuttgart

Du, X.Y., Zhang, Q.L. and Luo, Z-R. (2009) 'Comparison of four molecular markers for genetic analysis in *Diospyros* L. (Ebenaceae)', *Plant Systematics and Evolution,* vol 281, pp171–181

Elzinga, A.L., Salzer, D.W. and Willoughby, J.W. (1998) *Measuring and Monitoring Plant Populations,* Bureau of Land Management, Denver, CO, USA

Elzinga, C.L., Salzer, D.W., Willoughby, J.W. and Gibbs, D.P. (2001) *Monitoring Plant and Animal Populations,* Blackwell Scientific Publications, Abingdon, UK

Ervin, J., Mulongoy, K.J., Lawrence, K., Game, E., Sheppard, D., Bridgewater, P., Bennett, G., Gidda, S.B. and Bos, P. (2010) *Making Protected Areas Relevant: A Guide to Integrating Protected Areas into Wider Landscapes, Seascapes and Sectoral Plans and Strategies,* CBD Technical Series No. 44, Convention on Biological Diversity, Montreal, Canada

Given, D.R. (1994) *Principles and Practice of Plant Conservation,* Timber Press, Portland, OR, USA

Gottschalk, T.K., Huettmann, F. and Ehlers, M. (2005) 'Thirty years of analysing and modelling avian habitat relationships using satellite imagery data: A review', *International Journal of Remote Sensing,* vol 26, pp2631–2656, doi:10.1080/01431160512331338041

Hellawell, J.M. (1991) 'Development of a rationale for monitoring', in Goldsmith, F.B. (ed) *Monitoring for Conservation and Ecology,* pp1–14, Chapman and Hall, London, UK

Iriondo, J.M., Maxted, N. and Dulloo, M.E. (eds) (2008) *Conserving Plant Diversity in Protected Areas,* CAB International, Wallingford, UK

Joshi, S., Ranjekar, P. and Gupta, V. (1999) 'Molecular markers in plant genome analysis', *Current Science,* vol 77, pp230–240

Kent, M. and Coker, P. (1995) *Vegetation Description and Analysis: A Practical Approach,* John Wiley and Sons, New York, NY, USA

Kull, T., Sammul, M., Kull, K., Lanno, K., Tali, K., Gruber, B., Schmeller, D. and Henle, K. (2008) 'Necessity and reality of monitoring threatened European vascular plants', *Biodiversity and Conservation,* vol 17, pp3383–3402

Lepetz, V., Massot, M., Schmeller, D. and Clobert, J. (2009) 'Biodiversity monitoring: Some proposals to adequately study species' responses to climate change', *Biodiversity and Conservation,* vol 18, pp3185–3203

Marsh, D.M. and Trenham, P.C. (2008) 'Tracking current trends in plant and animal population monitoring', *Conservation Biology,* vol 22, pp647–655

Mueller-Dombois, D. and Ellenberg, H. (1974) *Aims and Methods of Vegetation Ecology,* Wiley, New York, NY, USA

Noon, B.R. (2003) 'Conceptual issues in monitoring ecological resources', in D.E. Busch and J.C. Trexler (eds) *Monitoring Ecosystems: Interdisciplinary Approaches for Evaluating Ecoregional Initiatives,* pp27–72, Island Press, Washington, DC

Peace, C.P., Vithanage, V., Neal, J., Turnbull, C.G.N. and Carroll, B.J. (2004) 'A comparison of molecular markers for genetic analysis of macadamia', *Journal of Horticultural Science and Biotechnology,* vol 79, pp965–970

Schmeller, D.S. (2008) 'European species and habitat monitoring: Where are we now?', *Biodiversity and Conservation,* vol 17, pp3321–3326

Schröder, S., Begemann, F. and Harrer, S. (2007) 'Agrobiodiversity monitoring – documentation at European level', *Journal für Verbraucherschutz und Lebensmittelsicherheit,* vol 1, pp29–32

Stork, N.E. and Samways, M.J. (1995) 'Section 7: Inventorying and monitoring', in V.H. Heywood (ed), *Global Biodiversity Assessment,* Cambridge University Press, Cambridge

Sutherland, W.J. (2000) *The Conservation Handbook: Techniques in Research, Management and Policy,* Blackwell Science Ltd, Oxford, UK

Sutherland, W.J. (2006) *Ecological Census Techniques – A Handbook,* Edition 2, Cambridge University Press, Cambridge

Tucker, G., Bubb, P., de Heer, M., Miles, L., Lawrence, A., Bajracharya, S.B., Nepal, R.C., Sherchan, R. and Chapagain, N.R. (2005) *Guidelines for Biodiversity Assessment and Monitoring for Protected Areas,* KMTNC, Kathmandu, Nepal

Turner, W., Spector, S., Gardiner, N., Fladeland, M., Sterling, E. and Steininger, M. (2003) 'Remote sensing for biodiversity science and conservation', *Trends in Ecology and Evolution,* vol 18, no 6, pp306–314, doi:10.1016/S0169-5347(03)00070-3

van der Maarel, E. (ed) (2005) *Vegetation Ecology,* Blackwell Science Ltd, Oxford, UK

Yoccoz, N.G., Hichols, J.D. and Boulinier, T. (2001) 'Monitoring of biological diversity in space and time', *Trends in Ecology and Evolution,* vol 16, pp 446–453.

Part IV

Other Major Issues

This part addresses the overarching issues of global change likely to affect the survival of many CWR by introducing new threats or altering the intensity of existing threats, with implications for conservation management. Also included here are the critical matters of capacity building and informing the public of the importance and significance of CWR and of the need to support their conservation.

Chapter 14

Adapting to Global Change

A profound transformation of Earth's environment is now apparent, owing not to the great forces of nature or to extraterrestrial sources but to the numbers and activities of people – the phenomenon of global change (Steffen et al, 2004).

The implications of climate change for the environment and society will depend not only on the response of the Earth system to changes in radiative forcings, but also on how humankind responds through changes in technology, economies, lifestyle and policy (Moss et al, 2010).

Until recently, the conservation of biodiversity has been undertaken based on the assumption that we live in a dynamic but slowly changing world. Such an assumption must be reconsidered in light of the rapid rate of change to which our planet is being subjected. The main components of this change are summarized in Box 14.1 and are collectively referred to as global change. Today, climate change is attracting a great deal of both scientific and public interest because of its implications for food security, health, global and national economies and our ways of life. It is important, however, to recognize that other components of global change, such as population growth, habitat change, deforestation and degradation, will also have major effects on the world and will interact with climate change as well. This chapter will first consider the impacts of climate change on biodiversity and, in particular, CWR and then the effects of the other aspects of global change.

Climate change and biodiversity conservation

In the last few years, accelerated climate change has attracted a great deal of attention, publicity and concern. This has been fuelled by a series of documents such as *The Economics of Climate Change* (Stern, 2007), the IPCC Reports (IPCC, 2007)

Box 14.1 The main components of global change

Population change

- human population movement/migrations;
- demographic growth;
- changes in population pattern.

Changes in land use and disturbance regimes

- deforestation;
- degradation, simplification or loss of habitats;
- loss of biodiversity.

Climate change – as defined by the Intergovernmental Panel on Climate Change (IPCC)

- temperature change;
- atmospheric change (greenhouse gases: carbon dioxide, methane, ozone and nitrous oxide).

Other climate-related factors

- distribution of nitrogen deposition;
- global dust deposition (including brown dust and yellow dust);
- ocean acidification;
- air pollution in mega-cities.

and *Confronting Climate Change: Avoiding the Unmanageable and Managing the Unavoidable* (Bierbaum et al, 2007) which, together with many other findings in the literature, present a picture of large, serious and damaging climatic impacts on our way of life and on biodiversity in the short, medium and long term. Current and predicted patterns of global climate change are a major cause of concern in many areas of biodiversity and agrobiodiversity, conservation planning, socio-economics, ecology and politics.

Although the evidence for climate change is overwhelming, there are still major uncertainties that need to be resolved and gaps in our knowledge (Schiermeier, 2010). While the general trends revealed by the use of general circulation models (GCMs) are evident, they are accurate only to a resolution of one to three degrees in latitude and longitude, and details are far from clear at the regional and local scale. There are also problems with the use of bioclimatic models for estimation of likely migrations of species as discussed below. This makes planning adaptation or mitigation strategies difficult. We need estimates of changes in biodiversity that are sufficiently accurate to allow us to make the necessary adjustments to population management and conservation. In response to these issues, a set of next-generation scenarios for climate change research and assessment has been developed by Moss et al (2010).

Another serious problem is that we do not know with any confidence how far we can allow global change to continue before reaching a tipping point; or as a recent study has termed it, transgressing planetary boundaries with unacceptable environmental change (Rockström et al, 2009a, 2009b).

We already have good evidence of recent phenological change – time of bud burst, flowering, fruiting, etc. – attributable to climate change (Cleland et al, 2007) and of shifts in altitudinal range of species and communities (e.g. Parolo and Rossi, 2007; Lenoir et al, 2008). If such trends continue or increase, the impacts on biodiversity will be significant.

Already countless studies of the impacts of global – and more specifically climate – change have been published at global, regional and national levels. The impacts on plant life have been particularly well studied in parts of Europe (e.g. Thuiller et al, 2005; MACIS, 2008; EEA/JRC/WHO, 2008; Berry, 2008; Araújo, 2009; Heywood, 2009) where it has been estimated that up to half of plant species may be at risk because of climate change. As noted below, very few studies have been carried out on the possible fate of CWR.

Changes in both temperature and precipitation regimes over the coming decades are likely to affect many biological processes, including the distribution of species. Observational and empirical data attest to recent shifts in the distributions and altitudinal range of species and changes in phenology and disturbance regimes that can be attributed to climate change. These are predicted to continue or intensify over the coming decades and will require us to adapt our current biodiversity conservation strategies or adopt new ones. With regard to CWR, the impacts of climate (and other aspects of global) change on protected areas and on the distribution of species will be critical.

As regards the CWR Project countries, the expected consequences of climate change in Armenia are summarized in Box 14.2 while the projected climatic changes and responses for Madagascar are outlined by Hannah et al (2008). Strategies for maintaining biodiversity under global change in Madagascar are proposed by Virah-Sawmy (2009).

Climate change and protected areas

In situ conservation of CWR will mostly take place in some form of protected area, so the effects of global change on such areas are of major concern. It is clear that the projected impacts on protected areas in many parts of the world will force us to rethink their role in biodiversity conservation. The political boundaries of protected areas are fixed, but the biological landscape is not (Lovejoy, 2006). It is clearly difficult for a fixed system of protected areas to respond to global change and considerable rethinking in the design of such areas will be needed if they are to survive and remain effective. Climate change, therefore, has major implications not only for protected areas but for protected area management and managers (Schliep et al, 2008). Generally, protected area managers have tended to adopt minimum intervention procedures, but climate change will force them to reassess management objectives, paying attention to the maintenance of ecosystem health and the conservation needs of target species. They will need to be prepared for

Box 14.2 Consequences of climate change in Armenia

According to Armenia's draft Second National Communication to the United Nations Framework Convention on Climate Change (UNFCCC) in 2009, climate change models predict that annual temperatures in the country will increase by 1°C by 2030, 2°C by 2070 and 4°C by 2100. Precipitation is projected to decrease by 3 per cent, 6 per cent and 9 per cent, respectively. These consequences can essentially affect the climate-dependent branches of economy. Global climate change and internal micro-climatic changes on the territory of Armenia might have the following consequences:

- The modelling of the vulnerability of mountain ecosystems of Armenia with regard to the climate change for the next 100 years foresees a shift of the landscape-zone borders up the mountain for 100–150m. It is expected that the desert to semi-desert zone area will expand by 33 per cent. The steppe belt will be expanded by 4 per cent and shifted upwards by 150–200m, which will cause transformation of steppe vegetation communities. The lower border of the forest belt will move upward by 100–200m. The area of subalpine belt will be reduced by 21 per cent and that of the alpine belt by 22 per cent on average.
- An increase of climate aridity and intensification of desertification processes can be expected under the projected increase of temperature and precipitation reduction.
- In the case of the accepted scenario of climate change, reduction of annual river flow by 15 per cent and an increase in evaporation from the surface of Lake Sevan by 13–14 per cent is expected.
- Under the projected change of climatic characteristics, the efficiency of plant cultivation in Armenia will be reduced by 8–14 per cent. The productivity of cereals will be reduced on average by 9–13 per cent, vegetable cultures by 7–14 per cent, potatoes by 8–10 per cent and fruits by 5–8 per cent. The productivity of more heat-resistant grapes may grow by 8–10 per cent.

Source: Climate Change Information Centre of Armenia; http://www.nature-ic.am/ClimateChange/Env_NGO/EnvNGO.htm/

more frequent and sometimes intensive management interventions (Hagerman and Chan, 2009). A comprehensive strategy should include (Ervin et al, 2010):

- *Improved linkages between protected areas:* by creating biological corridors that allow species to move and genes to flow, from one protected or conserved area to another;
- *Improved protected area management:* by better managing existing protected areas to ensure species survival within these areas and other intact habitats and species persistence within intact habitats;
- *Improved protected area design:* by ensuring that the design, layout and configuration enhances species survival and enhances connectivity with the surrounding landscape;

- *Improved management of the surrounding matrix:* by encouraging natural resource sectors to adopt practices that either positively impact (or at least do not negatively impact) biodiversity conservation and connectivity;
- *Improved connectivity to allow species to migrate in the face of climate change:* by ensuring species have a wider range of options for movement and adaptation in the face of climate change.

> *Protected areas that were set up to safeguard biodiversity and ecological processes are likely to be affected by climate change in a number of ways. Climate change is expected to cause species to migrate to areas with more favourable temperature and precipitation. There is a high probability that competing, sometimes invasive species, more adapted to a new climate, will move in. Such movements could leave some protected areas with a different habitat and species assemblage than they were initially designed to protect* (Mansourian et al, 2009).

Various papers suggest that many protected areas will suffer moderate to substantial species loss and some may experience catastrophic species loss and cease to be functional. However, the evidence is still equivocal and is likely to remain so while there continues to be uncertainty as to the scale and extent of climatic and other change. For example, an assessment was undertaken by Araújo et al (2004) of the ability of existing reserve-selection methods to secure species in a climate-change context. It used the European distributions of 1200 plant species, considering two extreme scenarios of response to climate change: no dispersal and universal dispersal. The results indicated that 6–11 per cent of species modelled would potentially be lost from selected reserves in a 50-year period. A study by Hannah and Salm (2003) on protected area needs in a changing climate concluded that such areas can be an important conservation strategy under a moderate climate change scenario, and that early action may be both more effective and less costly than not taking or delaying action. In the three areas observed (Mexico, Cape Floristic Region of South Africa and Western Europe) the study showed that protected areas remain effective in the early stages of climate change, while adding new protected areas or expanding current ones would maintain species protection in future decades and centuries.

A report by the Secretariat of the CBD (2009) notes that 'an assessment of the ecological regions that are most at risk due to current and projected climate change trends might suggest that the conservation of 10 per cent of ecological regions could be too small a threshold to prevent further extinctions'.

The likely responses of species to climate change

A great deal of effort has gone into developing tools that will help us predict the impacts of climate change on the future distribution of plants. Among the questions we need to answer are (Heywood, 2009):

- Which species will be able to track their climate envelopes as they move?
- Which will not be able to migrate and why (lack of dispersal capacity or reproductive capacity, lack of suitable niches, etc.)?
- What will the physical (climate–soil) conditions in these new climate envelopes be?
- What are sources of potential immigrants (both native and non-native) for many regions, i.e. where will the species that occupy the new habitats come from?
- What will the biotic diversity be like, i.e. what combinations or assemblages of species (plants, animals, micro-organisms, pollinators etc.) will grow there?
- Will the novel (emerging) assemblages be able to provide similar values of ecosystem services (including pollinators) to those that they replace?

In response to climate change, plants have three possibilities: adapt, migrate or become extinct.

Bioclimatic modelling

The tool that is most frequently used in attempting to predict the impacts of climate change is *bioclimatic modelling*. Bioclimatic models (bioclimatic envelope models) are a special case of ecological niche or distribution models. Currently, most current predictions of the future migration of plants use the 'climate envelope' or bioclimatic modelling techniques (Nix, 1986; Guisan and Thuiller, 2005) in which projected future distributions are based on the current climate in the species' native range. But it should be noted that models are simplifications of reality and primarily important aids to research, as Thuiller et al (2008) point out. Bioclimatic modelling techniques combine computer-based models of the current climate with information on the current distribution of species to establish a bioclimatic (also known as edaphic, fundamental, environmental or Grinellian) niche model. This model of optimal environmental parameters is then fitted to a range of future climate scenarios to establish likely shifts in environmental optima for species. Although commonly referred to as predictions, their proper role is in providing part of the information base on which predictions of future change are made.

Bioclimatic modelling has been applied extensively in Europe and is also being applied in other parts of the world. There is no single standard approach and techniques are constantly being developed.

> *While we can use various types of model to predict the possible migrations of species into 'new' climatic envelopes, what we cannot do with existing modelling approaches is to predict what the new vegetation cover will be nor the overall environmental conditions, in areas impacted by climate change. This applies both to the move-out areas and the move-in areas, a distinction that is not often made but which may be critical in some parts of Europe such as the Mediterranean zone, as mentioned above. Since the likelihood of survival and multiplication of migrant species will depend*

> *on the environmental context into which they move, not to mention stochastic factors which may intervene, we have to accept that our present understanding of the consequences of climate change is severely limited and sometimes dependent on little more than intelligent speculation. If we add to this the level of uncertainty that still surrounds the details of the extent of climate change and their impact at a local level, much of our planning has to be broadly based rather than site-specific, such as modifying or enhancing our protected area systems, or precautionary such as employing* ex situ *complementarity* (Heywood, 2009).

In an agrobiodiversity context, it would obviously be of great importance to be able to predict the effects of climate change on the future distribution and survival of target species of economic importance such as wild relatives or crops. One of the few studies so far published (Lane and Jarvis, 2007; Jarvis et al, 2008) used current and projected future climate data for ~2055, and a climate envelope species' distribution model to predict the impact of climate change on the wild relatives of three of the world's major food crops: peanut (*Arachis*), potato (*Solanum*) and cowpea (*Vigna*). They considered three migrational scenarios for modelling the range shifts (unlimited, limited and no migration) and found that climate change strongly affected all taxa, with an estimated 16–22 per cent of these species predicted to become extinct and most species losing over 50 per cent of their range size.

Climate envelope modelling has been used to indicate possible shifts in the distribution of *Pinus kesiya* and *P. merkusii* in Southeast Asia, and their possible implications for the conservation and use of their genetic resources (van Zonneveld et al, 2009a). This showed that in the case of *P. kesiya*, in addition to the areas where natural populations of the species have been recorded, it could potentially occur in several other locations in Myanmar, north-eastern and southern Thailand, the Lao People's Democratic Republic and south-western Cambodia, where it now occurs naturally. In addition, the Indonesian provinces of Java and Nusa Tenggara, which are outside its recorded natural distribution range, appear to have a suitable climate for the species. In the case of *P. merkusii*, its climate envelope coincides with the observed distribution of the species in mainland Southeast Asia and in Sumatra, while suggesting that the climate in several parts of the Malay Archipelago and in northern Australia is suitable for *P. merkusii* outside its natural distributional range.

Another study by van Zonneveld et al (2009b) of climate change impact predictions on populations of two important forest plantation species, *Pinus patula* and *Pinus tecunumanii*, in Mexico and Central America, using climate envelope modelling (CEM) found that climate change significantly impacts on the natural species distribution of the two pine species. However, assessment of the adaptive ability of the these species based on the evaluation of provenance trials, undertaken to validate the CEM impact assessment studies, showed that they performed well in a wide range of climates, including conditions that were recorded by CEM as unsuitable for natural pine occurrence. They interpret these

Box 14.3 CWR and bioclimatic modelling in Mexico

Using bioclimatic modelling, two possible scenarios of climatic change in Mexico were used to analyse the distribution patterns of eight wild Cucurbitaceae closely related to cultivated plants, *Cucurbita argyrosperma* subsp. *sororia*, *C. lundelliana*, *C. pepo* subsp. *fraterna*, *C. okeechobeensis* subsp. *martinezii*, *Sechium chinantlense*, *S. compositum*, *S. edule* subsp. *sylvestre* and *S. hintonii*. Most of these taxa have restricted distributions. Many of them also show proven resistance to various diseases, which could be crucial for the improvement of their related cultivars. The possible role that the Mexican system of protected areas might have in the conservation of these taxa was also assessed. The results showed a marked contraction of the distributions of all eight taxa under both scenarios. It was also found that, under a drastic climatic change scenario, the eight taxa will be maintained in just 29 out of the 69 natural protected areas where they currently occur. Accordingly, it seems that most of the eight wild taxa will not have many opportunities to survive under climate change. However, the ability of these plants to maintain low-density isolated populations for long periods, as well as the low resolution of the bioclimatic models, are discussed as possible mitigators of these rather grim predictions.

Source: Lira et al, 2009

findings as suggesting that the pine species in their natural habitat are better adapted to climate change than is predicted from CEM and recommend caution in interpreting CEM climate change impact predictions.

Bioclimate envelope modelling analysed the distribution patterns of eight Cucurbit wild relatives and their survival prospects under climate change (Lira et al, 2009) (Box 14.3).

A recent study modelling the shifts in species' ranges in Madagascar in response to forthcoming climatic change predicts that the littoral forest will disappear (Hannah et al, 2008), although Virah-Sawmy (2009) notes that palaeoecological reconstructions show the littoral forest remaining stable throughout several pronounced arid intervals, lasting hundreds of years each, during the last 6500 years, as well as during past sea-level rises of 1–3m. Temperature rises were not accounted for in this timeframe.

Non-modelling approaches

Although bioclimatic modelling is the most common method of suggesting the likely response of species to climate change, other approaches can be used to assess species' vulnerability on the basis of their biological and ecological characteristics, and other factors, that determine their sensitivity, adaptive capacity and exposure to climate change (Gran Canaria Group, 2006; CBD/AHTEG, 2009) (see Box 14.4).

Box 14.4 Criteria for identifying taxa vulnerable to climate change

- taxa with nowhere to go, such as mountain tops, low-lying islands, high latitudes and edges of continents;
- plants with restricted ranges such as rare and endemic species;
- taxa with poor dispersal capacity and/or long generation times;
- species susceptible to extreme conditions such as flood or drought;
- plants with extreme habitat/niche specialization such as narrow tolerance to climate-sensitive variables;
- taxa with co-evolved or synchronous relationships with other species;
- species with inflexible physiological responses to climate variables;
- keystone taxa important in primary production or ecosystem processes and function;
- taxa with direct value for humans or with potential for future use.

Source: Gran Canaria Group, 2006

Indigenous peoples and climate change

> *Sustainable agricultural growth in developing countries is challenged as never before – by climate change, increasingly volatile food and energy markets, natural resource exploitation, and a growing population with aspirations for a better standard of living* (Mark Rosegrant, Director of Environment and Production Technology at the International Food Policy Research Institute (IFPRI), 2010).

Indigenous peoples relying on traditional agriculture will be among the most severely affected by climate change although their reliance on a diversity of local crops and traditional varieties may provide some insurance against major losses. Their possible role in adaptation to and mitigation of the effects of climate change are discussed in Box 14.5. Examples of the use of indigenous knowledge for climate change mitigation and adaptation strategies by tree planting, conservation measures, management of natural resources, better land-use practices in Kenya, South Africa, Botswana, Ghana and Nigeria are given in a report by the Bureau of Environmental Analysis (BEA) International (Karani et al, 2010). The conservation of CWR *in situ* as part of such measures would be a win–win situation.

REDD (Reducing Emissions from Deforestation and Forest Degradation)

Given that forest clearing and degradation is responsible for about 17 per cent of global greenhouse emissions, according to estimates by the IPCC, efforts to reduce such emissions are an essential component of climate change adaptation strategies. The United Nations Collaborative Programme on Reducing Emissions

Box 14.5 Indigenous people and addressing the climate change agenda

Indigenous peoples have played a key role in climate change mitigation and adaptation. The territories of indigenous groups who have been given the rights to their lands have been better conserved than the adjacent lands (i.e. Brazil, Colombia, Nicaragua, etc.). Preserving large extensions of forests would not only support the climate change objectives, but it would respect the rights of indigenous peoples and conserve biodiversity as well. A climate change agenda fully involving indigenous peoples has many more benefits than if only government and/or the private sector are involved. Indigenous peoples are some of the most vulnerable to the negative effects of climate change. Also, they are a source of knowledge to the many solutions needed to avoid or ameliorate those effects. For example, ancestral territories often provide excellent examples of a landscape design that can resist the negative effects of climate change. Over the millennia, indigenous peoples have developed adaptation models to climate change. They have also developed genetic varieties of medicinal and useful plants and animal breeds with a wider natural range of resistance to climatic and ecological variability.

Source: Sobrevila, 2008

from Deforestation and Forest Degradation in Developing Countries (UN-REDD) is a mechanism that creates incentives for developing forested countries to protect, and better manage their forest resources, thus contributing to the global fight against climate change. REDD+ goes beyond reducing deforestation and forest degradation solely for the purpose of emissions reductions, and its strategies include the role of conservation, sustainable management of forests and enhancement of forest carbon stocks. The aim is to make standing forest more valuable than the timber obtained from felling it by giving a financial value to the carbon stored in the standing trees (Katerere, 2010).

It has been suggested that indigenous lands protected areas (ILPAs) should form part of government REDD strategies (Ricketts et al, 2010). They suggest that the steps that national governments could take to include ILPAs effectively in their REDD strategies could consist of:

- identifying where establishing or strengthening ILPAs would most effectively reduce emissions;
- as a matter of urgency, the establishment of national monitoring schemes to measure deforestation rates and quantify carbon emissions reductions (cf. Brazil's system of remotely sensed monitoring); and
- establishing insurance mechanisms, pooling the risk that illegal logging or fires reverse gains in individual ILPAs.

Of course, as they point out, it is also essential to ensure that governments provide indigenous groups and local communities with the information and capacities they need to participate and that payments are distributed transparently to reward those responsible for reducing emissions.

Global change, agriculture and food security

> *Although considered by many to be a success story, the benefits of productivity increases in world agriculture are unevenly spread. Often the poorest of the poor have gained little or nothing; and 850 million people are still hungry or malnourished with an additional 4 million more joining their ranks annually. We are putting food that appears cheap on our tables; but it is food that is not always healthy and that costs us dearly in terms of water, soil and the biological diversity on which all our futures depend* (Watson, 2008).

It is obvious that substantial improvements are needed in current crops to achieve higher yields and sustainable farming and this should be done without a major expansion of agricultural land and in such a way that it does not exacerbate climate change. In achieving these aims, all possible means and techniques will be needed to streamline breeding programmes, including the more extensive use of the genetic diversity found in CWR. As the *World Development Report 2010: Development and Climate Change*[1] notes, the weedy and wild relatives of today's crops retain higher genetic diversity and may be a useful base for enhancing crops' plasticity and their adaptability to changing conditions – some weeds, for example, thrive in conditions of higher CO_2 and warmer temperature. One of the main reasons for conserving CWR is so that genetic variation will be available for plant breeders so as to be able to breed new cultivars for crops in response to the conditions under climate change. Material of traditional landraces will also be an important source of genes for breeding new cultivars adapted to the conditions of abiotic environmental stress that may be expected as a result of climate change. As Semenov and Halford (2009) note: 'Breeders select new cultivars of agricultural crops that are better suited to a specific environment utilizing available resources in the most optimal way. However, cultivars that are recommended for use at present might not be suitable if the climate changes. Breeding for a new cultivar usually takes 10–12 years, if the target traits are known and the environment in which to test new lines is available. Faced with the prospect of a rapidly changing climate, breeders do not have access to the climatic conditions of even the near future in which to carry out field trials, and they do not know which … traits might be important in 15–25 years time.'

> *We know that the main sources of agricultural growth in the 20th century are drying up. Theoretically the global agricultural area could still be expanded by 80% but most spare land is little suited for productive agriculture. Only Africa and Latin America have significant reserves of suitable land. In several grain belts, especially in Asia, freshwater supply for irrigation is running dry. And yield potentials of major food crops*

have stagnated, even though there might still be some room for lifting potential yields along conventional pathways' (Koning and van Ittersum, 2009).

Climate change and forestry genetic resources

The effects of climate change on forestry species and their CWR are likely to be significant, given that many of them are already impacted by non-climatic factors such as habitat loss or fragmentation with a consequent loss of genetic diversity in their populations (Bawa and Dayanandan, 1998). These effects will include rising temperatures, changes in precipitation patterns, extreme weather events, prolonged droughts leading to more frequent incidence of forest fires and changes in the physiology and reproductive success of tree species (Rimbawanto, 2010).

Strategic responses and new conservation strategies

As we have seen, conventional approaches to biodiversity conservation may not be a broad enough strategy to combat the effects of climate change and a number of novel approaches are being considered. These include the controversial approach known as *human-aided translocation of species*. Human-aided translocation of species' populations as a means of countering biodiversity loss from global change is a very recent approach and is being proposed for situations where the rate of change, the existence of obstacles or barriers or the lack of continuous suitable habitat is considered likely to prevent natural migration. Known as *assisted migration* (McLachlan et al, 2007) or *assisted colonization*[2] (Hunter, 2007; Hoegh-Guldberg et al, 2008), it is a complex and potentially costly venture and needs to be subject to careful cost–benefit analysis and perhaps used only in exceptional circumstances. Moving species into new environments is, as McLachlan et al (2007) say, a contentious issue and may involve considerable risks. It is a complex process involving not just scientific, technical and economic but also sociological and ethical considerations.

Seddon et al (2009), for example, state that 'calls to take proactive conservation measures need to consider that there are currently huge uncertainties involved, not only in climate change predictions and consequent species responses ... but also in our understanding of the habitat requirements of species ... and the effects of translocations on ecosystem function'. Ricciardi and Simberloff (2009) argue against assisted colonization as a viable conservation strategy on the grounds that: (1) species translocations can erode biodiversity and disrupt ecosystems; (2) planned introductions carry high risks; (3) risk assessments and decision frameworks are unreliable; and (4) the lack of power in predicting species invasiveness suggests that assisted colonization is ecological gambling and should be avoided as the precautionary principle.

On the other hand, human-assisted migration also has strong supporters: Richardson et al (2009), for example, believe that its importance as a conservation strategy will increase as global change takes hold and that it should not be considered *a priori* as a last resort approach but as one of a portfolio of options. It is evident that assisted migration requires a sound and well-thought-out policy framework before being widely undertaken as a management response to global change. It may be worth considering for CWR of particular importance but is unlikely to become a major component of CWR conservation strategies.

Other components of global change

Although the emphasis in recent years has been very much on the predicted impacts of climate change, it is important to recognize that the world is experiencing the effects of global change which, as Steffen et al (2004) observe, 'is much more than climate change. It is real, it is happening now and it is accelerating.'

Population change

Population change refers to both changes in the *pattern of distribution* of human populations and to *demographic growth*. Large-scale migrations of human populations can be caused by social, economic, political and health factors. The effects of war and civil conflict can leave large areas of land devastated or unusable and cause large human migrations, thus affecting the natural and agro-ecosystems involved and their biodiversity. In 2008, more than about half of the world's population (an estimated 3.3 billion people) lived in urban areas, and every day about 160,000 people move from rural areas to cities (United Nations, 2006; UNFPA, 2007). In comparison, the world's rural population is expected to *decrease* by some 28 million between 2005 and 2030, so that at the global level, *all* future population growth will thus be in towns and cities. Urbanization levels are rising, especially in less developed countries: in 2000, approximately 40 per cent of people living in less developed countries were in urban areas, but this proportion is anticipated to rise to 54 per cent by 2025.

Changes in land use and disturbance regimes

During the course of the past hundred years, changes in land cover and land use have accelerated largely in line with human demographic growth, as a result of industrialization, agricultural intensification, abandonment of traditional agricultural practices, population movements away from the land and many other factors.

Sometimes, land-use practices alter the natural disturbance regimes that generate the complex patterns of habitats that native plants and animals need for survival. If land-use practices change the frequency, size and intensity of natural disturbances, such as floods, fires, droughts and other extreme climatic events, then ecosystem functioning will be affected and communities with quite a different composition may develop. Deforestation and other forms of habitat destruction or degradation remain the major cause of biodiversity loss.

Tourism

Annual tourism is another form, albeit temporary, of population migration. The increase of tourism has led to massive urban and touristic development with accompanying infrastructural effects. It is estimated that carbon dioxide emissions from the tourism sector account for 4–6 per cent of total emissions and changing climate patterns might alter major tourism flows where climate is of paramount importance, such as southern Europe, the Mediterranean and the Caribbean. This will leave coastal and mountain-based destinations in least developed countries and small island developing states particularly vulnerable to direct and indirect impacts of climate change (such as storms and extreme climatic events, coastal erosion, physical damage to infrastructure, sea-level rise, flooding, water shortages and water contamination), given that most infrastructure is located within a short distance of the shoreline (UNWTO, 2008).

The number of environmental refugees – 'people who can no longer gain a secure livelihood in their homelands because of drought, soil erosion, desertification, deforestation and other environmental problems' (Myers, 1997) – is expected to increase by 200 million by the middle of this century. Their effects on biodiversity could be serious in that they will move into territories not able to support or feed them without large-scale disruption. Displaced people have to rely heavily on the surrounding environment for food and fuelwood, leading to forest and other vegetation degradation or loss.

Sources of further information

de Chazal, J. and Rounsevell, M. (2009) 'Land-use and climate change within assessments of biodiversity change: A review', *Global Environmental Change*, vol 19, pp306–315.

Heinz Center (2008) *Strategies for Managing the Effects of Climate Change on Wildlife and Ecosystems*, The H. John Heinz III Centre for Science, Economics and the Environment, Washington, DC

Hoegh-Guldberg, O., Hughes, L., McIntyre, S., Lindenmayer, D.B., Parmesan, C., Possingham, H.P. and Thomas, C.D. (2008) 'Assisted colonization and rapid climate change', *Science*, vol 321, pp345–346.

IPCC (2007) *Climate Change 2007: Impacts, Adaptation and Vulnerability*, Working Group II contribution to the Fourth Assessment Report of the Intergovernmental Panel on Climate Change (IPCC), Cambridge University Press, Cambridge.

Lovejoy, T.E. and Hannah, L. (eds) (2004) *Climate Change and Biodiversity*, Yale University Press, New Haven, CT and London, UK.

SEG (2007) *Confronting Climate Change: Avoiding the Unmanageable and Managing the Unavoidable*, Scientific Expert Group on Climate Change (SEG), [Rosina M. Bierbaum, John P. Holdren, Michael C. MacCracken, Richard H. Moss, and Peter H. Raven (eds)], report prepared for the United Nations Commission on Sustainable Development, Sigma Xi, Research Triangle Park, NC and the United Nations Foundation, Washington, DC.

Notes

1. WDR (2010), 'Chapter 3: Managing land and water to feed nine billion people and protect natural systems'.
2. Hunter uses the term *assisted colonization* in contrast to *assisted migration* 'because many animal ecologists reserve the word *migration* for the seasonal, round-trip movements of animals ... and because the real goal of translocation goes beyond assisting dispersal to assuring successful colonization, a step that will often require extended husbandry'.

References

Araújo, M.B. (2009) 'Protected areas and climate change in Europe', Report prepared by Professor Miguel B. Araújo, National Museum of Natural Sciences, CSIC, Madrid, Spain and 'Rui Nabeiro' Biodiversity Chair, CIBIO, University of Évora, Portugal, with contributions by Ms Raquel Garcia, Convention on the Conservation of European Wildlife and Natural Habitats, Standing Committee, Strasbourg, 25 June 2009, T-VS/Inf(2009)10

Araújo, M.B., Cabezas, M., Thuiller, W. and Hannah, L. (2004) 'Would climate change drive species out of reserves? An assessment of existing reserve selection methods', *Global Change Biology*, vol 10, pp1618–1626

Bawa, K. and Dayanandan, S. (1998) 'Global climate change and tropical forest genetic resources', *Climate Change*, vol 39, pp473–485

Berry, P. (2008) 'Climate change and the vulnerability of Bern Convention species and habitats', report for the Convention on the Conservation of European Wildlife and Natural Habitats, Standing Committee, Strasbourg, 16 June 2008, T-PVS/Inf(2008)6 rev

Bierbaum, R., Holdren, J.P., MacCracken, M., Moss, R.H. and Raven, P.H. (eds) (2007) *Confronting Climate Change: Avoiding the Unmanageable and Managing the Unavoidable*, Sigma Xi, Research Triangle Park, NC and the United Nations Foundation, Washington, DC

CBD (2009) *The Convention on Biological Diversity Plant Conservation Report: A Review of Progress in Implementing the Global Strategy of Plant Conservation (GSPC)*, Convention on Biological Diversity (CBD) Secretariat, Montreal, Canada

CBD/AHTEG (2009) 'Draft findings of the Ad Hoc Technical Expert Group on Biodiversity and Climate Change', Convention on Biological Diversity (CBD), http://www.cbd.int/Climate/Meetings/Ahteg-Bdcc-02-02/Ahteg-Bdcc-02-02-Findings-Review-En.Pdf, accessed 21 May 2010

Cleland, E.E., Chuine, I., Menzel, A., Mooney, H.A. and Schwartz, M.D. (2007) 'Shifting Alant phenology in response to global change', *Trends in Ecology and Evolution*, vol 22, pp357–365

EEA/JRC/WHO (2008) *Impacts of Europe's Changing Climate — 2008 Indicator-Based Assessment*, European Environment Agency (EEA) Report No 4/2008, Office for Official Publications of the European Communities, Luxembourg http://reports.eea.europa.eu/eea_report_2008_4/en, accessed 21 May 2010

Ervin, J., Mulongoy, K.J., Lawrence, K., Game, E., Sheppard, D., Bridgewater, P., Bennett, G., Gidda, S.B. and Bos, P. (2010) *Making Protected Areas Relevant: A Guide to Integrating*

Protected Areas into Wider Landscapes, Seascapes and Sectoral Plans and Strategies, CBD Technical Series No. 44, Convention on Biological Diversity, Montreal, Canada

Gran Canaria Group (2006) *The Gran Canaria Declaration II on Climate Change and Plant Conservation,* Cabildo de Gran Canaria, Jardín Botánico 'Viera y Clavijo' and Botanic Gardens Conservation International (BGCI)

Guisan, A. and Thuiller, W. (2005) 'Predicting species distribution: Offering more than simple habitat models', *Ecology Letters,* vol 8, pp993–1009

Hagerman, S.M. and Chan, K.M.A. (2009) 'Climate change and biodiversity conservation: Impacts, adaptation strategies and future research directions', F1000 Biology Reports 1:16, doi:10.3410/B1-16. The electronic version can be found at: http://F1000.com/Reports/Biology/content/1/16

Hannah, L. and Salm, R. (2003) 'Protected areas and climate change', in L. Hannah and T. Lovejoy (eds) *Climate Change and Biodiversity: Synergistic Impacts,* pp91–100, Conservation International, Washington, DC

Hannah, L., Dave, R., Lowry, P.P., Andelman, S., Andrianarisata, M., Andriamaro, L., Cameron, A., Hijmans, R., Kremen, C., MacKinnon, J., Randrianasolo, H.H., Andriambololonera, S., Razafimpahanana, A., Randriamahazo, H., Randrianarisoa, J., Razafinjatovo, P., Raxworthy, C., Schatz, G.E., Tadross, M. and Wilme, L. (2008) 'Climate change adaptation for conservation in Madagascar', *Biology Letters,* vol 4, pp590–594

Heywood, V.H. (2009) *The Impacts of Climate Change on Plant Species in Europe,* Final Version, Report prepared by Professor Vernon Heywood, School of Biological Sciences, University of Reading with contributions by Dr Alastair Culham. Convention on the Conservation of European Wildlife and Natural Habitats – 29th meeting of the Standing Committee – Bern, 23–26 November 2009, T-PVS/Inf(2009)9E

Hoegh-Guldberg, O., Hughes, L., McIntyre, S., Lindenmayer, D.B., Parmesan, C., Possingham, H.P. and Thomas, C.D. (2008) 'Assisted colonization and rapid climate change', *Science,* vol 321, pp345–346

Hunter, M.L. (2007) 'Climate change and moving species: Furthering the debate on assisted colonization', *Conservation Biology,* vol 21, pp1356–1358

IPCC (2007) *Climate Change 2007 – Impacts, Adaptation and Vulnerability,* Working Group II contribution to the Fourth Assessment Report of the IPCC, Intergovernmental Panel on Climate Change (IPCC), Cambridge University Press, Cambridge

Jarvis, A., Lane, A. and Hijmans, R. (2008) 'The effect of climate change on crop wild relatives', *Agriculture, Ecosystems and Environment,* vol 126, pp13–23

Karani, P., Ahwireng-Obeng, F., Kung'u, J. and Wafula, C. (2010) *Clean Development Mechanism (CDM) Carbon Markets Opportunities for Investments and Sustainable Development in Local Communities: The Application of Indigenous Knowledge Case Studies,* Prepared by the Bureau of Environmental Analysis (BEA) International, Nairobi

Katerere, Y. (2010) 'A climate change solution?' *World Finance,* May–June 2010, pp104–106

Koning, N. and van Ittersum, M.K. (2009) 'Will the world have enough to eat?', *Current Opinion in Environmental Sustainability,* vol 1, pp77–82

Lane, A. and Jarvis, A. (2007) 'Changes in climate will modify the geography of crop suitability: Agricultural biodiversity can help with adaptation', *SAT e-journal/e-journal.icrisat.org,* vol 4, no 1, pp1–12, http://www.icrisat.org/Journal/SpecialProject/sp2.pdf, accessed 27 May 2010

Lenoir, J., Gegout, J.C., Marquet, P.A., de Ruffray, P. and Brisse, H. (2008) 'A significant upward shift in plant species optimum elevation during the 20th Century', *Science,* vol 320, no 5884, pp1768–1771, doi:10.1126/science.1156831

Lira, R., Téllez, O. and Dávila, P. (2009) 'The effects of climate change on the geographic distribution of Mexican wild relatives of domesticated Cucurbitaceae', *Genetic Resources and Crop Evolution,* vol 56, no 5, pp691–703

Lovejoy, T.E. (2006) 'Protected areas: A prism for a changing world, *TREE,* vol 21, pp329–333

MACIS (2008) 'Deliverable 1.1: Climate change impacts on European biodiversity – observations and future projections', Jörgen Olofsson, Thomas Hickler, Martin T. Sykes, Miguel B. Araújo, Emilio Baletto, Pam M. Berry, Simona Bonelli, Mar Cabeza, Anne Dubuis, Antoine Guisan, Ingolf Kühn, Heini Kujala, Jake Piper, Mark Rounsevell, Josef Settele and Wilfried Thuiller and MACIS Co-ordination Team, Minimisation of and Adaptation to Climate Change Impacts on Biodiversity (MACIS), http://www.macis-project.net/pub.html, accessed 23 May 2010

Mansourian, S., Belokurov, A. and Stephenson, P.J. (2009) 'The role of forest protected areas in adaptation to climate change', *Unasylva,* vol 60, no 231/232, pp63–69

McLachlan, J.S., Hellmann, J.J. and Schwartz, M.W. (2007) 'A framework for debate of assisted migration in an era of climate change', *Conservation Biology,* vol 21, pp297–302

Moss, R.H., Edmonds, J.A., Hibbard, K.A., Manning, M.R., Rose, S.K., van Vuuren, D.P., Carter, T.R., Emori, S., Kainuma, M., Kram, T., Meehl. G.A., Mitchell, J.F.B., Nakicenovic, N., Riahi, K., Smith, S.J., Stouffer, R.J., Thomson, A.M., Weyant, J.P. and Wilbanks, T.J. (2010) 'The next generation of scenarios for climate change research and assessment', *Nature,* vol 463, pp747–756 (11 February 2010), doi:10.1038/nature08823

Myers, N. (1997) 'Environmental refugees', *Population and Environment,* vol 19, pp167–182

Nix, H.A. (1986) 'A biogeographic analysis of Australian elapid snakes', in R. Longmore, (ed) *Australian Flora and Fauna Series Number 7: Atlas of Elapid Snakes of Australia,* Australian Government Publishing Service, Canberra, pp4–15

Parolo, G. and Rossi, G. (2007) 'Upward migration of vascular plants following a climate warming trend in the Alps', *Basic and Applied Ecology,* doi:10.1016/j.baae.2007.01.005

Ricciardi, A. and Simberloff, D. (2009) 'Assisted colonization is not a viable conservation strategy', *TREE,* vol 24, pp248–253

Richardson, D.M., Hellmann, J.J., McLachlan, J.S., Sax, D.F, Schwartz, M.W., Gonzalez, P., Brennan, E.J., Camacho, A., Root, T.L., Sala, O.E., Schneider, S.H., Ashe, D.M., Clark, J.R., Early, R., Etterson, J.R., Fielder, E.D., Gill, J.L., Minteer, B.A., Polasky, S., Safford, H.D., Thompson, A.R. and Vellend, M. (2009) 'Multidimensional evaluation of managed relocation', *Proc. Natl. Acad. Sci. USA,* vol 106, pp9721–9724

Ricketts, T.H., Soares-Filho, B., da Fonseca, G.A.B, Nepstad, D., Pfaff, A., Petsonk, A., Anderson, A., Boucher, D., Cattaneo, A., Conte, M., Creighton, K., Linden, L., Maretti, C., Moutinho, P., Ullman, R. and Victurine, R. (2010) 'Indigenous lands, protected areas, and slowing climate change', *PLoS Biol,* vol 8, no 3, e1000331, doi:10.1371/journal.pbio.1000331

Rimbawanto, A. (2010) *Climate Change and the Potential Risk to Forest Genetic Resources,* Centre for Forest Biotechnology and Tree Improvement (CFBTI), http://www.apafri.org/FGR09%20CD%20final/Presentation%20Pdf/03_Anto%20Rimbawanto.pdf, accessed 3 April 2010

Rockström, J., Steffen, W., Noone, K., Persson, A., Chapin, F.S., Lambin, E.F., Lenton, T.M., Scheffer, M., Folke, C., Schellnhuber, H.J., Nykvist, B., de Wit, C.A., Hughes, T., van der Leeuw, S., Rodhe, H., Sörlin, S., Snyder, P.K., Costanza, R., Svedin, U., Falkenmark, M., Karlberg, L., Corell, R.W., Fabry, V.J., Hansen, J., Walker, B., Liverman, D., Richardson, K., Crutzen, P. and Foley, J.A. (2009a) 'A safe operating space for humanity', *Nature,* vol 461, pp472–475

Rockström, J., Steffen, W., Noone, K. (2009b) 'Planetary boundaries: Exploring the safe operating space for humanity', *Ecology and Society*, vol 14, no 2, p32, http://www.ecologyandsociety.org/vol14/iss2/art32/, accessed 27 May 2010

Schiermeier, Q. (2010) 'Climate: The real holes in climate science', *Nature* (London), vol 463, p284

Schleip, R., Bertzky, M., Hirschnitz, M. and Stoll-Kleemann, S. (2008) 'Changing climate in protected areas: Risk perception of climate changed by biosphere reserve managers', *GAIA*, vol 17/S1, pp116–124

Seddon, P.J., Armstrong, D.P., Soorae, P., Launay, F. and Walker, S. (2009) 'The risks of assisted colonization', *Conservation Biology*, vol 23, pp788–789

Semenov, M.A. and Halford, N.G. (2009) 'Identifying target traits and molecular mechanisms for wheat breeding under a changing climate', *Journal of Experimental Botany*, vol 60, pp2791–2804, doi:10.1093/jxb/erp164

Sobrevila, C. (2008) *The Role of Indigenous Peoples in Biodiversity Conservation: The Natural but Often Forgotten Partners*, The World Bank, Washington, DC

Steffen, W., Sanderson, A., Jäger, J., Tyson, P.D., Moore III, B., Matson, P.A., Richardson, K., Oldfield, F., Schellnhuber, H.J., Turner II, B.L. and Wasson, R.J. (2004) *Global Change and the Earth System: A Planet Under Pressure*, Springer Verlag, Heidelberg, Germany

Stern, N. (2007) *The Economics of Climate Change (The Stern Review)*, Cambridge University Press, Cambridge

Thuiller, W., Lavorel, S., Araújo, M.B., Sykes, M.T. and Prentice, I.C. (2005) 'Climate change threats to plant diversity in Europe', *PNAS USA*, vol 102, pp8245–8250

Thuiller, W., Albert, C., Araújo, M.B., Berry, P.M., Guisan, A., Hickler, T., Midgley, G.F., Paterson, J., Schurr, F.M., Sykes, M.T. and Zimmermann, N.E. (2008) 'Predicting climate change impacts on plant diversity: Where to go from here?', *Perspectives in Plant Ecology, Evolution and Systematics*, vol 9, pp137–152

United Nations (2006) *World Urbanization Prospects: The 2005 Revision*, Population Division, Department of Economic and Social Affairs, United Nations, New York, USA

UNFPA (2007) *State of the World Population 2007: Unleashing the Potential of Urban Growth*, United Nations Population Fund (UNFPA), New York, USA

UNWTO (2008) *Climate Change and Tourism – Responding to Global Challenge*, United Nations World Tourism Organization (UNWTO) and the United Nations Environment Programme (UNEP), Madrid, Spain

van Zonneveld, M., Koskela, J., Vinceti, B. and Jarvis, A. (2009a) 'Impact of climate change on the distribution of tropical pines in Southeast Asia', *Unasylva*, no 231/232, vol 60/1–2, pp24–29.

van Zonneveld, M., Jarvis, A., Koskela, J., Dvorak, W., Lema, G., Vinceti, B. and Leibing, C. (2009b) 'Climate change impact predictions on *Pinus patula* and *Pinus tecunumanii* populations in Mexico and Central America', *Forest Ecology and Management*, vol 257, pp1566–1576

Virah-Sawmy, M. (2009) 'Ecosystem management in Madagascar during global change', *Conservation Letters*, vol 2, pp163–170

Watson, R. (2008) *Inter-Governmental Report Aims to Set New Agenda for Global Food Production*, www.iaastd.com/docs/IAASTD_backgroundpaper_280308.doc

WDR (2010) *World Development Report 2010: Development and Climate Change*, The World Bank, Washington, DC

Chapter 15

Capacity Building

> *Developing capacity is about facilitating and encouraging a process of transformation or change by which individuals, organizations and societies develop their abilities, both individually and collectively, to perform functions, solve problems, and set and achieve their own goals* (Hough, 2006).

Aim of the chapter

The success of a project or initiative on CWR *in situ* conservation depends, to a large extent, on the capacity of the individuals and organizations involved. This chapter provides guidance on how the capacity of individuals, and to some extent organizations, can be strengthened to better undertake key activities for CWR *in situ* conservation as described in detail elsewhere in this manual – planning, team building, prioritizing, data collection and analysis, developing plans and strategies, monitoring, communicating and raising awareness, and so forth. While issues related to organizational and societal transformation are beyond the scope of this manual, the chapter does stress that all CWR *in situ* conservation activities take place in particular institutional and societal contexts that will have significant influence on how both individuals and organizations perform and, ultimately, on how successful conservation initiatives are.

We suggest that capacity building should be an integrated element of CWR initiatives, as formal qualifications in this area tend to be weak among key stakeholders. The primary audience for this chapter is a project manager of a CWR *in situ* conservation project or intervention. The chapter may also be of interest to institutional leaders and policy-makers who have a stake in such projects. Tertiary education institutions might also find the chapter useful as a reference in their curriculum review processes.

The aim is to raise awareness of the role of capacity building in a CWR initiative and to support capacity development processes linked to such initiatives. The chapter provides a quick guide on how to analyse capacity needs and how to plan, implement and evaluate capacity building – principally, capacity building of individuals. The text focuses mainly on the process of education and training, with a particular emphasis on participatory methodologies. A reference section at the end of the chapter offers suggestions on further reading and internet resources.

Capacity for CWR *in situ* conservation

Regions with the richest biodiversity, including genetic diversity of CWR, also tend to have the lowest levels of skilled specialists and the most fragile institutions. Hence, capacity building must be a major component of the process of CWR *in situ* conservation.

Capacity building is the process of developing competencies in individuals, groups or organizations, which will contribute to their sustained improved performance. It is much more than training of individuals; it is about equipping individuals *and* organizations with abilities, resources and opportunities to solve problems and with the confidence to influence others. The capacity of the individual is thus important, but the ability of the individual to apply the knowledge and influence the institution depends on his or her institutional context: the institution's programme and strategies, facilities and resources, leadership and the external environment such as access to networks. A broader view of capacity development relates to theories on systems thinking, societal change and complexity. Although such processes are also relevant to CWR *in situ* conservation, they involve quite different actors and fall outside the scope of this brief chapter. Approaches to develop capacity thus need to be situated in wider efforts to support the strengthening of capacity at other levels, as Figure 15.1 illustrates.

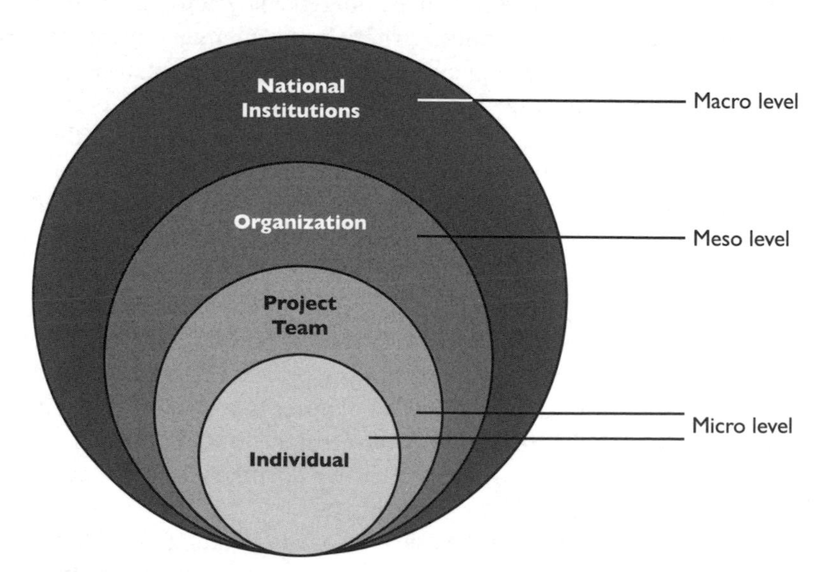

Figure 15.1 *Capacity development needs to be considered at different levels*

Source: Horton et al, 2003

Chapter 1, as well as later chapters of this manual, highlight the complexity and multidisciplinary nature of CWR *in situ* conservation. This creates many challenges. It is a process that addresses actions covering planning, data gathering, information management and analysis which lead to on-the-ground conservation actions and which touch upon a range of technical, political and institutional issues. Capacity building is a cross-cutting issue central to the success of this process. This manual has already highlighted some capacity building-related issues that individuals and organizations will face when undertaking CWR *in situ* conservation including:

- limited understanding and awareness of the importance of CWR at all levels of society;
- poor enabling environments created by inappropriate or lack of policy and legislation;
- national strategies and programmes that give no, or only token, consideration to CWR *in situ* conservation;
- no allocated funds in national annual budgets to sustain or initiate new activities commenced under donor-funded CWR projects;
- lack of cross-sectoral approaches – agricultural, forestry and environmental agencies that lack a tradition of collaboration;
- no generally agreed procedures or protocols to follow;
- limited understanding of conservation components or the sequence in which they need to be carried out and what *in situ* conservation of target species actually entails;
- limited practical experience of CWR *in situ* conservation both inside and outside of protected areas, especially the development and implementation of management plans and monitoring;
- limited capacity for data collection and information management;
- little understanding of the benefits of involving stakeholders, especially local and indigenous communities, in conservation approaches and how to facilitate their participation; and
- complexity of national political, institutional and administrative structures, making it difficult to implement a common strategy.

These all present major challenges for CWR conservation and highlight the role of capacity building at all levels in helping to overcome them. An individual's (or organization's) ability to solve a particular problem will not depend on his or her skills and training alone. It will also depend on the support, resources and equipment at their disposal within their own organization and that of their partners and networks. Ideally, developing capacity must focus on the entire conservation chain and facilitate the necessary process of transformation in individuals, organizations and society (see Figure 15.1) to enhance CWR *in situ* conservation.

Successful implementation of the many steps identified in the process of CWR *in situ* conservation will require that attention is given to capacity building at the outset. This cross-cutting issue is all too often neglected in the early stages of implementation, whether at the project or national level. As a result, training is often undertaken on an *arranged* basis or not given the consideration it requires until implementation is well underway. Failure to address capacity development needs might result in delays or reduced efficiency and impact.

As pointed out in Chapter 6, very few countries have ever developed national CWR strategies or action plans. Chapter 4 highlights that the majority of CWR initiatives that have been implemented to date have been sponsored by grants from agencies such as the Global Environment Facility (GEF). It follows that most capacity building to support CWR conservation takes place in a context that is largely project-driven and time-bound. General longer-term CWR capacity building efforts or commitments at the national level are rare. While the two are obviously related, there are significant differences regarding the scale, time and approaches to address these issues. There are also important implications regarding the sustainability and impact of capacity building initiatives that take place in a project-driven context as compared to a capacity building programme which might be part of a national programme or strategy. This chapter primarily seeks to explore options for capacity building for CWR *in situ* conservation at the project level.

It is beyond the scope of this book to focus on capacity development at the macro or societal level, as illustrated in Figure 15.1. However, that is not to imply that the need for such capacity development or targeted efforts is not necessary or possible. These are certainly needed, and it is important to keep in mind that much of the focus of national communication strategies is about making these connections and creating awareness among the wider societal actors. For example, Chapter 16 briefly describes the need for communication and advocacy strategies and activities that specifically target groups such as senior policy-makers who can make changes at this level. In fact, this could be a main thrust of an awareness campaign with limited resources as highlighted in that chapter.

> Conservation managers or practitioners, despite having the skills and the best of intentions, must operate in an environment that is largely outside of their control. Such an environment is often characterized by competing and conflicting organizations working within defined legal and regulatory frameworks and national committees and decision-making processes within a broader policy environment moulded by local, national and international contexts. Ultimately, we must look beyond individual skills to the ability of organizations as a whole to achieve the goal of CWR *in situ* conservation so that capacity building also contributes to institutional building and learning which brings about the needed organizational transformation in structures, cultures and procedures that help facilitate much more conducive environments for professionals as well as collaborations between relevant agencies and organizations. Daunting as this may seem, there is much that practitioners can do to bring about change in attitudes and behaviours of actors at this higher level, including targeted awareness and education campaigns as well as high-level lobbying and negotiation.
>
> *Source:* adapted from Hough (2006)

Developing a capacity building strategy

A capacity building strategy for supporting CWR *in situ* conservation at the macro and meso levels (see Figure 15.1) would require broad and long-term efforts that involve many stakeholders, their institutions and the policy environment they operate in, as Chapter 6 discusses. A capacity building strategy for the micro level – the focus of this chapter – would aim more specifically on developing competent project teams that are able to work effectively and efficiently with key stakeholders and in participation with local communities.

The first step in developing a capacity building strategy is to determine the competencies required for a successful intervention. Next, one will need to establish the current capacity of the stakeholders of the project. A training needs assessment will give a sense of the gaps in knowledge, skills or attitudes – competencies – that need to be addressed. One can then plan and implement the capacity building actions. Finally, monitoring and evaluation will give you valuable feedback for continuous improvement. As illustrated in Figure 15.2, this process involves:

1. reviewing the tasks involved in CWR *in situ* conservation;
2. a stakeholder analysis including assessment of stakeholders' roles in relation to the project;
3. establishing the competencies required in stakeholders to carry out or facilitate the tasks involved;
4. assessing training needs and conducting a situation analysis;
5. developing a capacity building plan;
6. monitoring and evaluation.

Step 1: Reviewing the tasks involved in CWR *in situ* conservation

A quick glance at the different chapters of this manual will give an idea of the types of activities that will be required if CWR *in situ* conservation is to be successful. The scheme presented in Chapter 1 as Table 1.3, 'The process of *in situ* conservation of CWR' provides a more detailed and clear picture of the steps and actions involved.

Step 2: Capacity building for whom? – Stakeholder analysis

The next step or question to ask is 'Capacity building for whom?'. Earlier chapters of this manual will give some idea of the answer to this. Chapter 4, which focuses on planning for CWR conservation and partnership building, provides guidance on identifying the main stakeholders and, therefore, who might need to be considered for capacity building. Chapter 5 focuses on participatory approaches and guides one in working with stakeholders and communities. There is now considerable evidence of the benefits of including indigenous and local communities in biodiversity management. Therefore, developing community-

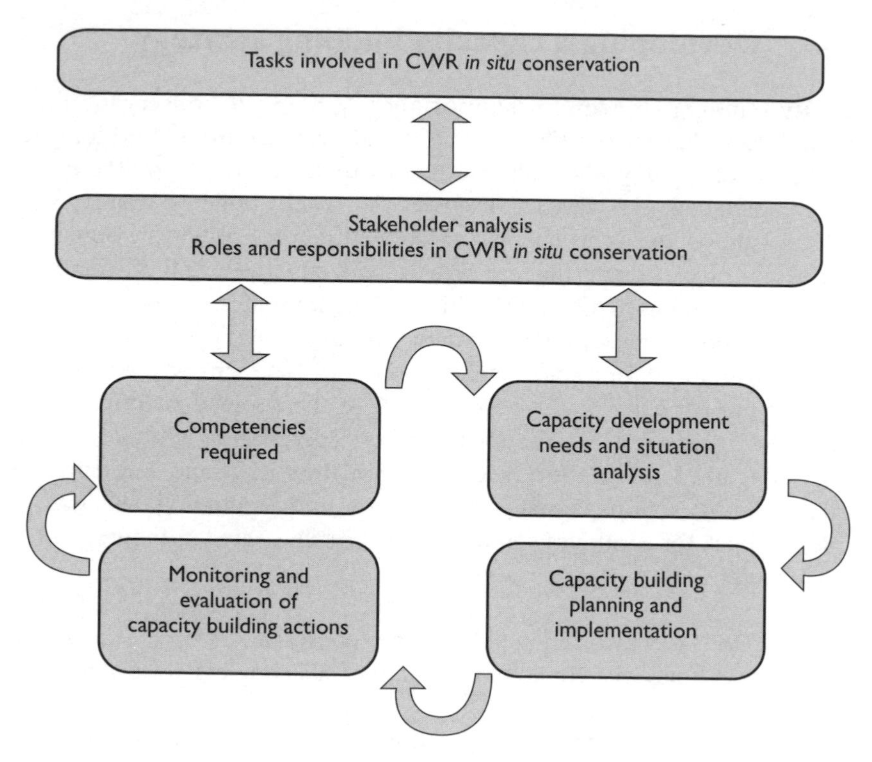

Figure 15.2 *Steps involved in developing a capacity building strategy*

based capacity is important for enhancing CWR *in situ* conservation. Special skills are required to facilitate this but, more often than not, such skills are lacking in those individuals and organizations involved in CWR conservation. More specifically, Table 9.2 and the 'stakeholders' subsection of Chapter 10 provide information on stakeholders involved in developing and implementing management plans.

The clear message is that there are a diverse range of individuals, groups and organizations that might require some level of capacity building if they are to make a successful contribution to CWR conservation. A list of stakeholders might include:

- political leaders and senior policy-makers;
- senior biodiversity, environment and agriculture decision-makers;
- heads of relevant organizations and institutes;
- national and local planners;
- scientists and researchers;
- protected area managers;
- project management staff;
- field technicians;
- university lecturers and postgraduate students;

	Low influence	High influence
High importance	Community groups	Protected area managers
Low importance	General public	Policy-makers

Figure 15.3 *An example of a stakeholder matrix*

Source: Rudebjer et al, 2001

- communications and public awareness specialists;
- extension and outreach specialists;
- information analysts and managers; and
- community leaders and groups.

Since time and funding for capacity building will always be limited you will need to set priorities and determine where and how to focus efforts. A simple method to aid priority setting is to group the stakeholders as 'insiders' and 'outsiders'. Insiders will be directly involved in the project team. They would need to be able to play their role in the various tasks involved in CWR *in situ* conservation. Outsiders, on the other hand, may provide an enabling environment that is critical for success and impact. For example, senior policy-makers might need to be sensitized to pave the way for the work at community level.

A 'stakeholder matrix' describing stakeholders' importance and influence in relation to the project can further deepen the analysis. Positioning each stakeholder in the grid (Figure 15.3) can aid in priority setting and might reveal important power relations or conflicts of interest that may be critical to your intervention's success.

Step 3: What capacity building is needed? – Establishing the competencies required

A variety of competencies are required covering technical aspects of CWR conservation, as well as process-oriented competencies – 'soft skills' – such as facilitation or leadership skills. A brief analysis of the UNEP/GEF CWR Project highlights that the following list of competencies were required for enhanced CWR *in situ* conservation:

Process-oriented competencies

- partnership building;
- facilitation;
- stakeholder analysis;
- leadership;
- team building;
- participatory approaches and community development;
- conflict, negotiation and advocacy skills.

Project management competencies

- project development and management;
- project monitoring and evaluation;
- budget preparation and financial management;
- resource mobilization;
- communications, public awareness and outreach.

Technical competencies

- Red Listing;
- ecogeographical surveying;
- conservation status and threat assessment;
- geographic information systems (GIS);
- preparing national CWR action plans and strategies;
- preparing species management and monitoring plans;
- monitoring and surveillance;
- data gathering, analysis and management;
- report and proposal writing;
- scientific and technical writing and communicating;
- educational and capacity strengthening strategies and methods; and
- training of trainers.

As this list indicates, professionals and organizations involved in CWR conservation require a balance of both technical and 'soft' skills. They also need to be able to apply those skills in a multidisciplinary environment, using participatory approaches. With a history of little collaboration between relevant agencies or organizations and minimal efforts to involve indigenous and local communities in CWR conservation, this poses a particular challenge that needs to be addressed in a capacity building strategy. Obviously, what will be required in terms of capacity building will depend on the local context and a range of other factors and must be defined on a specific case-by-case basis. For example, it is worth doing an inventory of related projects that require similar multidisciplinary approaches, such as community-based forest management, agroforestry or buffer-zone management projects. These may have relevant tools, capacity and expertise that the CWR projects could draw upon.

Step 4: Assessing capacity building needs and conducting a situation analysis

When starting a CWR *in situ* conservation programme it can be assumed that there will be a considerable gap between current capacity and skills, and the level that is actually required or desired. Assessing this gap between 'what is' and 'what should be' is known as a capacity building needs assessment; this will help define the type of training required and who it is required for. *A needs assessment should be done at the outset of a programme or project.*

There is a wide range of methods and tools that can be used to carry out a needs assessment. The tools highlighted in Chapter 5 (participatory approaches) can be used both to facilitate stakeholder participation and data collection on training history, strengths, gaps and future needs. A needs assessment may use a combination of individual questionnaires or more qualitative approaches like brainstorming, focus group discussions or other tools (see Table 15.1).

A broader situation analysis would complement the direct needs assessment. The situation analysis may cover new research results, relevant policies and processes, and other external factors that may trigger or influence capacity needs. If resources are available, specific studies can be commissioned to map out such aspects. The information and feedback from these consultations and studies can be presented in a workshop involving all stakeholders for review, priority setting and participatory planning. Needs must be prioritized in consultation and in an open and transparent manner.

The situational analysis also involves reviewing the resources available for capacity building. Limited resources should be allocated so as to make the most impact, and investing in capacity development early can pay off later. Allocating funds for capacity building is especially important in a project context where other components and activities will have a strong demand on resources and funds. It is the job of the project manager or the focal point for CWR conservation to balance these varying demands in light of available resources and diverse needs. This balancing act will be considered in the development of the capacity building plan, the next step of the process.

For those CWR initiatives which fall within a project context, it is advisable that a comprehensive inception workshop is held at a very early stage. A very clear message that emerged from the UNEP/GEF CWR Project was that it is essential to ensure from the outset that there is a clear understanding of the aims and purpose of the project, the different project components and the sequence of steps and activities necessary to achieve such aims and purposes. Such a workshop can help ensure that all those involved start with a common and basic understanding of the project and its various technical components and activities. It provides an opportunity for participants to seek clarification on issues, identify and fill in gaps in knowledge and understanding of the sequence and scheduling of implementation of activities. It also provides an early point in the project that participants can use to identify training needs from which the capacity building plan will be developed.

Table 15.1 *Tools for assessing capacity needs and related capacity levels*

Tools	*Societal levels*	*Organizational levels*	*Individual levels*
Brainstorming	X	X	
Case study analysis	X	X	X
Concept mapping	X	X	
Consensus-building discussions	X	X	
Delphi process	X		
Direct observation	X		
Document reviews	X	X	X
Expert panels	X	X	
Focus groups	X		
Force field analysis	X	X	
Gap analysis	X	X	X
Informant interviews	X	X	X
Job analysis	X	X	
Logical framework analysis	X	X	
Nominal group techniques	X	X	X
Organizational audits	X		
Participatory appraisals	X	X	
Prioritization matrix	X	X	X
Problem tree/root cause analysis	X	X	
Questionnaires and surveys	X	X	
Site visits	X	X	
Stakeholder analysis	X	X	
Staff audits	X		
SWOT (strengths, weaknesses, opportunities and threats) analysis	X	X	
Systems analysis	X	X	
Terms of reference	X	X	
Testing	X	X	X
Work plans	X	X	
Workshops/working groups	X	X	

Source: Lockwood et al, 2006

Probably the most serious failing of the UNEP/GEF CWR Project was the failure to appreciate, until rather late, the importance of the conservation components or the sequence in which they needed to be carried out and what *in situ* conservation of target species (as opposed to area conservation) entailed. As a result, Red Listing, extensive ecogeographic surveys and data management were, in some cases, carried out almost as an end in itself rather than as a means to establish the necessary background information for undertaking species conservation.[1] Had a capacity building plan been in place to link the training needs to the overall aims, learning objectives and outcomes, there might have been a different outcome. This is an easy trap to fall into, and this research-implementation gap is commonly found in conservation projects.

Step 5: Developing a capacity building plan

Building on the results of the stakeholder analysis and capacity needs assessment, and considering broader project/programme objectives and resource availability, a capacity building plan can then be developed. Such a plan can be part of a broader national CWR strategy and action plan (see Chapter 6).

The capacity building plan may take rather different formats depending on the level of intervention (local, project, national, etc.) but would generally include:

- aims – the broad purpose of the capacity building actions;
- learning objectives or outcomes; the acronym 'SMART' can help in formulating effective objectives:
  - specific;
  - measurable;
  - achievable;
  - relevant;
  - time-bound;
- contents – topics to be covered to address the competence gaps identified;
- implementation plan, including: selection of tools and methods for capacity building; time allocation; identification of trainers, facilitators, mentors, etc., including external resource persons (also consider using a training of trainers approach, for more impact); resources required; logistic considerations, etc.;
- monitoring and evaluation of training.

In developing the capacity building plan, a wide range of tools, methods and approaches will need to be considered, often in combination, to achieve the learning objectives. Further in this chapter a list of options for capacity building through education and training with a focus on individual training is provided and the lessons arising from capacity building under the UNEP/GEF CWR Project are described.

In addition, many of the examples and case studies on raising awareness and understanding of CWR provided in Chapter 16 could contribute to developing capacity in certain stakeholders. A good example is to sensitize policy- and decision-makers who play such an important role in determining the enabling environment for CWR conservation.

Step 6: Monitoring and evaluation of the capacity building plan

Monitoring and evaluation should be part of the capacity building plan as it will provide important feedback for continuous improvement. Well-planned and carefully implemented monitoring and evaluation will reveal if the capacity building plan is on track and highlight where it may need to be adjusted. It will show if learning objectives are achieved and if resources are well spent – which is of high interest to those investing in CWR conservation.

Monitoring could provide an 'early warning', which might help adjust an ongoing course or other capacity development activity to better meet the aims and objectives. Or, it could involve post-course feedback, which will help improve the capacity building activity the next time around.

The methods, criteria and indicators for evaluation need to be formulated early in the process. Decisions should be taken on what information should be collected and analysed throughout the capacity building activity and by whom. Evaluation assesses the achievement of learning objectives of the capacity building, that is, the knowledge, skills and attitudes acquired by the learner, the relevance of the content of capacity building and effectiveness of the learning processes. Both internal evaluation (by those involved in the intervention) and external evaluation (undertaken by independent evaluators) should be planned for – they provide different types of feedback for different purposes. Evaluation data – for example baseline data on existing capacity – can also be valuable inputs to future impact assessments.

Participatory approaches to evaluation of capacity building are useful to consider for CWR conservation, particularly when multiple stakeholders have been involved in the design of the capacity development plan. If stakeholders are involved in ongoing participatory evaluation and subsequent improvements to the capacity building plan, the project outcomes are likely to be more useful.

Further reading is suggested at the end of the chapter on carrying out monitoring and evaluation of capacity building.

Tools, methods and approaches for education and training

A subset of capacity building, *education and training*, is central to developing individual capacity within a CWR initiative. There are many ways and approaches available and it is important to pick the most appropriate approach, or combination of approaches, for addressing the identified capacity development needs. Capacity can be developed formally through training courses and other activities that are planned and implemented for that purpose. Informal learning that occurs without the presence of a curriculum, for example mentoring, collaborative research, networking or learning-by-doing, can also be important. The most common education and training approaches that may be considered are briefly presented here, each having its advantages and disadvantages:

- formal education;
- short courses;
- training workshops;
- internships, mentoring and study exchanges;
- fellowships;
- para-professional training.

Box 15.1 Capacity building and mainstreaming CWR information and knowledge into formal university courses

There are many reasons for considering partnerships with universities when it comes to CWR conservation. Universities and their teaching staff will be important custodians of knowledge on specific CWR species and on processes for their conservation. They provide opportunities for young graduates to pursue supervised postgraduate programmes in CWR conservation. Collaboration with universities within a project also influences curricula review and can strengthen course content in relation to CWR conservation and use. During the course of the UNEP/GEF CWR Project, participating countries were able to support students to undertake Masters and PhD programmes, which also resulted in important research and data outputs for the project. For example, in Madagascar, research was undertaken in ethnobotanical, biological and ecogeographical studies of wild *Dioscorea* spp. and wild species of *Coffea*. Many university courses in agriculture and conservation in participating countries lacked sufficient and up-to-date information on CWR. During project implementation, partners worked closely with relevant universities and staff to ensure that information generated from the project was mainstreamed into relevant university-level courses and programmes. In Armenia, one of the achievements of the partnership with the Armenian State Agrarian University was the establishment of a special course on agrobiodiversity, which addresses CWR conservation and utilization. The course was included in the Bachelors' and Masters' programme curricula of agronomy, crop selection and genetics of the university's Agrarian Department.

The question of how individual capacity developed through such approaches translates into institutional capacity lies beyond the scope of this chapter. For guidance on institutional and societal capacity, the reader is referred to the section on further reading. In addition, Chapter 16 adds information on awareness raising approaches and tools, which is an important complement as it generates support for interventions from responsible policy- or decision-makers.

Formal education

Tertiary education is society's fundamental approach to capacity development of individuals, leading to formal qualifications in subjects of relevant specialization at technical, undergraduate and postgraduate levels. However, agrobiodiversity in general, not to mention CWR conservation, is rarely a stand-alone course or full programme. Consequently, project staff and partners in a CWR initiative would rarely have a formal qualification in the subject.

The Masters of Research in Conservation and Use of Plant Genetic Resources offered by the University of Birmingham is one of the few programmes available that covers a range of topics pertinent to the process of CWR conservation and utilization. Although few other universities offer formal courses or programmes on CWR, many do provide opportunities to undertake such thesis

Box 15.2 Development of certificate-level modules targeting policy-makers, researchers and NGO staff – Sri Lanka

Addressing a major gap in capacity in Sri Lanka, staff at the Faculty of Agriculture at the University of Peradeniya, in collaboration with other organizations, developed three course modules on wild relatives of crops and their conservation. These short courses are aimed at policy-makers, researchers and NGO staff, but are also offered to graduate students. The Agriculture Education Unit of the university worked closely with national partners involved in the UNEP/GEF CWR Project to develop the curricula and the educational materials. The general content of the courses was reviewed and relevant stakeholders, who collaborated in curriculum development, were identified. One aspect of the review involved identifying earlier weaknesses of undergraduate and postgraduate courses in relation to wild relatives of crops. A stakeholder workshop took place in 2008 to finalize the curricula, brochures and other teaching materials. The courses commenced in September 2008.

research at the Masters and PhD level. Thesis research students could be a great resource for a CWR initiative, and it is worth considering budgeting for this at the onset of a project. During the course of the UNEP/GEF CWR Project, countries took advantage of such opportunities to build capacity, facilitate data collection and analysis, and implement conservation actions by enrolling staff and students in postgraduate programmes. At the same time, these experiences resulted in knowledge flowing back to the universities and contributed to mainstreaming the knowledge of CWR conservation and utilization into existing or new courses (see Box 15.1). In another instance, Sri Lanka addressed identified gaps in capacity by developing certificate courses targeting specific stakeholder groups during the UNEP/GEF CWR Project (see Box 15.2). These are good examples of how locating energy and commitment to educational change can be very useful beyond the project itself.

Short courses

A great many of the competencies outlined earlier are suited to short training courses. Short courses (one to a few weeks) can quickly develop new knowledge and skills, while ensuring that individuals are not away from their workplace for too long. There has been considerable growth in short-term training providers in recent years and it is possible to find some sort of training in most areas relevant to CWR conservation. For example, staff of the Royal Botanic Gardens, Kew, regularly offer short courses in topics such as conservation assessment techniques, organized with counterparts in herbaria and museums in many countries and regions of the world. Short-term courses, both face-to-face courses and online self-learning courses, can also be found for most of the process and project management competencies noted in Step 3 above.

Training workshops within the project

Training workshops are one of the most common ways of providing short-term training and capacity building for staff and partners of a CWR conservation project. Designed and implemented within the context of the project, they can target project goals with precision. In addition to developing technical and process-oriented knowledge and skills, they also help in building the project team. This approach was often used at the country level in the UNEP/GEF CWR Project and included training in the application and interpretation of IUCN Red Listing categories and criteria, basic GIS tools and information management.

Suitable resource persons can often be found in-country. On certain occasions expertise will have to be sought from outside. The UNEP/GEF CWR Project was able to source expertise and resource persons from its international partners such as the IUCN (Red Listing), BGCI (public awareness and outreach), WCMC (biodiversity monitoring) and FAO (legislation and policy review) for capacity building. This is an important role for international partners involved in such projects.

The advantage of organizing project training workshops is that the content is very focused and context sensitive. Training workshops use relevant examples from real life and allow the participants to share and learn from each others' experiences. They are also suited to developing skills in using participatory approaches. Depending on the situation, participants can sometimes bring their data to work on or to receive feedback from expert resource persons or from other participants. Box 15.3 illustrates how Bolivia effectively used regionally available expertise to address a major capacity gap and which eventually led to the implementation of extensive Red Listing and the publication of the first Red Book of CWR Plants in the region.

Internships, mentoring and study exchanges

Internships, mentoring and study exchanges can be put in place to develop capacity in project staff. Alternatively, the project can host interns and receive study visits, which aids in knowledge exchange and contributes to building capacity outside of the project.

Junior staff can undergo extended placements of work with more senior and experienced professionals. Placements can occur within the individual's organization or at another organization. Occasionally, there are opportunities for internships in international organizations such as CGIAR centres, botanic gardens, conservation organizations, and so forth; these opportunities are well worth exploring. Box 15.4 describes a study exchange that took place between Bioversity International's Regional Americas Office and a national Bolivian partner organization of the UNEP-GEF CWR Project to strengthen understanding of conservation assessment.

Box 15.3 Seeing red: Building national capacity to assess threat status

At the beginning of the UNEP/GEF CWR Project there were few experts in Bolivia with knowledge and experience in implementing the IUCN Red List categories and criteria. Fortunately, IUCN, as an international partner in the project, was well placed to help address this capacity gap. Bolivia made a direct request to the IUCN Regional Office in Ecuador to assist with the identification of an expert to train Bolivian researchers in the process of assessing the status of threatened species. The Bolivian partners identified Dr Gloria Galeano (from the National University of Colombia) because of her involvement in the development of Colombian Flora Red Books and also as a way of facilitating South-to-South cooperation. Dr Galeano, together with Arturo Mora from IUCN, trained Bolivian researchers through two workshops. The first workshop, aimed to familiarize researchers with the terminology, methodology and concepts of IUCN Red Listing and the application of the criteria and categories of IUCN Red List for species assessments, was held in La Paz in February 2006. Sixty-five researchers from national partner institutions and herbaria were trained during this workshop. The second workshop consisted of training on the technical review of CWR that were threatened according to IUCN categories and was held in La Paz in October 2007. Twenty-five researchers who attended the first workshop reviewed the categories given to the assessed species and the contents of technical sheets, under the supervision and guidance of Dr Gloria Galeano. Fourteen of the researchers who participated in the second workshop then applied the criteria and categories of IUCN as authors of the technical sheets contained in the Red Book of CWR Plants, the first of its kind in Bolivia.

Figure 15.4 *The Red Book of CWR Plants – Bolivia*

Source: Beatriz Zapata Ferrufino, UNEP/GEF CWR Project National Coordinator, Bolivia

Box 15.4 Providing mentoring in conservation assessment tools

The close collaboration between Bolivia and the Regional Office for the Americas of Bioversity International in Cali, Colombia, led to the short-term placement of a young researcher from one of the national partner institutions of the Bolivian CWR Project. During a month-long, hands-on internship, the Bioversity Regional Office provided training and mentoring in the use of conservation assessment tools (CATs), including ArcView for developing area of occupancy (AOO) and extent of occurrence (EOO), as a basis for analysing the degree of threat, based on the IUCN criteria. Upon return to Bolivia, the researcher was able to replicate the training to other authors of the technical sheets of the planned CWR Red Book. The tool was used to standardize the calculation of AOO and EOO and to apply the IUCN criteria and determine the category of threat to the species included in the Red Book of CWR in Bolivia.

Fellowships

Some organizations and funding agencies may offer scholarship or fellowship opportunities for individuals to undertake thesis or postdoctoral research in an area of importance to CWR conservation. The Vavilov-Frankel Fellowship (see Box 15.5) administered by Bioversity International is a good example. It has allowed individuals from developing countries to conduct research on plant genetic resources, including CWR, at advanced research institutes. Organizations providing research grants, such as the International Foundation for Science (IFS), can also be a source for funding CWR research projects for scientists at the beginning of their career. Many universities with conservation-related graduate programmes offer scholarships and fellowships in conjunction with their study programmes. There are many directories and websites that list such research fellowship and scholarship opportunities.

Para-professional training

Para-professional training can be used to build capacity in key individuals in local communities involved with a CWR conservation programme. The approach can develop their conservation skills through participation in workshops, training courses and seminars, or attachments to a conservation project or national programme. This can expose these key individuals to a range of skills and provide local communities with an enhanced capacity to implement, monitor and evaluate conservation actions. A good example is the training and deployment of para-taxonomists by the Instituto Nacional de Biodiversidad (INBio), Costa Rica – the first programme of this type (Basset et al, 2004). We are not aware of any such training approach being used in the area of CWR conservation, but there seems no reason why such approaches should not be used.

Box 15.5 The Vavilov-Frankel Fellowship

Dr Nicolai I. Vavilov was one of the first scientists to appreciate the importance of CWR. In his honour, and that of another important scientist, Sir Otto Frankel, Bioversity International set up a fellowship fund to encourage the conservation and utilization of plant genetic resources by enabling outstanding young scientists to carry out innovative research internationally. To date, fellowships have been awarded to 33 scientists from 22 countries. Topics relevant to CWR conservation have included work on: morphological and systematic characterization of diversity of the wild potato *Solanum brevicaule* complex; simple sequence repeat (SSR) evaluation of population genetic structure of common wild rice *Oryza rufipogon* for developing *in situ* conservation in China; analysis of genetic diversity and classification of wild and cultivated Iranian pistachio (*Pistacia* L.) using molecular markers; genetic structure and gene flow between wild and domesticated populations of *Polaskia chichipe* (Cactaceae) in the Tehuacán Valley, Mexico; structural and functional genomics of drought resistance in the progenitors of wheat and barley for crop improvement; and analysis of the gene genealogies and population structure in *Citrullus lanatus* L. and its wild relative, *Citrullus colocynthis* L. (Cucurbitacease) and the implications for genetic resources conservation.

Conclusion

CWR *in situ* conservation is rarely well-covered in educational programmes. As a result, formal qualifications in this area tend to be lacking among project staff and key partners of such initiatives. Hence, this chapter argues that capacity building should be an integrated element of CWR initiatives to ensure a project's success. With a focus on education and training of individuals, the chapter provides a quick guide on how to determine capacity building needs, planning for capacity building actions and evaluating the results. However, the ability of those individuals to apply their new competencies also depends on the institutional and societal context in which they operate. Ultimately, such organizational capacity will come down to issues of power, leadership, culture and belief systems, and control of resources and decision-making processes, as much as specific competencies on CWR *in situ* conservation.

Further reading

Baser, H. and Morgan, P. (2008) *Capacity, Change and Performance,* Study report, Discussion Paper No 59B, European Centre for Development Policy Management; www.ecdpm.org/capacitystudy.

Bioversity International has a list of fellowship and scholarship opportunities that can be accessed on their website: www.bioversityinternational.org.

Capacity.org is a web-based magazine and portal for practitioners and policy-makers who work in or on capacity development in international cooperation in the South. It includes a quarterly journal and sections on tools and methods, and practice reports. See: www.capacity.org.

The Centre for Forests and People (formerly Regional Community Forestry Training Center, RECOFTC) website maintains an excellent range of modules and training guides. There are three downloadable modules covering capacity building and training needs assessment. See: www.recoftc.org/site/index.php?id=432.

Horton et al (2003) *Evaluating Capacity Development: Experiences from Research and Development Organizations Around the World*, ISNAR/CTA/IDRC, www.idrc.ca/en/ev-31556-201-1-DO_TOPIC.html#begining.

The Institutional Learning and Change Initiative has a range of resources and tools focusing on areas relevant to capacity development and communications and knowledge sharing. See: www.cgiar-ilac.org.

Lockwood, M., Worboys, G.L. and Kothari, A. (2006) *Managing Protected Areas: A Global Guide*, Earthscan, London, UK. Chapter 7 has useful information on capacity development and training in the context of protected area management.

Rudebjer, P., Taylor, P. and Del Castillo, R.A. (eds) (2001) *A Guide to Learning Agroforestry – A Framework for Developing Agroforestry Curricula in Southeast Asia,* Training and Education Report No 51, ICRAF, Bogor, Indonesia, www.worldagroforestry.org/Sea/networks/Seanafe/Books/GLearnAF-Part1.pdf.

Taylor, P. (2003) *How to Design a Training Course*. A guide to participatory curriculum, which integrates the philosophy and orientation of a training programme, expected learning outcomes, key content, methodology and evaluation for the teaching and learning process. London: VSO/Continuum.

Taylor, P. and Clarke, P. (2008) *Capacity for a Change*. A document based on outcomes of the 'Capacity Collective' workshop, Dunford House, 25–27 September, 2007, Institute for Developing Studies, Sussex, www.ids.ac.uk/go/idsproject/capacity-collective.

United Nations Development Programme (UNDP) publishes a selection of publications relating to capacity development, which can be found on its website: www.undp.org/capacity/recommended_reading.shtml.

The World Agroforestry Centre has developed 'Training in Agroforestry', a toolkit for trainers to facilitate the planning, organization and implementation of training and education activities. It focuses on the design of a training progamme using a participatory approach. See: www.worldagroforestry.org/downloads/publications/PDFS/b12460.pdf.

Note

1. As stated in the Technical Advisory Committee Report (2009). Report of the Sixth Meeting of the International Steering Committee for the UNEP/GEF supported project "*In situ* conservation of crop wild relatives through enhanced information management and field application'. Bioversity International, Rome, Italy. pp. 55.

References

Basset, Y., Novotny, V., Miller, S.E., Weiblen, G.D. and Stewart, A.J. (2004) 'Conservation and biological monitoring of tropical forests: The role of parataxonomists' *Journal of Applied Ecology,* vol 41, pp163–174, http://geo.cbs.umn.edu/BassetEtAl2004.pdf, accessed 24 May 2010

Horton, D., Alexaki, A., Bennett-Lartey, S., Brice, K.N., Campilan, D., Carden, F., de Souza Silva, J., Duong, L.T., Khadar, I., Maestrey Boza, A., Kayes Muniruzzaman, I., Perez, J., Somarriba Chang, M., Vernooy, R. and Watts, J. (2003) 'Evaluating capacity development: Experiences from research and development organizations around the world', The Netherlands: International Service for National Agricultural Research (ISNAR); Canada: International Development Research Centre (IDRC), The Netherlands: ACP-EU Technical Centre for Agricultural and Rural Cooperation (CTA), www.idrc.ca/en/ev-31556-201-1-DO_TOPIC.html#begining.

Hough, J. (2006) 'Developing capacity', in M. Lockwood, G. Worboys and A. Kothari (eds) *Managing Protected Areas: A Global Guide,* Chapter 7, pp164–192, Earthscan, London, UK

Lockwood, M., Worboys, G.L. and Kothari, A. (eds) (2006) *Managing Protected Areas: A Global Guide*, Earthscan, UK

Rudebjer, P., Taylor, P. and Del Castillo R.A. (eds) (2001) *A Guide to Learning Agroforestry – A Framework for Developing Agroforestry Curricula in Southeast Asia,* Training and Education Report, no 51, World Agroforestry Centre (ICRAF), Bogor, Indonesia www.worldagroforestry.org/Sea/networks/Seanafe/Books/GLearnAF-Part1.pdf

Chapter 16

Communication, Public Awareness and Outreach

Action to foster biodiversity is urgently needed, and that requires politicians – and thus the wider public – to understand the significance of the changes taking place. This can be a complex message to communicate. The issue is not whether it is worth conserving a charismatic mammal or whether it matters if a few nematodes become extinct: it needs to be far more widely understood that declines in individual species herald the decline of diversity in whole ecosystems, which, in turn, has implications for human survival (Richard Lane, *New Scientist*, September 2009).

Aim of this chapter

Crop wild relatives (CWR) represent a significant body of neglected and threatened species whose importance is, with rare exception, poorly appreciated. This lack of appreciation of the value of CWR, the threats they face and their critical role in food security and ecosystem health is one of the greatest challenges facing their *in situ* conservation. This has resulted in a general lack of interest by the public, low commitment and political will by policy-makers, which translates into low priority and minimal conservation action at the country level.

Clearly, we are at a critical crossroads for CWR. We know the threats undermining their survival are intensifying and from limited studies we know that a significant number of CWR species are threatened with extinction as a consequence of changing climate. The outlook for CWR as the bedrock for agriculture in securing and sustaining food production and security is bleak if action is not taken soon. While the future may be hazardous, it is important to understand that these scenarios present key opportunities for the CWR community to make the case for increased attention to be given to CWR conservation through strengthened advocacy and communication efforts.

Effective communication strategies must play a part in changing the attitudes of key audiences about CWR and are critical to the overall success and sustainability of conservation efforts. In developing such strategies, clarity of message and clear definition of target audiences are essential. A well-planned communication strategy can ensure that the right messages and results reach those people and institutions that are in a position to influence the conservation policies and practices around CWR. It is the aim of this chapter to help practitioners think more strategically about communications, to introduce the range of available communications tools, and to explore the means for measuring communications impact. The body of knowledge on communications is substantial and the reader is directed to relevant sources. Recognizing the limitations of what can be presented in a single chapter of this nature, the authors hope that, at the very least, it presents some 'food for thought' on a critical, but often neglected, cross-cutting area vital for successful conservation.

The importance of communication

Taking action to change attitudes is probably the most reliable way to influence a change in behaviour over the long term. If the goal is the *in situ* conservation of CWR, the behaviour we wish to change is anything that prevents this goal from being reached. It might be that policies are in place that prevent – or at least do not support – the conservation and use of CWR in a given country or locality. It might be that people do not value CWR, viewing them as weeds or animal fodder. In this case, people probably do not know the role that wild relatives can play in improving agricultural productivity and food security or the functioning of the habitat in which they live.

These are only some of the possible constraints to conservation. Almost certainly there are others and it is probable that these will vary from place to place. But taking the above examples as indicative, and assuming that attitude change does in fact influence behavioural change, at least two things need to happen before these constraints can be removed:

1. Policy-makers and the people and institutions that influence policy (the so-called 'agents of change') must be convinced of the need to put into place policies, strategies and incentives to support the conservation of CWR.
2. Scientific institutions need to be convinced of the value of putting measures into place to conserve CWR.

Changing attitudes is not a quick or easy business. It is not likely to be accomplished with a single conversation, let alone a fact sheet, poster or even media mention. Changing attitudes on the scale necessary to achieve an impact that will ensure the conservation of CWR wherever they are at risk requires capacity, resources and a long-term institutional commitment. It will also require a comprehensive profile of the people who hold the key to ensuring that we meet our strategic goals, the best way to approach such people, and the means and messages most likely to compel them to change their attitudes. These factors will

Rising to the challenge

In many biodiversity-rich countries, the forces promoting biodiversity conservation are rarely consolidated and powerful enough to influence major policy decisions in favour of effective conservation policies. In most cases, government agencies do not play an effective enough leading role for biodiversity conservation for reasons already highlighted in this manual, including lack of political will, inadequate funding, low technical capacity, inappropriate policies and mismanagement of available resources. This gap in effective leadership means governments remain a significant impediment to achieving real progress in the implementation of international agreements such as the CBD and ITPGRFA, including promoting and enhancing CWR *in situ* conservation. Another issue in some countries is that the responsibility for biodiversity conservation within the context of the CBD lies with the Ministries of Environment, which sometimes tend to consider agriculture as detrimental to biodiversity rather than highly dependent on (agricultural) biodiversity. The result is that agricultural biodiversity does not receive the attention it merits at the CBD level. In addition, Ministries of Agriculture, which do have responsibility for agricultural biodiversity, do not always communicate well with their counterparts in the environmental sector (and vice versa); consequently, important opportunities may be lost.

However, even with limited resources, governments can support community education and awareness initiatives by making use of networks and organizations in their countries, as well as those existing regionally and globally. Carefully targeted awareness and education programmes can enable communities to protect and conserve the natural heritage in their local environment on which their cultures and livelihoods depend.

Source: adapted from *Communication, Education and Public Awareness: A Toolkit for National Focal Points and NBSAP Coordinators*, http://69.90.183.227/cepa/toolkit/2008/doc/CBD-Toolkit-Complete.pdf

tend to vary from place to place. In the case of CWR, it is likely that the individuals and institutions that can influence their conservation status will be relatively limited in number in each country. It makes sense to focus efforts on reaching this small audience rather than undertaking a broad-based campaign targeting the general public, whose support would be hard won, expensive and, in the end, probably not all that helpful.

At the global level, communicating information about CWR might help to achieve the recognition they merit in the global policy arena and also the financial support required from donors and relevant agencies. Organizations such as Bioversity International, the Food and Agricultural Organization of the United Nations (FAO) and the International Union for the Conservation of Nature (IUCN) routinely work in global forums and with international agreements such as the International Treaty on Plant Genetic Resources for Food and Agriculture (ITPGRFA) and the Convention on Biological Diversity (CBD) which are policy instruments that address CWR. A visible presence in global forums where relevant issues are addressed will help to ensure that CWR receive due considera-

The Climate Project

A recent study of The Climate Project (TCP) concluded that TCP presentations have had marked effects on public attitudes about climate change. The report found that those who previously did not identify as 'environmentalists' underwent the greatest mental shift, becoming more likely to support emissions reduction and to reduce their carbon footprints. Moreover, the evaluation suggested that TCP, an international non-profit organization founded by former Vice President Al Gore, has created a new, unique environmental movement by customizing its message to the region and community.

People who attend TCP talks were found to be more likely to change their behaviour on behalf of the environment after watching the presentation, based on the slideshow presented by Al Gore in the film *An Inconvenient Truth*. According to the study, if this intention translates into simple actions with households, such as changing incandescent light bulbs to energy-efficient bulbs, presentation attendees would reduce carbon emissions by 569,755 tons annually – the approximate equivalent of taking 109,702 passenger cars off the road each year.

The Climate Project's efforts have not only affected audiences, but also the presenters themselves. As a result of their work with TCP, presenters committed to changing their lifestyles to conserve energy and reduce their environmental impacts. Collectively, presenters cut their personal carbon emissions by an estimated 30 per cent. The presenters also reported that climate change became an important factor when voting and making investment decisions, a direct result of their work with TCP.

Source: TCP News; http://www.theclimateprojectus.org/tcpnews.php?id=1249

tion; however, interventions must be strategic and innovative if they are to successfully compete for attention with a long list of other conservation needs and priorities.

One arena where people have been effective in changing attitudes and shaping actions is that of climate change. There is much that the biodiversity community can learn from the climate change arena about how to communicate the right messages to the right audiences. The Climate Project, highlighted in the box above, demonstrates particularly well the power of communications to influence both attitudes and actions.

Developing a communications strategy

In a world where more and more people are experiencing information overload, it is especially important to understand how to communicate effectively. Policy-makers and other influential people receive a constant stream of information on various subjects from many different sources. Spending large amounts of money on a glossy brochure is not sensible if the brochure is immediately placed in the trash or sits unread on a shelf. Having more information products does not neces-

Box 16.1 Find a helping hand

Agriculture and biodiversity specialists frequently find it difficult to move out of their scientific mindset, which is required in order to understand the diversity of perceptions and opinions that exist among different stakeholders. For this reason, it is good practice to seek professional help and guidance from communication specialists when developing a communications strategy. The expertise of communication and social science professionals is increasingly available through networks that share and exchange expertise across various sectors. Examine other projects or initiatives in your country that have resulted in significant attitudinal and behavioural change. How was this achieved? What approach was used? How did they plan and organize?

Source: adapted from *Communication, Education and Public Awareness: A Toolkit for National Focal Points and NBSAP Coordinators*, http://69.90.183.227/cepa/toolkit/2008/doc/CBD-Toolkit-Complete.pdf

sarily translate into more action, outcomes or results. A better strategy might be to engineer a face-to-face encounter with an important and influential individual. The key word is strategy. No communications intervention should ever be undertaken without serious consideration of objectives, targets and audiences. *It is good practice to seek professional advice from a communications expert when planning your intervention* (see Box 16.1).

An effective communications strategy should be based on two major assumptions:

1 Public awareness can be used to change behaviour by influencing changes in attitudes.
2 Influencing profound changes in attitudes will require sustained, long-term effort.

The objective of a communications strategy is to provide a road map for convincing individuals and institutions whose actions – or inaction – are so impeding the conservation and use of CWR that any constraints to such activities should be removed.

A well-developed communications strategy must start by describing the communications objective, the target audience, the audience's current attitude towards the issue, the messages that need to be communicated to change that attitude, and the best ways to reach the target audience. The more you engage and consult with your target audiences about their information and communication needs, the communications tools that they prefer to receive, and the messages and arguments they find convincing (and those they do not), the more likely it is that your communication activities will have a positive impact. Therefore, a communications strategy should be developed at the beginning of a project and refined in the light of feedback during the project's lifetime. To re-emphasize a previous point: *it is strongly recommended to include a communications specialist in the development of the strategy.*

A recently concluded review in Sri Lanka highlighted there was a poor understanding among non-conservation sectors (including both state agencies and the business sector) and provincial, regional and municipal authorities concerning biodiversity and other environmental plans and policies. There was also a low level of awareness about the responsibilities of these sectors to implement such plans and policies. Among the requirements identified by stakeholders in Sri Lanka, was the need to develop a well-planned communications strategy to map out continuous dialogue and communications with relevant sector agencies, business and policy-makers and to provide capacity building to conservation agencies to enable them to better communicate, promote and 'sell' their image and work plans.

Source: adapted from *Communication, Education and Public Awareness: A Toolkit for National Focal Points and NBSAP Coordinators*, http://69.90.183.227/cepa/toolkit/2008/doc/CBD-Toolkit-Complete.pdf

Box 16.2 provides a checklist of things to consider when developing a communications strategy.

Identifying and shaping your key messages is critical. While the example presented in Box 16.3 is a good message, it is not the only one. Clearly, it is important to communicate the many benefits of CWR in terms of how they have been used to underpin food production and security, but the rate and consequence of the destruction of CWR and their habitats is an equally important key message. Chapter 14 highlights studies which indicate that by as early as 2055 an estimated 16–22 per cent of wild relatives species of peanut (*Arachis*), potato (*Solanum*) and cowpea (*Vigna*) may become extinct. Many more will lose their range size and the current system of protected areas will only maintain and protect a reduced proportion of CWR species. Also, what we know about future climate change scenarios indicates that many of the characteristics that our agricultural crops will require in future, such as resistance to new pests, increased drought and salinity, will most likely be found among the genetic traits CWR have to offer. The implications and importance of CWR for future food production and well-being are clear. Furthermore, the influential journal, *Science*, recently published a special Food Security issue (12 February 2010) and two papers in particular draw attention to the future importance of CWR for food security, presenting clear opportunities to piggy-back key messages to a wider audience. Given the importance of CWR for keeping agriculture safe and productive, these resources simply must be conserved. The arguments must be made and the case must be built. There is no way around it.

Many of the themes mentioned thus far – biodiversity, climate change, food security and food crises – are all newsworthy and draw considerable attention and interest from the media. The key messages for CWR can clearly be aligned with these themes but will still need to compete with everyone. The publication that recently highlighted the potential impact of climate change on the wild relatives of peanut, potato and cowpea (Jarvis et al, 2008) is probably the best example of a CWR key message being picked up by the media (see Chapter 14). This article

Box 16.2 Developing a communications strategy

There is plenty of information and help available on developing a communications strategy; the majority is freely accessible over the internet. There may be a communications specialist in your organization or at a partner agency. Be sure to make use of such expertise when developing a strategy. As a general rule a communications strategy should determine the following components and in the order presented below:

Objectives

The very first step is to determine the objective of the communications intervention. What do you hope to accomplish? Is the objective to bring about policy change? To raise funds? To inspire a change in priorities among research institutions? The strategy must be driven by the overall objectives of the project or organization.

Target audience

Identify the audience that you must influence in order to meet your objectives. Define all relevant audiences and target groups clearly. Some of your target audience will be broad and will need to be addressed using far-reaching tools (e.g. the internet) whereas some will be highly defined and may be best addressed through face-to-face contact.

Key messages

These should be strategic, targeted and consistent. Different audiences will respond to different messages. No matter which audience you are addressing, the case should be summarized in no more than three key points that can be constantly repeated. Box 16.3 provides an example of how to shape key messages.

Communication tools and activities

Different audiences will warrant different tools. Be aware of which tools the audiences find useful and those which they do not. For example, using the internet to reach a target audience in a country with low bandwidth will not get you very far. The examples in this chapter illustrate the variety of available tools.

Budgets and resources

The budget must be sufficient to support plans and activities or else the strategy should include a well-articulated case for more resources.

Timeline

This will include a phasing of activities and actions that might start with undertaking a needs assessment of target audiences, capacity building and so forth.

Evaluation and refinement of the strategy

This is important for monitoring and evaluating success. Such an evaluation can gather information from both internal and external audiences. Adjustments to the strategy should be made where necessary.

Source: adapted from Media Trust http://www.odi.org.uk/rapid/tools/toolkits/Communication/Communications_strategy.html

Box 16.3 Selling crop wild relatives

One major selling point to policy- and decision-makers is the contribution that CWR can make as gene donors to increase crop yields and quality. Globally this contribution has been estimated at about US$115 billion annually worldwide (Pimentel et al, 1997). Genes from wild plants have provided cultivars of food crops with resistance to pests and diseases, improved tolerance to abiotic stresses, tolerance to extreme temperatures, salinity and resistance to drought, as well as enhanced nutritional quality (see Chapter 1). To take a very specific example, a single wild tomato has contributed to a 2.4 per cent increase in solids contents worth an estimated US$250 million. Furthermore, with climate change, the demand for such genetic traits will rise.

certainly created a sense of urgency regarding the need to collect CWR for *ex situ* conservation, which led the Global Crop Diversity Trust to take this issue seriously, and donors have reacted positively as well. The lesson from the work itself is the importance of generating accurate numbers that capture the scale of the problem and that can be used in the media: 16–22 per cent under threat of extinction from climate change (A. Jarvis, personal communication).

At a more local level, the example of media tours in Uzbekistan highlights how greater attention can be drawn to CWR through the media. In pitching stories to the media, it is important to start with what people know and care about. Very few people know and care about biodiversity. In practice, the best thing to do is to find out what the *media* know and care about. This can be done by reading newspapers and blogs and by asking friendly journalists about their interests. We know that people – including the media – care about climate change, and everyone cares about food. CWR stories that relate to climate change or food issues might be easier to 'sell' than abstract stories or those that are overly technical. Linking your story to something already in the media is always a good strategy, but be sure to have facts and figures; otherwise the story may be vague and will not feel like news. Do not just contact the media periodically; instead, build relationships with them, checking in with the 'friendlies' regularly to bring them up to speed. If these journalists like and trust you, they are far more likely to cover your stories.

Achieving a major goal – such as influencing national CWR policy – is best done in partnership with like-minded individuals and organizations. Partnerships must be cultivated and this can take a significant length of time. All partners need to understand exactly what is expected of them and what they will gain from the partnership. Partnerships require effort, but will give greater weight to your message (if the partners are reputable) and may be able to open doors for you, helping you to get your messages to places that you may not be able to reach on your own, that is, the offices of those key individuals in strategic organizations, agencies and communities.

A communication strategy must also consider that the most effective communications are not merely one-way affairs that consist of bombarding audiences

Biodiversity loss matters and communication is crucial

Communicating the reason why biodiversity loss matters for people is essential if we are to reverse this trend. Like climate change, the threat of large-scale biodiversity loss – and the need for global political commitment and action to halt it – is growing daily. Persuading political leaders and the public of the urgent need to take action is both a complex and formidable challenge. Part of the answer lies in enhancing the media's ability to communicate messages emerging from underlying science, so these accurately reflect both the urgency of the situation and how the lives of ordinary people may be affected. Getting these messages across is not an easy task. So far, in the case of biodiversity, efforts have largely failed and, as a result, CBD targets have not been reached. The scientific community has not been able to effectively communicate its concerns to decision-makers. Often, issues scientists believe are most important do not resonate with the day-to-day concerns of the public, let alone policy-makers. New approaches must address weaknesses apparent in current efforts and be accompanied by more innovative communication strategies.

Source: David Dickson, 5 February 2010, www.scidev.net

with messages and materials. Communications as dialogue and communications for building and maintaining good relations with partners must be part of the strategy. Chapters 4 and 5 provide a context where communication is seen as vital for the development of effective partnerships and successfully engaging with stakeholders. While many communication interventions are clearly aimed at fairly broad audiences (whose influence may be limited), in many instances the most effective communications approach or strategy will consist largely of targeted face-to-face contacts with a few key individuals in strategic organizations, agencies and communities.

Case studies highlighted earlier in this manual clearly illustrate this point. Good examples are the processes of consultation and engagement that were necessary for the establishment of the Sierra de Manantlán Biosphere Reserve in Mexico (Box 1.6) and the development of wild yam species management plans for the National Park of Ankarafantsika, Madagascar (Box 5.5), as are the examples of collaborative work in other protected areas in the UNEP/GEF CWR Project countries highlighted in Chapter 9. The Sierra de Manantlán Biosphere Reserve, established by presidential decree in 1987, was created to protect wild species related to maize. Prior to the establishment of the reserve, local indigenous communities were often in conflict with private logging companies for control of land. This led to the emergence, in the late 1970s, of a strong peasant alliance against the timber companies. Around this time, the discovery of the endemic maize relative *Zea diploperennis*, in its natural habitat in Jalisco, and the interest it attracted from many scientists, was seen by local communities as an opportunity to establish communication with government agencies that had ignored them in the past. Direct communication between groups and institutions together with effective advocacy with state and national government agencies were instrumental in making the reserve a reality, despite considerable opposition from powerful groups with vested economic interests in the area. Continuing difficulties and conflicts in terms of developing and implementing management strategies in the reserve draws attention to the importance of ongoing communications as dialogue (Nathan Russell, personal communication).[1]

The above example of the Sierra de Manantlán Biosphere Reserve, as well as that of the development of wild yam species management plans for the National Park of Ankarafantsika, also highlight the need – depending on the audience – to give due consideration to community-based communications as opposed to more formal tools when developing a communications strategy (see Chapter 5). When trying to sensitize rural communities to the importance of CWR through awareness programmes or general consultation, it is important to consider approaches that are embedded in local culture and appropriate to local contexts and norms. Among various tools that might be considered are biodiversity fairs, folksong competitions, rural poetry journeys and rural roadside drama.

Communication and public awareness tools

There are many communication and public awareness tools to select from. It is beyond the scope of this manual to describe them fully. The reader is therefore referred to some useful sources listed at the end of this chapter, which include techniques, tools, guidelines, case studies and information on networks and sources of experts. The list that follows (see Box 16.4) is extensive but by no means exhaustive; it will serve as a guide to selecting the appropriate tools. The case studies that have been selected should also stimulate thinking about innovative ways of communicating and creating awareness about CWR (see Boxes 16.5 to 16.9).

A distinction should be made between external and internal audiences for communications. Internal audiences comprise organizational or project staff and partners that are involved directly in project planning and implementation, and other current or potential collaborators and relevant donors. These actors are described in Chapter 4. External audiences include the general public and policy-makers, and special consideration would have to be given to the communication tools used to target the two groups.

A special mention should be made about the growing importance of weblogs, wikis, listserves and other social networking tools. These are an efficient and effective way of sharing current information on CWR. Options for using such tools to disseminate newsworthy stories on CWR were recently reviewed by Guarino (2008). Such networking tools play an important role in facilitating information sharing within a globally distributed community of CWR conservation specialists, but their role is limited in that they usually are 'preaching to the converted'.

Box 16.4 Communication and public awareness tools

External communications tools

Print/radio

- media press release;
- radio programmes;
- feature articles.

TV

- news;
- biodiversity, agriculture, science programmes;
- videos/CDs/DVDs of interesting activities and outcomes.

Advertising and feature stories

- print;
- radio;
- television.

Publishing

- brochures;
- posters;
- bill boards;
- letters;
- leaflets/flyers;
- technical reports;
- websites;
- blogging, listserves, wikis.

Public relations

- biodiversity, science and agriculture shows;
- T-shirts, bags, stickers;
- telephone calls;
- side events;
- conferences;
- networking.

Other tools

- policy papers;
- lobbying;
- role plays and drama;
- educational materials for schools and universities;
- making use of special occasions such as International Day for Biological Diversity (22 May) and World Food Day (16 October);
- special exhibits in botanic gardens;
- school painting, poetry, essay and quiz contests to target young generations.

Box 16.4 continued

Internal communications tools

- phone calls;
- country visits;
- face-to-face meetings with partners/stakeholders;
- email;
- progress reports;
- project newsletters;
- training workshops;
- international and national meetings;
- short-term attachments for information officers and research staff;
- study tours for project staff and other stakeholders;
- intranets;
- travelling seminar to bring together multidisciplinary groups and policy-makers.

Source: Bernadette Masianini, Communication Officer for the Development of Sustainable Agriculture in the Pacific Project (DSAP), http://wwwx.spc.int/dsap/about_dsap.htm (last accessed 14 October 2010)

A good guiding principle is to 'communicate internally before communicating externally'. Make sure the entire organization knows the plan and how they are expected to contribute to it.

Source: adapted from *Communication, Education and Public Awareness: A Toolkit for National Focal Points and NBSAP Coordinators* (http://www.cbd.int/cepa/toolkit/2008/doc/CBD-Toolkit-Complete.pdf)

The most important thing to know about any communications tool is whether it is meaningful to the target audience. Some people are impressed by things they read in the media – being in the media can add credibility to an initiative. Some like glossy publications or websites; others feel these are a waste of time. Remember that you are using a public awareness tool to reach a target audience. You can find out what your target will respond to by seeing what works or by asking them directly what they need. You will also get an idea of the kinds of tools that work best by asking communication specialists in your organization or locality.

Evaluating success

Earlier in this chapter the importance of evaluation and refining strategies for communication and public awareness was mentioned. This aspect of communication is often neglected. Communication is often seen as a one-way process of reaching or telling others, but communication is also a process whereby the 'communicator' can learn from the needs and interests of the target groups. Such

Box 16.5 The power of art in conveying the conservation message

There are many unconventional ways to communicate a conservation message. Perhaps one of the most effective and appealing ways is art. Japanese artist Mitsuaki Tanabe has chosen sculpture as his preferred means of expression and has shared his concern for conservation by blending art and science. Since the late 1970s, Tanabe has concentrated on creating nature-based sculptures and is passionate about the importance of his work in promoting the conservation of endangered species and the importance of biodiversity. In recent years, the leitmotif of his work has been wild rice and the plight of its conservation. Wild rice, whose natural distribution around the globe has been slowly declining due to habitat loss and degradation, is essential for food security and an important source of breeding traits for cultivated rice varieties. Tanabe has a number of artworks displayed in museums and agricultural research centres around the world.

Figure 16.1 *Mitsuaki Tanabe with one of his sculptures*

Source: Teresa Borelli, adapted from *Geneflow*, Bioversity International

an evaluation can only help to increase the impact of your communications strategy. *As this chapter highlights, communication is a long-term undertaking, so we need to continually reflect and ask questions* such as:

- Have we achieved our objectives?
- Did we reach the right audience?

Box 16.6 Development of CWR Information Parks – Sri Lanka

The Sri Lankan Department of Agriculture is taking full advantage of its beautiful setting to bring the story of agriculture – including the role played by wild relatives of crops – directly to the public. Inspired by the Department's attractive location in the central hills of Sri Lanka along both banks of the river Mahaweli, Rohan Wijekoon decided to give the public an opportunity to witness first-hand new agricultural technologies and research. This led to the establishment of the Department's first Agriculture Information Park, which now attracts about 30,000 people annually. Visitors to the park learn about important conventional crops in Sri Lanka as well as home gardens, paddy cultivation and traditional farming systems. There is also the national genebank and an agriculture museum. Importantly, the Department of Agriculture is using the park to raise public awareness about the importance of CWR. So far, wild relatives of pepper, bean, okra, banana and rice have been established along the banks of the Mahaweli River. The Department recently established its second Agriculture Information Park at Bataata, in the southern part of Sri Lanka. This park is on the way to one of the most venerable religious places in the country, which many Sri Lankans visit all the year round. The park, which features CWR prominently, was opened by the President of Sri Lanka in January 2008 and is already attracting 8000 to 10,000 visitors per month.

Botanic gardens can also be locations to showcase nationally important wild relatives. For example, in Sri Lanka, Uzbekistan, Armenia and Madagascar, national botanic gardens have dedicated sites for locally important CWR to educate visitors. The Royal Botanic Gardens in Peradeniya, Sri Lanka, receives over a million visitors a year including 250,000 school students.

Figure 16.2 *Entrance to the CWR Information Park*

Box 16.7 Organizing a media tour to promote and raise awareness of CWR conservation in Uzbekistan

In 2008, Uzbekistan hosted a national media tour involving more than 30 journalists from various national mass-media organizations. The event provided an opportunity for professional ecologists and journalists to come together and discuss the importance of CWR and ways to increase public awareness. The tour then provided an opportunity for the journalists to visit Ugam-Chatkal State National Natural Park, where various specialists working on CWR demonstrated conserved populations of wild relatives of pistachio, apple, almond and walnut to the journalists. Journalists also observed the devastating impact of threats such as water erosion, livestock grazing and tree cutting on CWR. Four television programmes, 10 radio programmes and 18 stories in the national press resulted from the tour.

Figure 16.3 *Journalists on the CWR media tour*

Source: Sativaldi Djataev and Feruza Mustafa

- Did they understand what the message was – did they do what had to be done?
- Did we reach the right people within the organization?
- Did we use the right tools?
- Were decisions taken as a result?
- Did this result in concrete actions?
- Did we meet our budget? If not, why not?

Box 16.8 Creating awareness of CWR inside protected areas

Protected areas are one of the most important locations for *in situ* conservation of CWR. They also receive large numbers of visitors annually. Most often, these visitors have little or no understanding of the kinds of wild relatives in the protected area, or their importance. This presents a useful opportunity for public awareness activities. In Sri Lanka, public awareness work was undertaken in the Kanneliya Forest Reserve with the aim to help visitors learn about the biodiversity of wild cinnamon in the park and efforts to enhance *in situ* conservation. Signboards were placed throughout the park and posters hung in visitors' dormitories, which explained the role and importance of CWR. There are also plans to create a display focusing on CWR at the entrance to the Forest Reserve.

Figure 16.4 *Signboards raise awareness on the importance of CWR in the Kanneliya Forest Reserve, Sri Lanka*

Source: text and photo, Anura Wijesekara

In this regard it is worth considering holding focus group discussions with your target audience to clarify:

- What do they read/see/hear?
- What works/does not work?
- What do they want to see more of?
- What information do they need that you do not currently have?
- How often do they want us to communicate with them?

Box 16.9 Rural poetry journey and rural drama in Nepal and Sri Lanka to raise public awareness of conservation of wild rice

The Nepal On-farm Project mobilized local cultural groups and rural poets to sensitize the community awareness programme with multiple approaches embedded in local culture and taste. Among various tools used, biodiversity fairs, folksong (teej geet) competitions, rural poetry journeys and rural roadside drama were found most popular and effective in communicating messages to a wide range of rural audiences. Rural drama was also effectively employed in Sri Lanka as part of the UNEP/GEF CWR Project.

Rural poetry journeys are a kind of participatory travelling seminar; in this project selected teams of national and local poets visited diversity-rich areas, including wild rice (*Oryza rufipogon L.*) habitats in the Begnas Rupa Lake watershed in Nepal. The teams spent time with farmers, learning of the value of wild rice and reciting poems and songs in the evening in the village. The impact of the poetic pilgrimages was encouraging and proved to be effective in generating awareness among a large number of farming communities. The poets recited their 'odes to wild biodiversity' to the community before moving on to the next village. At the end of journey, the poems were compiled and published as a book. Selected poems were regularly cited by rural radio to sensitize the community about environmental issues.

Figures 16.5 and 16.6
Teej song competition

Source: LI_BIRD/NARC 2000 Teejgeet Pratiyogita – Contributed by Bhuwon Sthapit

Figure 16.7 *Using rural drama to raise awareness*

Source: Mr R. Vijekoon, Sri Lanka

While it is easy to keep track of the number of public awareness materials produced and distributed, or the number of visitors to a website and files downloaded, it is more important, but more challenging, to measure the actual impact of these materials on your target group (see Box 16.10). Is your target audience more aware of CWR than before your intervention? Has the communications intervention changed the way they behave? What is the long-term impact of these changed behaviours? Has your intervention contributed to a better enabling environment for CWR conservation? Is there evidence that governments or other agencies are allocating more funds and resources to CWR *in situ* conservation as a result? The further you travel along these evaluation steps the more difficult and costly it is to measure impact, and the harder it is to clearly demonstrate a causal link to the initial intervention. Having said this, there are ample evaluation tools available for use, such as questionnaires, focus group discussions, case studies and participatory evaluation approaches, that can be found in the further information sources listed at the end of the chapter as well as on the internet.

Measuring impact is relatively easy if your audience is small and your objectives are measurable. If your objective is to influence policy and your target audience is key parliamentarians, you can be sure you have had some impact if policies do change as a result of consistently targeting these individuals and sharing the information they need. With a larger audience such as the general public, it is more difficult to judge the impact, even with an unlimited budget. Nonetheless, baseline attitude surveys are always a good way to start.

Before and after surveys to assess the impact of a CWR public awareness campaign as part of the UNEP/GEF CWR Project in Sri Lanka highlighted that greatest impact was among protected area managers and extension workers, but among policy-makers the impact was low (Figure 16.8).

Sources of further information

Bioversity International's *Geneflow* is an annual magazine that contributes to promoting awareness of the importance of the earth's agricultural biodiversity and the role it plays in improving people's lives and livelihoods. Website: http://www.bioversityinternational.org/publications/publications/geneflow/2008.html

The Communication Initiative Network is an excellent general website on communication, with extensive resources, tools, examples, funding sources, etc. In many cases it is searchable by country/region, issue and communication tool. It also has a site maintained in Spanish. Website: www.comminit.com

Hamu, D., Auchincloss, E. and Goldstein, W. (2004) *Communicating Protected Areas*, IUCN Commission on Education and Communication. This has useful information on strategic communications in the context of protected areas. Much of the information is highly relevant to professionals involved in CWR conservation. The book is illustrated with a number of case studies describing communication tools and approaches that could be easily adapted to build support for CWR *in situ* conservation.

Box 16.10 Measuring the success of public awareness

How can we measure the impact of public awareness activities or campaigns to promote understanding and conservation of CWR? All countries participating in the UNEP/GEF CWR Project undertook significant public awareness activities using a number of the tools listed in Box 16.4 and described in this chapter. Hardly surprising, before and after assessments demonstrate that such activities and campaigns do seem to contribute to better awareness and understanding of CWR among a wide range of target groups – the general public, policy-makers, scientists, protected area managers, NGO staff and so forth. In Armenia, for example, in 2005, before such activities commenced, 23 per cent of people in urban areas, including scientists, were able to name some CWR species and 36 per cent had some general understanding or knowledge of CWR (compared to 10 per cent and 17 per cent for rural areas). By 2009, following countrywide public awareness activities, these figures had increased to 37 per cent and 43 per cent, respectively (compared to 30 per cent and 35 per cent for rural areas). But the goal is not simply awareness for the sake of awareness alone. To what degree do such efforts translate into more support for actual conservation actions? Without long-term sustained and targeted interventions and the refinement of communication strategies over time, it is hard to know. While there are some indications that countries are committed to sustaining activities and implementing management plans and strategies, it is too early to say how likely this is to become a reality. We already have too many examples of how such initiatives fall by the wayside after a project stops. A real indication of success would be the allocation of dedicated budgets for *in situ* CWR conservation in national programmes, as well as greater financial commitments from donors.

Source: Armen Danielyan, National Project Coordinator, UNEP/GEF CWR Project, Armenia

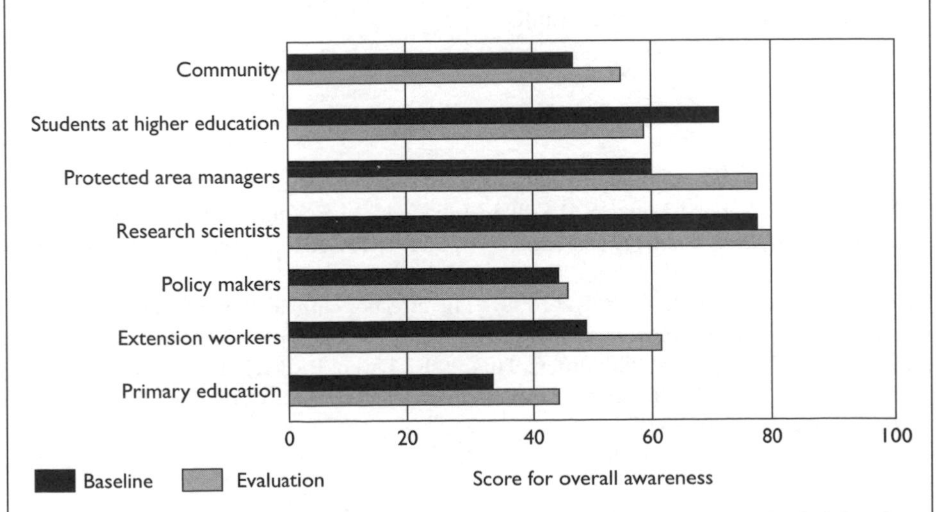

Figure 16.8 *General awareness of CWR among stakeholder groups in Sri Lanka following a project-led public awareness campaign*

Source: Mr Kamal Karunagoda, Sri Lanka

Hesselink et al (2007) *Communication, Education and Public Awareness: A Toolkit for NSBAP Coordinators*, CBD/IUCN. Website: www.cepatoolkit.org

Hovland, I. (2005) *Successful Communication: A Toolkit for Researchers and Civil Society Organisations*, Overseas Development Institute, London. Website: http://www.odi.org.uk/resources/download/155.pdf

The IUCN Commission on Education and Communication is a network that drives change for sustainability. More than 600 members volunteer their professional expertise in learning, knowledge management and strategic communication to achieve IUCN goals. It provides access to experts, thematic issues, networks and downloadable resources. Website: http://www.iucn.org/about/union/commissions/cec/

Lockwood, M., Worboys, G.L. and Kothari, A. (2006) *Managing Protected Areas: A Global Guide*, Earthscan. Chapter 10: 'Obtaining, managing and communicating information' has useful information on the principles of good communication and methods and approaches for communicating with local communities. The information is written in a protected area management context and therefore highly relevant to CWR conservation.

The Media Trust has a wealth of communications and publicity-related information on how to improve outreach including training resources and online guides for public relations, communications and dealing with the media. Website: www.mediatrust.org

Roots, *Cuttings* and *BGjournal* are magazines published by Botanic Gardens Conservation International, which contain stories and case studies about successful public awareness and education activities as well as regular listings of resources. Website: http://www.bgci.org/resources/publications/

Note

1. http://river.unu.edu/e-archive/14.pdf.

References

Guarino, L. (2008) 'Some thoughts on sources of news about crop wild relatives,' in N. Maxted, B.V. Ford-Lloyd, S.P. Kell, J.M. Iriondo, M.E. Dulloo and J. Turok(eds) *Crop Wild Relatives Conservation and Use,* pp521–531, CAB International, Wallingford, UK

Jarvis, A., Lane, A. and Hijmans, R. (2008) 'The effect of climate change on crop wild relatives', *Agriculture, Ecosystems and Environment,* vol 126, pp13–23

Pimentel, D., Wilson, C., McCullum, C. Huang, R., Dwen, P., Flack, J., Tran, Q., Saltmna, T. and Cliff, B. (1997) 'Economic and environmental benefits of biodiversity', *BioScience,* vol 47, pp747–757

Annexes

Annex I

CWR Species for which Field Data were Collected in Bolivia during 2006–2009, by Institution

Genus	*Common name*	*Cultivated species*	*Wild relatives of crop*
Herbario Nacional de Bolivia (LPB) – Universidad Mayor de San Andres[1]			
Theobroma	Cacao	*Theobroma cacao*	1 *Theobroma cacao*
			2 *Theobroma obovatum*
			3 *Theobroma speciosum*
			4 *Theobroma subincanum*
Anacardium	Cayú	*Anacardium occidentale*	5 *Anacardium giganteum*
			6 *Anacardium humile*
			7 *Anacardium spruceanum*
Centro de Biodiversidad y Genética/Herbario Nacional Forestal Martin Cardenas (BOLV) – Universidad Mayor de San Simon[2]			
Annona	Chirimoya	*Annona cherimola*	8 *Annona amazonica*
			9 *Annona ambotay*
			10 *Annona cordifolia*
			11 *Annona cornifolia*
			12 *Annona coriacea*
			13 *Annona dioica*
			14 *Annona excellens*
			15 *Annona foetida*
			16 *Annona hypoglauca*
			17 *Annona macrocalyx*
			18 *Annona Montana*
			19 *Annona montícola*
			20 *Annona nutans*
			21 *Annona paludosa*
			22 *Annona scandens*
			23 *Annona sericea*
			24 *Annona tomentosa*
			25 *Annona* sp.
Vasconcellea	Papaya	*Carica papaya*	26 *Vasconcellea cundinamarcensis*
			27 *Vasconcellea glandulosa*
			28 *Vasconcellea microcarpa*

Genus	*Common name*	*Cultivated species*	*Wild relatives of crop*
			29 *Vasconcellea monoica*
			30 *Vasconcellea parviflora*
			31 *Vasconcellea quercifolia*
Rubus	Mora, Zarzamora	*Rubus procerus*	32 *Rubus adenothallus*
		Rubus rosifolius	33 *Rubus betonicifolius*
			34 *Rubus bogotensis*
			35 *Rubus boliviensis*
			36 *Rubus briareus**
			37 *Rubus bullatus**
			38 *Rubus coriaceus*
			39 *Rubus floribundus*
			40 *Rubus glabratus*
			41 *Rubus glaucus*
			42 *Rubus loxensis*
			43 *Rubus mandonii*
			44 *Rubus megalococcus*
			45 *Rubus nubigenus*
			46 *Rubus penduliflorus*
			47 *Rubus peruvianus*
			48 *Rubus rigidifolius*
			49 *Rubus robustus*
			50 *Rubus roseus*
			51 *Rubus ulmifolius*
			52 *Rubus urticifolius*
			53 *Rubus weberbaueri*
*Cyphomandra***	Tomate de árbol	*Cyphomandra betacea*	54 *Cyphomandra acuminata**
			55 *Cyphomandra benensis**
			56 *Cyphomandra maternum**
			57 *Cyphomandra oblongifolia*
			58 *Cyphomandra pendula*
			59 *Cyphomandra pilosa*
			60 *Cyphomandra tenuisetosa*
			61 *Cyphomandra uniloba**
Museo de Historia Natural Noel Kempff Mercado / Herbario del Oriente (USZ) – Universidad Autonoma Gabriel Rene Moreno & Instituto de Investigaciones Agricolas 'El Vallecito' – Universidad Autonoma Gabriel Rene Moreno[3]			
Manihot	Yuca	*Manihot esculenta*	62 *Manihot anisophylla* Müll Crantz
			63 *Manihot anomala* Pohl
			64 *Manihot brachyloba* Müll Arg.
			65 *Manihot condensata* Rogers & Appan
			66 *Manihot guaranitica* Chodate & Hassler.
			67 *Manihot grahamii Hooker*
			68 *Manihot quinquepartita* Huber ex Rogers & Appan
			69 *Manihot tripartita* Müll Arg.
			70 *Manihot tristis* Müll Arg.
			71 *Manihot violacea* Pohl

Genus	Common name	Cultivated species	Wild relatives of crop
			72 *Manihot* sp1
			73 *Manihot* sp2 – monte yucca
			74 *Manihot* sp3 – amazonia
			75 *Manihot* sp4 – chaco
			76 *Manihot* sp5 – cerrado-saxicola
			77 *Manihot* sp6 – cerrado-arenoso
			78 *Manihot* sp7 – cerrado-rocoso
			79 *Manihot* sp8 – 5-foliadas
			80 *Manihot* sp9 – chiquitania
Ipomoea	Camote	*Ipomoea batata*	81 *Ipomoea amnicola*
			82 *Ipomoea aquatica*
			83 *Ipomoea argentea*
			84 *Ipomoea bonariensis*
			85 *Ipomoea cheirophylla*
			86 *Ipomoea cuneifolia*
			87 *Ipomoea cynanchifolia*
			88 *Ipomoea decora*
			89 *Ipomoea descolei*
			90 *Ipomoea echinocalyx*
			91 *Ipomoea grandiflora*
			92 *Ipomoea haenkeana*
			93 *Ipomoea hieronymi*
			94 *Ipomoea magniflora*
			95 *Ipomoea martii*
			96 *Ipomoea maurandioides*
			97 *Ipomoea minuta*
			98 *Ipomoea neurocephala*
			99 *Ipomoea paludosa*
			100 *Ipomoea peredoi*
			101 *Ipomoea philomega*
			102 *Ipomoea procumbens*
			103 *Ipomoea schomburgkii*
			104 *Ipomoea schulziana*
			105 *Ipomoea sericophylla*
			106 *Ipomoea squamisepala*
			107 *Ipomoea tenera*
			108 *Ipomoea* spp. *1* (espécie no identificada)
			109 *Ipomoea* spp. *2* (e)spécie no identificada)
			110 *Ipomoea* spp. *3* (espécie no identificada)
			111 *Ipomoea* spp. *4* (espécie no identificada)
			112 *Ipomoea* spp. *5* (espécie no identificada)
			113 *Ipomoea* spp. *6* (espécie no identificada)
			114 *Ipomoea* spp. *7* (espécie no identificada)

Genus	Common name	Cultivated species	Wild relatives of crop
			115 *Ipomoea* spp. *8* (espécie no identificada)
			116 *Ipomoea* spp. *9* (espécie no identificada)
			117 *Ipomoea* spp. *10* (espécie no identificada)
			118 *Ipomoea* spp. *11* (espécie no identificada)
			119 *Ipomoea* spp. *12* (espécie no identificada)
			120 *Ipomoea* spp. 13 (espécie no identificada)
			121 *Ipomoea* spp. 14 (espécie no identificada)
			122 *Ipomoea* spp. *15* (espécie no identificada)
Ananas y Pseudananas	Piña	*Ananas comosus*	123 *Ananas ananassoides*
			124 *Ananas nanus*
			125 *Ananas paraguazensis*
			126 *Pseudananas sagenarius*
Fundación PROINPA La Paz[4]			
Chenopodium	Quinua	*Chenopodium quinoa*	127 *Chenopodium album*
			128 *Chenopodium hircinum*
			129 *Chenopodium hircinum* subsp. *catamarcensis*
			130 *Chenopodium hircinum* subsp. *eu-hircinum*
			131 *Chenopodium hircinum* subsp. *hircinum* var. *andinum*
			132 *Chenopodium quinoa* subsp. *melanospermum*
			133 *Chenopodium quinoa* var. *quinoa*
			134 *Chenopodium quinoa* subsp. *milleanum*
			135 *Chenopodium quinoa* var. *melanospermum*
Fundación PROINPA Cochabamba[5]			
Solanum	Papa	*Solanum tuberosum*	136 *Solanum acaule*
			137 *Solanum achacachense**
			138 *Solanum alandiae**
			139 *Solanum arnezii**
			140 *Solanum avilesii**
			141 *Solanum berthaultii**
			142 *Solanum boliviense* subsp. *astleyi*
			143 *Solanum bombycinum*
			144 *Solanum brevicaule*
			145 *Solanum candolleanum* f. *sihuanpampinum*
			146 *Solanum chacoense*

Genus	Common name	Cultivated species	Wild relatives of crop
			147 *Solanum circaefolium* var. *capsicibaccatum**
			148 *Solanum xdoddsii**
			149 *Solanum flavoviridens**
			150 *Solanum gandarillasii**
			151 *Solanum hoopesii**
			152 *Solanum infundibuliforme*
			153 *Solanum leptophyes*
			154 *Solanum xlitusinum*
			155 *Solanum megistacrolobum* subsp. *toralapanum*
			156 *Solanum microdontum* var. *montepuncoense*
			157 *Solanum neocardenasii**
			158 *Solanum neovavilovii*
			159 *Solanum okadae*
			160 *Solanum oplocense*
			161 *Solanum soestii*
			162 *Solanum sparsipilum*
			163 *Solanum xsucrense**
			164 *Solanum tarijense*
			165 *Solanum ugentii*
			166 *Solanum vidaurrei*
			167 *Solanum violaceimarmoratum**
			168 *Solanum virgultorum*
			169 *Solanum yungasense*
Centro de Investigaciones Fitoecogenéticas de Pairumani[6]			
Arachis	Maní	*Arachis hypogaea*	170 *Arachis batizocoi*
			171 *Arachis benensis**
			172 *Arachis chiquitana**
			173 *Arachis cruziana**
			174 *Arachis* cf. *cardenasii**
			175 *Arachis* aff. *diogoi*
			176 *Arachis duranensis*
			177 *Arachis glandulifera**
			178 *Arachis herzogii**
			179 *Arachis ipaensis**
			180 *Arachis kempff-mercadoi**
			181 *Arachis krapovickasii**
			182 *Arachis* cf. *magna**
			183 *Arachis matiensis*
			184 *Arachis rigonii**
			185 *Arachis simpsonii*
			186 *Arachis* cf. *trinitensis**
			187 *Arachis williamsii**
Phaseolus	Frijol	*Phaseolus vulgaris*	188 *Phaseolus augusti*
			189 *Phaseolus vulgaris* f. *silvestre*

(*) Species endemic to Bolivia; (**) According to molecular studies they form part of the genus *Solanum* and, in order not to confuse them with the wild relatives of potatoes, they are maintained in *Cyphomandra*.

Notes

1. Researchers from LPB that identified the species of *Theobroma* and *Anacardium:* Renate Seidel and Prem Jai Vidaurr; *Bactris* and *Euterpe:* Mónica Moraes
2. Researchers from CBG-BOLV that identifies the species of *Annona* and *Vasconcellea*: Nelly De la Barra; *Rubus* and *Cyphomandra*: Saúl Altamirano
3. Researchers from MHNNKM and IIA 'El Vallecito' that identified the species of *Manihot, Ipomoea, Ananas* and *Pseudananas:* Moisés Mendoza, Carlos Rivadeniera and Rolando Bustillos
4. Researchers from PROINPA La Paz that identified the species of *Chenopodium*: Wilfredo Rojas, Milton Pinto and Eliseo Mamani
5. Researchers of PROINPA Cochabamba that identified the species of *Solanum:* Fernando Patiño and Ximena Cadima
6. Researchers of the CIFP that identified the species of *Arachis, Phaseolus* and *Capsicum*: Margoth Atahuachi and Lorena Guzmán

Annex II

Monitoring Plan for Cereal Crop Wild Relatives in Erebuni State Reserve

Climate: Average daily temperature, air humidity and precipitation data will be provided by the meteorological centre. Data will be input into a database on a weekly basis.

Natural disasters: These will be recorded as they occur. Estimates of the affected area will be done by on-site evaluation. More accurate calculations will be done by the software tool after inputting coordinates using GPS. For the coordinates of the affected area, care should be taken to capture as many points as possible (but no more than the evaluator finds reasonable) for recording the outline of the area affected.

Daily observations of plants: This will be captured by the botanist assigned. Identification of the first shoots is considered the commencement of germination. If new shoots are no longer identified, the germination phase can be considered to have ended. The same approach is applied to other phenological phases.

Population size and distribution: This should be measured once per year in late May or early June, during the stage of early spike formation when identification of plants is easy (there can be variations depending on whether spring is early or late: appropriate time to be decided by the botanist assigned). Average number of plants will be determined by simple calculation of the mean in 10 selected experimental unit square plots (coordinates are available). In the journal of observations, the area occupied by each target species has to be assessed by simple on-site evaluation. More accurate calculations will be performed by the software tool after inputting GPS data. Again, care should be taken to capture as many points as possible. Mean number of seeds in spikes is calculated by averaging number of seeds in seven (7) randomly picked spikes. The procedure is performed by the botanist assigned.

Monitoring of pests and diseases: Inspection of plants for diseases/pathogens should be done 3 times during the season (late spring, summer and early autumn). Appropriate time to be chosen by the respective botanist/phytopathologist assigned. If not easily identified under filed conditions, samples should be collected for further identification in laboratories. Standard methodologies should be followed. Assessment of the damage caused should be done using a 1–5 scale with the value of 1 being least damaged and 5 being the most severely damaged. The latter assumes loss of viability and/or reproductive abilities. Here, the specialist assigned should use his/her subjective judgement to assess the average damage caused to affected plants. For calculating the percentage of infested plants the proportion of infested plants should be calculated in random unit (1m x 1m) squares by dividing the number of infested plants into the total number of plants belonging to the taxon of interest.

Monitoring of plant invaders: If the species is not easily identified in the field, this will be done in a laboratory. For the journal of observation records, area occupied is assessed by on-site evaluation. GPS data should be captured to outline the distribution area and to make more accurate calculations of the affected surface.

Data input into the software database is done on a weekly basis by assigned trained personnel.

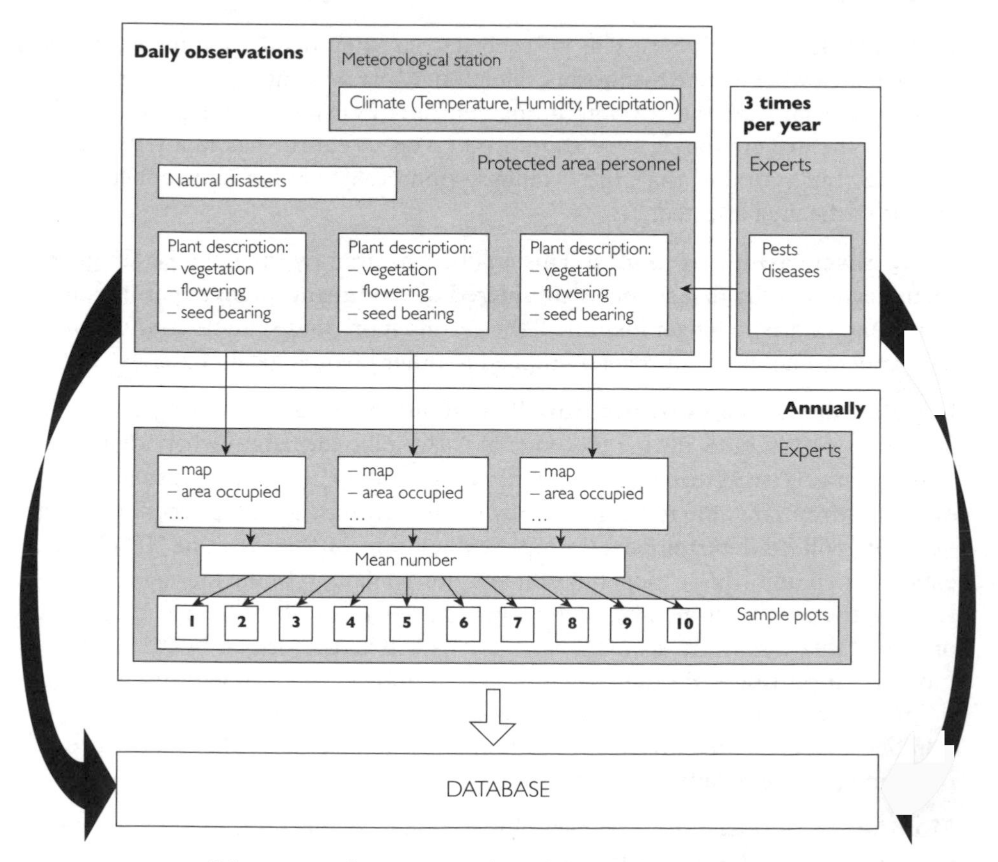

Diagrammatic representation of the monitoring process

JOURNAL OF OBSERVATIONS

1. Climate and landscape: Records of natural disasters

(Completed by the protected area personnel when the event occurs)

Date	Description of the event, consequences	Duration	Affected area (m^2)	Coordinates and description of the site affected	Type of natural disaster, e.g. flood, landslide, avalanche, sinkholes, thunder, storm, fire, drought, other: specify

2. Daily observations of plants

(Completed by protected area personnel when the event occurs)

Date	Plant species (botanical name)	Phenological phase		
		Germination (beginning and end)	Bush formation (beginning and end)	Spike formation (beginning and end)

3. Monitoring of priority plant population size and distribution

(Completed by botanist assigned, once during the season)

Date	Plant species (botanical name)	Number of plants in experimental plots											Average number of seeds in a spike	Area occupied by the species (Ha)
		1	2	3	4	5	6	7	8	9	10	average		
	Triticum boeoticum Boiss.													
	Triticum urartu Thum. ex Gandil.													
	Triticum araraticum Jakubz.													
	Aegilops tauschii Cosson.													

Distribution maps should be attached

4. Monitoring of plant pests and diseases

(Completed by assigned scientists 3 times during the season)

Date	Infested plant species	Name of the pathogen (in Latin)			Infested organs	Assessment of the damage caused to a single plant (1 to 5 scale)	Percent-age of plants damaged (%)	Notes
		Micro-organism	Fungi	Animal				

Distribution maps should be attached

5. Plant invaders

(Completed by botanist assigned, once during the season)

Date	Invaded plant species (indicate scientific name)	Area occupied (Ha)	Coverage % of a unit area	Notes

Distribution maps should be attached

Annex III

Management Plan for *Amygdalus bucharica* in the Chatkal Biosphere State Reserve, Uzbekistan

Description of *Amygdalus bucharica* L. (Rosaceae)

Amygdalus bucharica is distributed throughout the mountain chains of Uzbekistan in the lower and middle belts, from western Tien Shan in the north, (where it is very rare) on the Kuratin, Turkistan, Malguzar and Nurata chains, to the Gissar and Kuchitang chains in the south. *A. bucharica* occupies almond–pistachio populations on the southern slopes of the Gissar chain and the western slopes of Babataga chain. *A. bucharica* is more common in the south-western parts of the Chatkal chain within the Chatkal Reserve territory, on the northern slopes of the Nurata chain within the Nurata Reserve territory and in the Sangardak and Tupaland river basins in the Southern Gissar, where it comprises pure populations on the southern slopes. The species grows in small populations in the Gissar and Surhan reserves (Kuchitang chains), as well as in unprotected territories as small or very rare populations; *A. bucharica* was more widespread in the past. The species is distributed within the lowlands and hills from 500 to 1500m.

A. bucharica is a xerophyte and grows well in the hot summer conditions with precipitation of 300 to 700mm. It prefers rocky grey soil but also tolerates less rocky soil, and it comprises rare populations which are not very productive. *A. bucharica* is a very polymorphic species which is present in various forms, depending on the ecological conditions. It might be found as a many-stemmed bush, 1.5 to 2m high with a compact crown, in harsh conditions with a low amount of precipitation, high summer temperatures and on soil of poor fertility. In areas where the soil is fertile and precipitation levels are higher (700 to 800mm/year), the plant occurs as a tall (8m) tree with a single stem of 30cm or more in diameter and with a high and broad crown. In very hot summers, the plant drops almost all its leaves in August and spends the winter without damage. In this climate, the plant is characterized by drying out, but it recovers its crown very quickly through its young shoots.

Almond is a summer, green deciduous plant. It begins its vegetative growth early in the spring. It flowers early in summer, or even late in winter during warm periods, and its flowers might be damaged by spring frosts; the buds usually die. This is the reason for its irregular fruiting – almost once in 3–4 years. The fruits ripen early in July/August and then dry and remain in the crown for a long period. The fruits then fall as a result of wind or other factors. Vegetative growth finishes in September. It is resistant to winter frosts when the temperature drops to −15°C. If the temperature drops even lower, the species becomes frosted and dies, but due to resprouting, it might recover very quickly.

Almond flowers are well pollinated by insects. Reasonable yields can be generated (up to 15kg from one tree) under good conditions, if they are not damaged by frosts. Almond usually generates 2 to 3kg of fruits from one tree. The fruit is a drupe with a dry pericarp that ruptures at the sutures and opens when the fruit is ripe and thereby releases the stone fruit. The shell of the fruit is very thick and hard, but sometimes thin-shelled almond fruits can also be formed. The size of stone fruit varies from 2 to 4cm, but larger ones sometimes occur. The seed of the fruit is bitter, not edible for humans, but sometimes sweet seeds are produced. These forms have been the basis for breeding in the past. Sweet almond is a major fruit and widely consumed. Bitter almonds are in great demand by the perfume and by pharmaceutical industries, and the species is also harvested by local communities.

The main dispersers of the seed are large birds like magpies and crows which carry the seeds for long distances. Murine rodents also disperse some seeds by storing fruits for the winter. Almond possesses a strong, pink timber that is valued in joinery but is also often used as high-energy firewood; this is the reason it is gathered by the local community.

Evaluation

Importance

From early times, almonds have been used as a source of food and gathered in the forests by local communities and then later cultivated near their homes. Today, almond trees are cultivated in plantations not only in the mountains but also in the valleys under irrigation. In Uzbekistan, almonds are used as food for guests, along with walnut and pistachio. The wood has been used as fuel since ancient times. The branches of *A. spinosissima* are also used for fencing.

Potential importance of the almond for research and breeding

Almond shows great polymorphism, indicating its value as a genetic resource for use in breeding programmes to develop new forms with important features: high yielding, thin shell, various ripening timetable, late flowering (to avoid damage by early spring frosts), regular yielding, disease resistant and highly resistant to aridity, as well as ornamental and other features. All these characteristics are found in wild almond populations in Uzbekistan.

Threats to the species

A. bucharica is included in the IUCN conservation status Vulnerable [VU B2ab (iii,v)] (Eastwood et al, 2009). The main threat to the species is anthropogenic. Previously, the main factor leading to its extinction was cutting, but today it is cattle grazing. A high rate of population increase in recent times (during the last 50 years, the population grew threefold), especially in rural areas, resulted in an increase of grazing by cattle in almond plantations along with the almond tree's self-sown seedlings and undergrowth. All shrub layers are usually eaten, with the exception of inedible plants. The lower part of the almond's crown is also usually damaged by the cattle and fructification decreases as a result. Cattle grazing has completely destroyed almond communities and thus considerably decreased its resistance to disease, pathogens and various abiotic factors.

One of the main biotic factors is fungal disease which weakens the trees to the point of complete desiccation. Damage of almond communities by cattle considerably lowers their resistance to various kinds of diseases. Among natural abiotic factors that threaten the almond populations are early spring frosts, which affect the flowers and damage the buds. The assimilation apparatus of the plants is usually affected by high summer temperatures and hot winds. Extremely low winter temperatures cause the shoots to freeze, leading to the death of the trees. The most dangerous of the abiotic factors caused by human action is soil erosion, leading to landslides caused by torrential rain.

Key partners

The key partners in undertaking the conservation actions on almond populations are the authorities of Ugam-Chatkal National Park, Chatkal Reserve, authorities of Nurata Reserve, Main Management Department of Forestry and the State Committee on Nature Protection.

Analysis of the situation

The territories occupied by almond in the reserves are not sufficient for their effective conservation (the most significant conserved populations are found only in the Chatkal and Nurata Reserves). Thus, it is very important to develop measures for almond conservation in the unprotected areas. According to the law on protected areas of the Uzbekistan Republic, no human activity is permitted within such areas (especially on the territory of Nurata Reserve). This is why it is important to increase the number and extent of areas under protection actions.

Unprotected areas containing almond populations are found in the State Forest Fund, related to the Main Department of Forestry of the Ministry of Agriculture. In these areas, harmful anthropogenic actions such as unregulated cattle grazing, tree felling, overharvesting of fruits and destruction of plant communities often take place. Key partners who carry out measures to reduce the threats are the Forestry authorities, Main Department of Forestry Management, the State Committee on Nature Protection and the local community authorities.

Target and tasks

The target is to conserve the most important almond populations that exist today in the country's protected areas, as well as in the unprotected areas, and to create conditions for the restoration of populations in areas where the species previously existed.

The conservation plan for *A. bucharica* in the protected areas should be directed at strengthening the level of protection in these locations and excluding human activities that may influence the natural processes. This is especially important in the Nurata Reserve where cattle grazing and other human activities still take place.

The management plan for almond conservation includes:

1 improving the legal system of CWR conservation, developing amendments which reflect CWR conservation actions for the Forestry Codex and for the National Strategy and Action Plan on Biodiversity Conservation, developing amendments for the Forestry Law, and the Law on Protection of Plants;
2 fulfilling all measures according to existing laws;
3 restriction of cattle grazing and the harvesting of CWR fruits;
4 developing rental agreements (to make the areas with almond wild relatives available for renting) with an obligation on the renters to conserve almond populations in order to support the transfer to public management of forestry (PMF);
5 including facultative programmes in the curriculum of schools, higher educational organizations and colleges concerning CWR conservation actions;
6 carrying out research programmes on the selection of the most valuable almond forms in order to create *ex situ* collections and collect genetic material for breeding programmes;
7 increasing of the level of public awareness in the regions through mass media, booklets and posters publications, roadside posters and also through public organizations in rural places near almond communities.

Tasks and responsibilities

Actions	*Persons responsible and implementing organizations*	*Timetable*	*Budget*
Task 1: To conserve almond populations in protected areas			
1.1. Strengthening of protection of the territories with almond populations from cattle grazing, fruit picking and other threats	Authorities of Ugam Chatkal National Park, Chatkal and Nurata Reserves	Start in 2010	
Task 2: To improve the legal system on CWR conservation			
2.1. Making amendments to the Forestry Codex, National Strategy and Action Plans on Biodiversity Conservation	Project partners, Main Department of Forestry Management, State Committee of Nature Protection	Start in 2009	
2.2. Development of amendments into the Law on Forests and Nature Protection	N.K. Skripnikov, State Committee of Nature Protection	2009–2010	
Task 3: To implement the current laws on CWR conservation and monitor their implementation			
3.1. Restriction of cattle grazing and fruit harvesting	Forestry authorities, State Committee of Nature Protection, Main Department of Forestry Management	Start in 2010	
Task 4: To support the regeneration of almond populations throughout its natural habitat and in protected areas within the country			
4.1. To carry out measures on natural regeneration: planting almond seed, creating mineralized areas and others	Forestry authorities, Main Department of Forestry Management	Start in 2010	
Task 5: To involve local people in CWR conservation actions and widen the range of actions			
To adopt rental agreements in the areas of almond distribution	Forestry authorities, Main Department of Forestry Management	Start in 2010	
To develop public types of forestry management (PTFM)	Main Department of Forestry Management, Forestry authorities, local authorities	Start in 2011	

Actions	Persons responsible and implementing organizations	Timetable	Budget
Task 6: To increase the level of awareness of CWR importance in educational establishments			
6.1. Creation of curriculum programmes for schools, colleges and higher educational institutions	Scientific Plant Production Centre 'Botanica'	2009	
6.2. Agreement and introduction of these programmes in higher educational institutions	Scientific Plant Production Centre 'Botanica', Ministry of Education	Start in 2011	
Task 7: To carry out research programmes on CWR conservation			
7.1. Selection of the most valuable almond forms in nature for creation and maintenance of *ex situ* collections	Institute of Plant Industry, Shreder Institute, Authorities of Ugam-Chatkal National Park	Start in 2011	
7.2. Selection of gene material for breeding processes	Institute of Plant Industry, Shreder Institute	Start in 2010	
Task 8: To increase the level of public awareness			
8.1. Publication of booklets, posters and calendars on CWR conservation	National project partners	2009	
8.2. Conducting training courses for local authorities, forestry authorities and renters	National project partners	2009	

Monitoring strategy

Methodology

The methodology was agreed in advance and pilot areas in Chatkal Reserve have been identified and a monitoring strategy has been defined. Monitoring will be conducted every five years in both spring and summer. The results of monitoring have been prepared in Russian and are available on the website: www.cwr.uz. Currently, there are plans to translate these monitoring reports into English.

For further information on monitoring species/populations to assess the effectiveness of conservation actions and management plans, see Chapter 13.

References

Eastwood, A., Lazkov, G. and Newton, A. (2009) *The Red List of trees of Central Asia*, Fauna and Flora International, Cambridge, UK

Index of Organisms

General Index